COMPRESSIBLE FLUID FLOW

B. W. IMRIE
B.Sc.(Eng), M.Sc., C.Eng., M.I.Mech.E., M.Inst.M.C.
Lecturer, Department of Engineering, University of Leeds

LONDON
BUTTERWORTHS

THE BUTTERWORTH GROUP

ENGLAND
Butterworth & Co (Publishers) Ltd
London: 88 Kingsway, WC2B 6AB

AUSTRALIA
Butterworths Pty Ltd
Sydney: 586 Pacific Highway, NSW 2067
Melbourne: 343 Little Collins Street, 3000
Brisbane: 240 Queen Street, 4000

CANADA
Butterworth & Co (Canada) Ltd
Toronto: 14 Curity Avenue, 374

NEW ZEALAND
Butterworths of New Zealand Ltd
Wellington: 26-28 Waring Taylor Street, 1

SOUTH AFRICA
Butterworth & Co (South Africa) (Pty) Ltd
Durban: 152-154 Gale Street

(i) Reprinted by the National Book Foundation under the provision of the Copyright (Amendment) Act 1972 for the Government of Pakistan.

(ii) Reproduced by M/s. Nawa-i-Waqt Publications Limited Lahore

Code No. TBRP/B-208/NBF/377/1000

For sale and use in Pakistan.

First published 1973

ISBN 0 408 70434 9 Standard
0 408 70435 7 Limp

Set in Hungary
Printed in England by J. W. Arrowsmith Ltd, Bristol
and bound by R. J. Acford Ltd, Chichester

COMPRESSIBLE
FLUID
FLOW

ACKNOWLEDGEMENTS

I welcome this opportunity of acknowledging the excellent introduction to compressible fluid flow which I obtained in the Graduate School of Thermodynamics, University of Birmingham, directed by Professor F. K. Bannister.

Most of the examples in this book are taken from the degree examinations of Birmingham, Leeds and London Universities and I am indebted to the authorities of these universities for permission to use questions from their papers. I make the customary affirmation that I accept full responsibility for accuracy of solutions and answers.

I am most grateful to Dr. Paul Castle and Mr. Graham Jones for their work in preparing the computer programs and to Mr. Geoffrey Dean for his fine work in preparing the diagrams. A special word of appreciation is due to Miss Pat Kimber for coping so well with the typing of the manuscript and the presentation of the complicated algebraic formulae.

Finally, I am aware of the demands which the preparation of this book has made on my family and gratefully acknowledge the intangible contributions of my wife and children (who wanted to improve the graphs by colouring them!)

B. W. IMRIE

PREFACE

The lecture notes and experience from which this book has been compiled, have been developed by teaching and examining final year degree courses in the Department of Mechanical Engineering at Leeds University and by teaching introductory courses on compressible flow to postgraduates on the M.Sc. Course in Combustion and Explosion Studies.

The purpose of this book is to present the fundamentals of steady compressible fluid flow of a perfect gas in such a way that the student may appreciate the assumptions of theoretical analysis, the derivation and application of explicit parameter relationships and the use of tabulated data as solutions of such relationships. The presentation of Theory, Worked Examples and Tables is integrated to enable the student to consider the interaction of Theory and Tables in solving compressible flow problems which are taken mainly from university examination papers. In this respect it may be considered that the book will also help students to pass examinations in this subject.

While all theoretical equations are presented in a general form, the final parameter relationships are also presented for air considered as a perfect gas with a constant ratio of specific heats ($\gamma = 1.4$) and a gas constant, $R = 0.2871$ kJ/kgK. This value of γ for air has been chosen for convenience in manipulating the equations to avoid unnecessary elaboration and, in this form, the equations have been incorporated in computer programs to produce the tables of data. Each equation thus used is also presented graphically to enable the student to appreciate the variations of the parameters involved. All such parameters are non-dimensional but elsewhere Système Internationale (S.I.) Units have been used.

Worked examples are used to provide relevant application of fundamental principles and to develop particular aspects of theoretical analysis. The worked examples are set out with comprehensive numerical and algebraic detail and with appropriate cross-references to theoretical equations and to tables.

The effects of variable area (Chapter 2), friction (Chapter 5) and of heat energy transfer (Chapter 6) on one-dimensional steady flow in ducts of a compressible fluid have been considered separately and, in this respect, the presentation of the tabulated data differs markedly from that of Benedict and Steltz. While real situations may require consideration of interactions of these different effects, the uniform approach used in each Chapter should allow the student to develop a fundamental understanding of the physical processes, which he can then apply to more complicated problems.

Two-dimensional steady flow of a compressible fluid is developed in a similar manner for Prandtl-Meyer analysis (Chapter 3) of isentropic expansion and compression processes and for oblique shock wave analysis (Chapter 7) which includes the limiting conditions of one-dimensional isentropic flow (Chapter 2) and flow through the plane normal shock wave (Chapter 4). The inclusion of Chapter 8 as an introduction to wave propagation in Gas Dynamics is appropriate as a complement to the stationary wave concepts of Chapters 2 and 4.

It is not in the best interests of the student to leave his introduction to compressible fluid flow until his final year after a lengthy involvement with incompressible fluid mechanics and it is therefore hoped that this book will be of use in second and final year courses in fluid mechanics and in some postgraduate courses.

B. W. IMRIE

CONTENTS

1 SOME FUNDAMENTAL CONCEPTS OF FLUID MECHANICS

1.1 ONE-DIMENSIONAL FLOW CONSIDERATIONS

For gas flow in general and duct flow (Figure 1.1) in particular, it will be assumed that variations in boundary or cross-sectional area, will be gradual so that the rate of change of fluid properties normal to streamlines will be negligibly small. All flow properties such as density (ϱ), pressure (p), velocity (u) etc., will be assumed to be uniform across any section of the duct. Unless specifically excluded

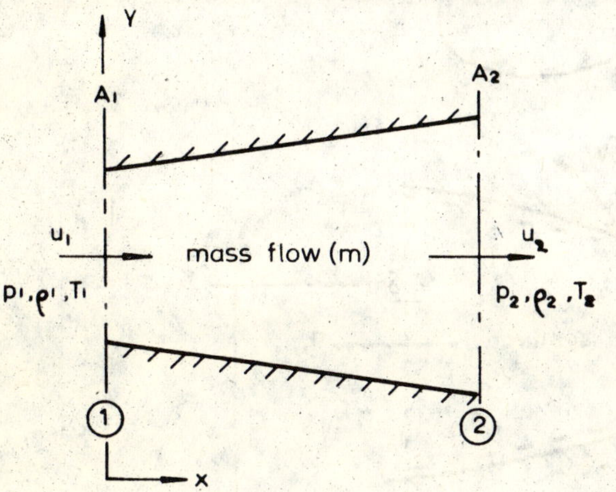

Fig. 1·1 uniform duct flow values

it will be assumed that the effects of compressibility, heat energy transfer and viscosity will be included in the general description of fluid flow. Only steady flow applications will be considered.

1.2 CONTINUITY EQUATION (CONSERVATION OF MASS)

For one-dimensional steady flow in a rigid duct (Figure 1.1) the mass flow is a constant at any section and is given by

$$m = \varrho uA = \varrho_1 u_1 A_1 = \varrho_2 u_2 A_2 = \text{constant} \quad (1.1)$$

thus, in differential form $d(\varrho uA) = 0$ (1.2)

or, $$\frac{d\rho}{\rho} + \frac{du}{u} + \frac{dA}{A} = 0 \quad (1.3)$$

1.3 MOMENTUM EQUATION (CONSERVATION OF MOMENTUM)

For one-dimensional steady flow in a rigid duct consider the elemental control volume shown in Figure 1.2. The net change between the outgoing and incoming momentum

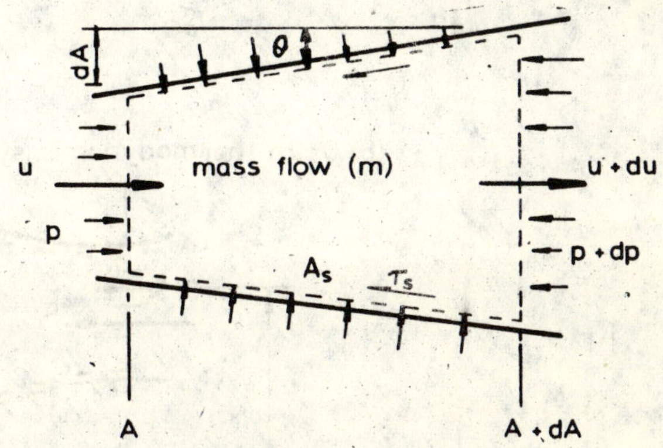

Fig.1.2 elemental control volume

fluxes is equal to the forces exerted on the fluid within the control volume. If the mean pressure exerted by the solid boundary on the control volume is $(p+dp/2)$ and the viscous shear stress at the boundary is τ_s acting on the surface area A_s then, (X-direction positive),

$$m((u+du)-u) = pA-(p+dp)(A+dA) + (p+dp/2)dA_s \sin\theta - \tau_s A_s \quad (1.4)$$

neglecting second order small terms, e.g., $(dp \times dA)$ and putting $dA = dA_s \sin\theta$,

from (1.1), $$(\varrho uA)\, du = -A dp - \tau_s A_s \quad (1.5)$$

for a parallel-sided duct (without viscous effects),

$$\varrho u\, du = -dp \quad (1.6)$$

Between any two sections in a duct (Figure 1.1) (without viscous effects) the momentum equation becomes,

$$m(u_2-u_1) = p_1A_1-p_2A_2+\int_1^2 p\,\mathrm{d}A \qquad (1.7)$$

where $\int_1^2 p\,\mathrm{d}A$ is the net force (usually unknown) exerted *on* the fluid inside the control volume *by* the pressure distribution on *all* solid surfaces (including a model) of the control volume (X-direction).

Note: For 'vertical' Y-direction changes, the 'body force' exerted by the gravitational field on the mass of fluid inside the control volume, must also be considered.

1.4 SUMMARY OF THERMODYNAMIC PRINCIPLES AND EQUATIONS

For the discussion of thermodynamic concepts and for the development of theoretical relationships in subsequent sections, it will be assumed that the working fluid is dry air which will be considered as a perfect gas. The specific heats of air at constant pressure (c_p) and at constant volume (c_v) will therefore be independent of temperature and the compressibility factor $Z = p/(\varrho RT)$ will be equal to unity for all temperatures and pressures.

The ratio of specific heats $\gamma = c_p/c_v = 1{\cdot}40$, and is the value used (with R) for evaluation of data in the compressible flow tables. ($c_p - c_v = R$, $c_p = \dfrac{\gamma}{\gamma-1} R$).

1.5 ENERGY EQUATION (CONSERVATION OF ENERGY)

A thermodynamic system may have some solid boundaries or may be considered more generally as any fixed control volume containing the working fluid which is part of the energy balance evaluation of the system. Figures 1.3a and 1.3b illustrate two different systems to which the First Law of Thermodynamics applies in the following form:

$$\delta Q = \mathrm{d}E+\delta W \qquad (1.9)$$

where δQ (kJ) and δW (kJ) are respectively transfers of heat energy and work energy across the system boundaries while $\mathrm{d}E$ (kJ) is the change in E, the total energy of the working fluid in the fixed control volume and closed system (Figure 1.3a). Electrical, chemical, nuclear, etc., contributions to this expression of energy conservation will not be considered and E includes the internal energy of the fluid, U (molecular activity), the kinetic energy of the fluid KE (for a fluid in motion—open system), and the potential energy PE (for the position of the fluid mass in the system). For unit mass of the fixed system,

$$e = u+q^2/2+gy \text{ kJ/kg} \qquad (1.10)$$

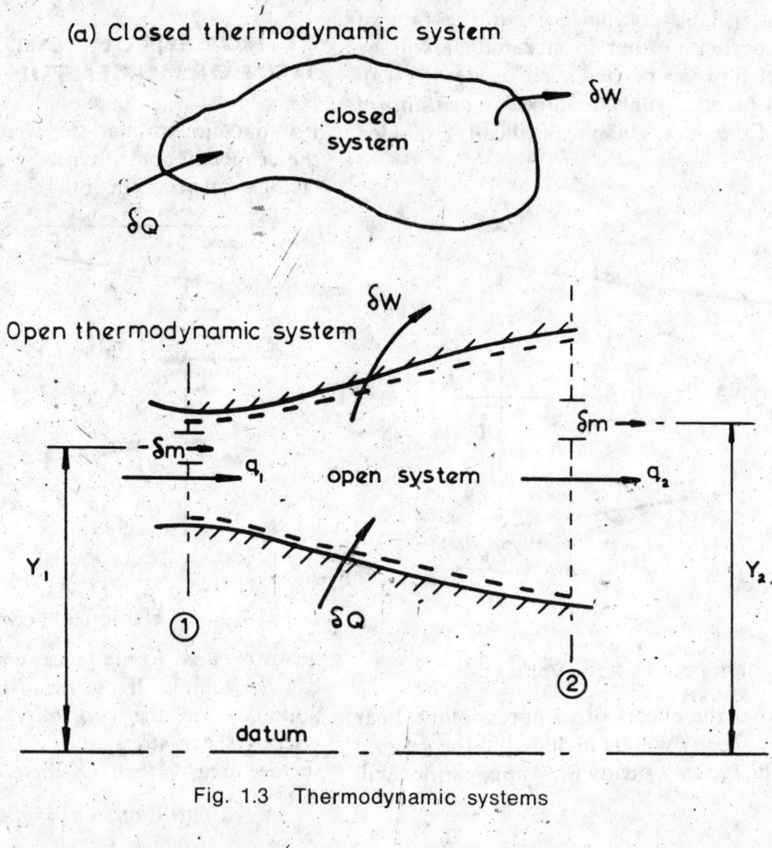

Fig. 1.3 Thermodynamic systems

Thus, for any two stations (1) and (2) in a homogeneous flow field the equation of state is

$$R = \frac{p}{\rho T} = \frac{p_1}{\rho_1 T_1} = \frac{p_2}{\rho_2 T_2} = \text{constant} \qquad (1.8)$$

The gas constant for air, $R = R_0/M_0 = 8{\cdot}3143/28{\cdot}96 = 0{\cdot}2871$ kJ/kg K where R_0 and M_0 are the universal gas constant and mean molecular weight respectively.

For steady flow of fluid between positions (1) and (2), Figure (1.3b), an elemental mass of fluid, δm, entering at state (1) will add energy $\delta m(u_1+q_1^2/2+gy_1)$ to the system, and an equal mass of fluid leaving at state (2) will remove energy $\delta m(u_2+q_2^2/2+gy_2)$ from the system; y_1 and y_2 are the mean positions of δm at (1) at (2) with respect to a common datum.

On entering the control volume, the mass δm will displace fluid inside the control volume thus adding energy to the system, at constant pressure, p_1, equal to $p_1 \delta m v_1$ or $\delta m p_1/\varrho_1$. Similarly, displacement energy will be transferred, from the system, at constant pressure, p_2, equal to $p_2\ \delta\ m v_2$ or $\delta m p_2/\varrho_2$.

If heat energy per unit mass, ${}_1Q_2$ (kJ/kg) is considered (conventionally) to be transferred to the working fluid across the system boundary between states (1) and (2), and work energy per unit mass, ${}_1W_2$ (kJ/kg) is transferred from the working fluid, then the principle of conservation of energy may be expressed as an equilibrium equation. The steady flow of energy into the system will be equal to the steady flow of energy out of the system, thus,

$$\begin{aligned} &{}_1Q_2+(p_1v_1+u_1)+q_1^2/2+gy_1 \\ &= {}_1W_2+(p_2v_2+u_2)+q_2^2/2+gy_2 \end{aligned} \qquad (1.11)$$

If no heat energy is transferred across the system boundary the flow is termed adiabatic, ${}_1Q_2 = 0$. For steady flow in duct systems no work energy will be transferred, ${}_1W_2 = 0$ and the energy equation becomes, with enthalpy, $h = u+pv$,

$$h_1+q_1^2/2+gy_1 = h_2+q_2^2/2+gy_2 \qquad (1.12)$$

Usually, for flow of compressible fluids which have very low densities compared with incompressible fluids, the change in $PE = (gy_2-gy_1)$ is negligible,

$$\therefore\ h_0 = h+q^2/2 = h_1+q_1^2/2 = h_2+q_2^2/2 = \text{constant} \qquad (1.13)$$

where h_0 is the stagnation enthalpy corresponding to a reservoir condition ($q = 0$) supplying fluid to the steady flow adiabatic system.

also $$c_pT_o = c_pT + q^2/2 = \text{constant} \qquad (1.13a)$$

or $$\frac{\gamma}{\gamma-1}RT_o = \frac{\gamma}{\gamma-1}RT + q^2/2 = \text{constant} \qquad (1.13b)$$

or $$\frac{\gamma}{\gamma-1}\frac{p_o}{\rho_o} = \frac{\gamma}{\gamma-1}\frac{p}{\rho} + q^2/2 = \text{constant} \qquad (1.13c)$$

These are different forms of the *steady flow adiabatic energy equation.*

1.6 ENTROPY

The Second Law of Thermodynamics has philosophical significance as well as being fundamental to the understanding of which processes are possible and of why the effects of such processes, with real fluids, differ from the same effects of the processes with ideal fluids.

With an ideal fluid (or perfect gas) having no viscous dissipation of energy and no random molecular activity, flow processes are assumed to be completely reversible without degradation of energy. With real fluids, flow processes are always irreversible and corollaries of the Second Law identify a property of a fluid which decribes the irreversibility of real processes. The property is entropy, s (per unit mass), and for a reversible process, the change of entropy is given by,

$$ds = \frac{\delta Q}{T}\ \text{kJ/kg K} \qquad (1.14a)$$

and, for an irreversible process,

$$ds > \frac{\delta Q}{T}\ \text{kJ/kg K} \qquad (1.14b)$$

For an adiabatic process ($\delta Q = 0$), $ds \geq 0$ and, if the process is also reversible, $ds = 0$. A reversible adiabatic processes is thus described as isentropic (constant entropy).

From the First Law (1.9),

$$\delta Q = de+\delta W\ \text{kJ/kg} \qquad (1.15)$$

For a non-flow process (closed system) in which kinetic, potential and displacement energies are zero,

$$\delta Q = du+\delta W\ \text{kJ/kg} \qquad (1.16)$$

For such a process, the work energy transferred at the system boundary $\delta W = p\,dv$.

δQ and δW depend on the nature or path of the process while properties (by definition) of the working fluid have unique values at any point in the process and property changes can be written as exact differentials dp, du, dh, ds, dv or $d\varrho$, etc., and may be integrated directly to give the net change for the process. For $\delta W = p\,dv$ the integration requires some functional relationship between p and v ($= 1/\varrho$) and has the general form pv^n or p/ϱ^n = constant; (n) is a polytropic index for all processes which may or may not be reversible.

Substituting from (1.14a) changes (1.16) to the form

$$T\,ds = du+p\,dv \qquad (1.17)$$

since $h = u + pv = c_vT + RT = c_pT$

(thus $c_p - c_v = R$)

$$dh = du+p\,dv+v\,dp$$

in (1.17), $$T\,ds = dh - v\,dp \qquad (1.18)$$

or (since $pv = RT$) $$ds = C_p\frac{dT}{T} - R\frac{dp}{p}$$

$$= -c_p\frac{d\rho}{\rho} + c_v\frac{dp}{p} \qquad (1.19)$$

whence, $$ds = s_2-s_1 = c_p \log_e\frac{T_2}{T_1} - R\log_e\frac{p_2}{p_1} \qquad (1.20a)$$

or $$s_2-s_1 = c_p\log_e\frac{v_2}{v_1}+c_v\log_e\frac{p_2}{p_1} \qquad (1.20b)$$

(which is conveniently symmetrical)

For isentropic flow, $ds = 0$,

(from 1,19), $$\therefore\ \frac{dp}{p} = \frac{c_p}{c_v}\frac{d\varrho}{\varrho} \qquad (1.21)$$

whence, integrating, $\log_e p = \gamma\log_e\varrho$+constant ($k$)

or $$p = k\varrho^\gamma \qquad (1.22)$$

Thus, for an isentropic process, the polytropic index, $n = \gamma$, the ratio of specific heats for a perfect gas.

Other processes may be identified by particular values of n and the differential form of $pv^n = k$ gives the result

$$\frac{dp}{dv} = -n\frac{p}{v} \qquad (1.23)$$

Note: Such processes may be plotted on a $p-v$ diagram and identified at any point by the slope of the curve determined by the value of n.

$n = \infty$ constant volume (isochoric)
$n = 0$ constant pressures (isobaric)
$n = 1$ constant temperature (isothermal)
$n = \gamma$ constant entropy (isentropic)

2 ONE-DIMENSIONAL ADIABATIC STEADY FLOW OF A COMPRESSIBLE FLUID

2.1 THE SPEED OF SOUND

If the medium is a compressible fluid considered as a perfect gas, the propagation of a pressure wave may be considered to be reversible for small amplitudes of pressure and adiabatic for rapid propagation of the wave through a stationary fluid. The basic fluid equations, however, have been developed by consideration of a fluid in motion with mass continuity. The pressure wave is a disturbance, without mass, which is transferred through the fluid. The resulting effect on fluid particles may be analysed by the flow equations only if the pressure wave is assumed to be stationary by considering the velocity of the fluid relative to the wave.

Figure 2.1a shows the effect of a small amplitude pressure wave ($\mathrm{d}p$) moving with speed (a) into stationary fluid. Figure 2.1b shows the stationary wave model with a control volume of fluid flow which can be considered as steady one-dimensional isentropic flow in a constant area duct. For analysis of such an isentropic flow process either the momentum or energy equations can be used in conjunction with the continuity equation and the isentropic relation, thus:

continuity, $\varrho a A = (\varrho + \mathrm{d}\varrho)(a - \mathrm{d}q)A$

for small amplitudes, $$\mathrm{d}q = a\frac{\mathrm{d}\varrho}{\varrho} \qquad (2.1)$$

momentum, $[p - (p+\mathrm{d}p)]A = \varrho a A\,[(a-\mathrm{d}q) - a]$

or $$\mathrm{d}q = \frac{\mathrm{d}p}{a\varrho} \qquad (2.2)$$

isentropic, $p = k\varrho^{\gamma}$

whence, $$\frac{dp}{d\varrho} = \frac{\gamma p}{\varrho} \qquad (2.3)$$

from (2.1) and (2.2), $$a^2 = \frac{\mathrm{d}p}{\mathrm{d}\varrho} \qquad (2.4)$$

from (2.3) and (2.4), $$a^2 = \frac{\gamma p}{\varrho} = \gamma RT \qquad (2.5)$$

Note: (1) For any gas for which γ and R kJ/kg K are assumed constant, $a = \text{constant}\,(T)^{0{\cdot}5}$ and the speed of sound may therefore be considered as a property of the gas,

(2) $a = \left(\gamma \dfrac{R_0}{M_0} T\right)^{0{\cdot}5}$ and will vary with the molecular weight of diatomic gases ($\gamma = 1.4$). For air, $a = (1{\cdot}4 \times 0{\cdot}2871 \times 10^3 \times T)^{0{\cdot}5}$ m/s

or $$a \approx 20(T)^{0{\cdot}5} \text{ m/s} \qquad (2.6)$$

at 16°C (air) $a \approx 340$ m/s

at 16°C (H_2) $a \approx 340(28{\cdot}96/2)^{0{\cdot}5} \approx 1294$ m/s

(3) The fluid particle velocity behind the wave moving into a stationary fluid is

$\mathrm{d}q\ (=q)$ and from (2.1), $$\frac{q}{a} = M = \frac{\mathrm{d}\varrho}{\varrho} \qquad (2.7)$$

The Mach number, M, is a measure of the rate of change of density, i.e., compressibility effect, and corresponds to the ratio of elasticity forces to convective inertia forces, F_E/F_{ic}. E, the bulk modulus of elasticity, is a measure of the ratio stress/strain for a homogeneous physical medium,

$$E = -\left(\frac{\Delta p}{\Delta v/v}\right) = \varrho\frac{\mathrm{d}p}{\mathrm{d}\varrho}$$

E may also be considered as the reciprocal of compressibility.

Thus $$\frac{F_E}{F_{ic}} \propto \phi\left(\frac{E}{\varrho q^2}\right)$$

Since $$\frac{E}{\varrho} = \frac{\mathrm{d}p}{\mathrm{d}\varrho} = a^2 \qquad \text{(from 2.4)}$$

Then, $$F_E/F_{ic} \propto \phi\left(\frac{a^2}{q^2}\right) \equiv \text{Mach Number, } M.$$

For water at 16°C, $a \approx 1473$ m/s

For steel at 16°C, $a \approx 5152$ m/s

(4) For a completely incompressible medium $d\varrho \to 0$, $a \to \infty$, $M \to 0$.

The Mach number may be less than, equal to, or greater than unity and the value of M is used to identify the characteristics of the different flow conditions (subsonic and supersonic) of a compressible fluid or of motion of a body through the fluid. These different conditions may be illustrated by considering the motion of a point disturbance relative to the propagation of the pressure waves from the position of the point at any given time and for successive equal time intervals. Three such intervals are considered in Figures 2.2, but an infinite number of pressure waves will be created by the disturbance:

(*a*) For a stationary point, $q = 0$, the spherical wave fronts after 1, 2 and 3 units of time with propagation speed (a) will be as shown in Figure 2.2a.

(b) The subsonic condition, $M < 1$, $q < a$, illustrated in Figure 2.2b, shows how the wave fronts move ahead of the disturbance (e.g. aircraft). The observer will hear the aircraft approaching before it reaches his position.

(c) The sonic condition, $M = 1$, $q = a$, is a transition condition, Figure 2.2c. The acoustic or small amplitude pressure waves coincide at a single point and merge into a finite amplitude wave. (A large amplitude pressure change or discontinuity is irreversible and is referred to as a shock wave.)

(d) The supersonic condition, $M > 1$, $q > a$ is shown in Figure 2.2d. The disturbance velocity now exceeds the speed of sound and the point coincidence (Figure 2.2c) of acoustic waves becomes a conical surface or envelope shown in section by lines inclined at angle (μ) to the horizontal as shown. Such lines or cone generators are referred to as Mach lines and may be considered as zero amplitude shock waves (Section 8). The effects of the disturbance are entirely contained by the cone.

For any tangent condition, $\sin u = \dfrac{a}{q} = \dfrac{1}{M}$

$$\text{or} \qquad u = \arc \sin (1/M) \qquad (2.8)$$

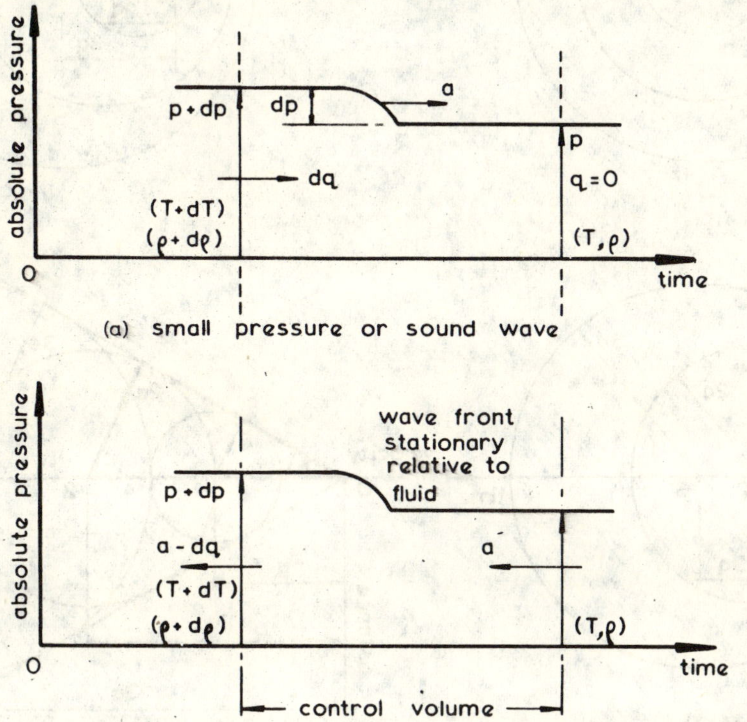

(a) small pressure or sound wave

Fig. 2.1 (b) control volume for stationary wave analysis

2.2 TOTAL HEAD CONDITIONS

The words 'total head', 'stagnation', 'Pitot-head' and 'reservoir' are frequently used to describe the fluid flow conditions for which the flow velocity is assumed to be zero. Indiscriminate use of such terms may lead to confusion and the 'reservoir' concept is considered separately.

In Figure 2.3, at any section *A-A* in a duct, the flow may be imagined to come from a reservoir of infinite capacity such that the reservoir conditions p_0 and T_0 remain constant and the fluid velocity $q = 0$.

The total head or stagnation pressure and temperature (p_t and T_t) at any station (1) or (2), in the steady flow are the pressure and temperature which would be measured if the flow velocity, at the same level, were reduced to zero isentropically. This is the effect which a Pitot-tube is designed to produce and to measure by insertion in a fluid flow, e.g., at station (2). The reservoir condition may then be compared with the total head or stagnation condition at (2) for steady adiabatic flow.

(a) If the fluid is considered to be incompressible, ($c_p = c_v$)

$$p_0 = p_1 + \tfrac{1}{2}\varrho q_1^2 = p_2 + \tfrac{1}{2}\varrho q_2^2 = p_t \qquad (2.9)$$

and the total head pressure is constant at all points in the flow.

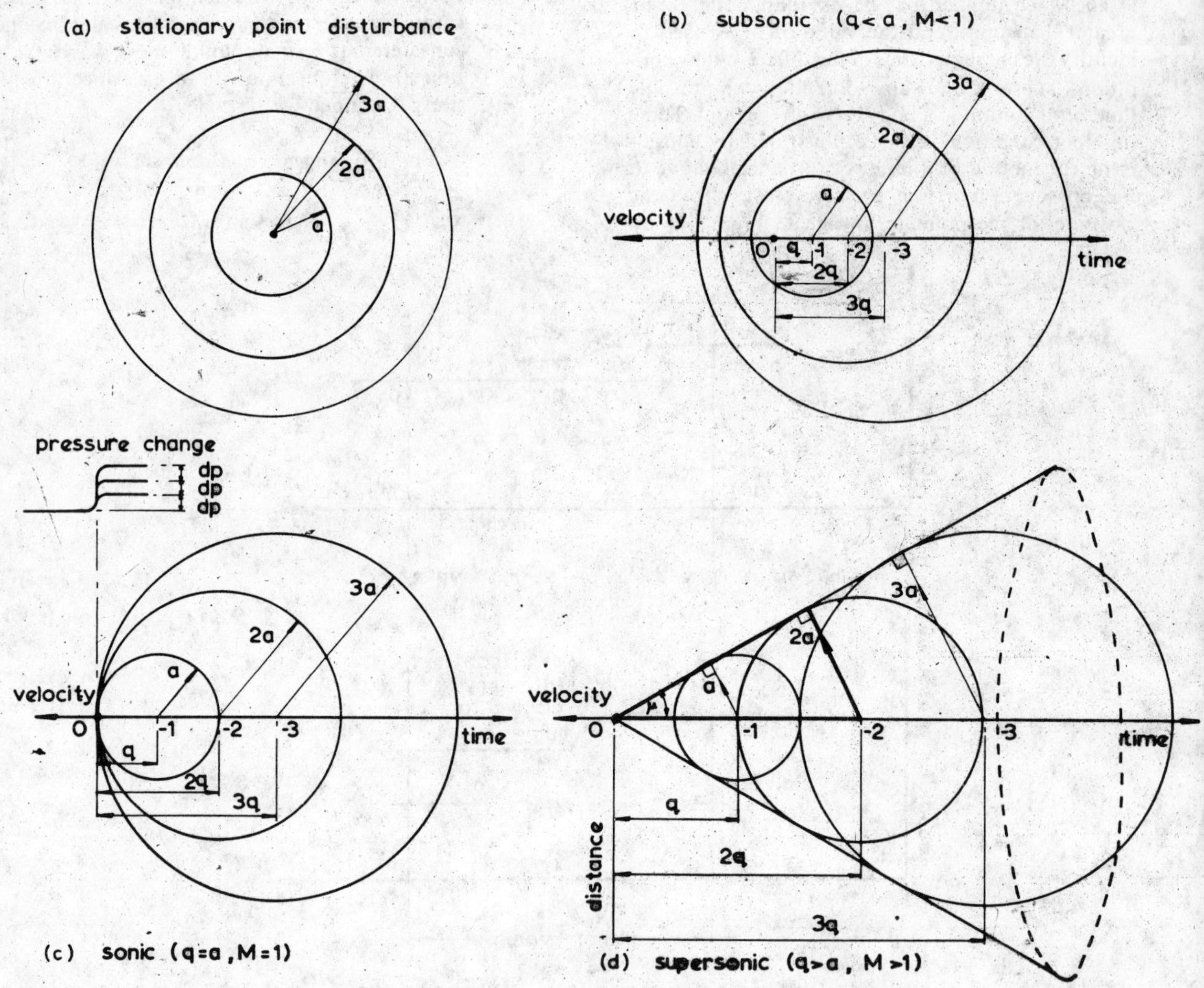

Fig. 2.2 Print disturbance in a stationary compressible fluid (unit time intervals)

(b) For a *compressible* fluid, $\left(c_p = \frac{\gamma}{\gamma - 1} R \text{ kJ/kg K}\right)$ from (1.13) etc.

$$\frac{\gamma}{\gamma-1}\frac{p_0}{\varrho_0} = \frac{\gamma}{\gamma-1}\frac{p_1}{\varrho_1} + \frac{q_1^2}{2} = \frac{\gamma}{\gamma-1}\frac{p_2}{\varrho_2} + \frac{q_2^2}{2} = \frac{\gamma}{\gamma-1}\frac{p_t}{\varrho_t} \quad (2.10a)$$

or,

$$\frac{\gamma}{\gamma-1} RT_0 = \frac{\gamma}{\gamma-1} RT_1 + \frac{q_1^2}{2} = \frac{\gamma}{\gamma-1} RT_2 + \frac{q_2^2}{2} = \frac{\gamma}{\gamma-1} RT_t \ldots \quad (2.10b)$$

or, $$T_0 = T_1 + \frac{q_1^2}{2c_p . 10^3} = T_2 + \frac{q_2^2}{2c_p . 10^3} = T_t \quad (2.10c)$$

For steady adiabatic flow of a compressible fluid, the total head temperature is constant at all points in the flow but (2.10a) does not permit a similar conclusion for total head pressure or total head density. If the flow is now considered to be isentropic (reversible adiabatic) then,

$$\frac{p_0}{\varrho_0^{\gamma}} = \frac{p_1}{\varrho_1^{\gamma}} = \frac{p_2}{\varrho_2^{\gamma}} = \frac{p_t}{\varrho_t^{\gamma}}$$

and comparison with (2.10a) justifies the conclusion that the total head pressure and total head density are now constant at all points in the flow *only if the flow is isentropic.* The suffix (0) will be used exclusively to denote total head conditions in the flow field.

Consider, $$s_2 - s_1 = c_p \log_e T_2/T_1 - R \log_e p_2/p_1 \quad (1.20a)$$

adiabatic, $T_{02} = T_{01} = T_0, \quad p_{02} \neq p_{01}$

isentropic, at the *same* station, $\frac{p_0}{p} = \left(\frac{T_0}{T}\right)^{\frac{\gamma}{\gamma-1}}$

rearranging (1.20a),

$$s_2 - s_1 = c_p \log_e \left[\frac{T_2}{T_{02}} \times \frac{T_{01}}{T_1}\right] - R \log_e \left[\frac{p_2}{p_{02}} \times \frac{p_{01}}{p_1} \times \frac{p_{02}}{p_{01}}\right]$$

substituting $\frac{p_{01}}{p_1} = \left(\frac{T_{01}}{T_1}\right)^{\frac{\gamma-1}{\gamma}}$ and $\frac{p_{02}}{p_2} = \left(\frac{T_{02}}{T_2}\right)^{\frac{\gamma-1}{\gamma}}$

$\left(c_p = \frac{\gamma}{\gamma-1} R\right)$ gives, $$\frac{s_2 - s_1}{R} = \log_e \frac{p_{01}}{p_{02}} \quad (2.11)$$

for *any adiabatic* process the non-dimensional change (increase) in entropy is a function of the change in total head pressure, i.e., the loss of total head pressure corresponds to the increase in entropy and is an indication of inefficiency or irreversibility.

2.3 DIFFERENT FORMS OF THE ENERGY EQUATION—REFERENCE SPEEDS

Following the development of an expression for the speed of sound in terms of fluid properties (2.5) the following reference speeds may usefully be considered:

(a) a_0—propagation (acoustic) speed at reservoir or total head condition,

(b) c—limit or maximum particle speed,

(c) $M = 1$, $q_* = a_*$—sonic condition; combination of particle speed and acoustic speed.

from (2.12b) and (2.13),

$$\frac{q}{c} = \left(1-\frac{a^2}{a_0^2}\right)^{0\cdot5} = \left[1-\left(\frac{1}{1+\frac{\gamma-1}{2}M^2}\right)\right]^{0\cdot5}$$

$$\therefore \frac{q}{c} = M\left[\frac{\gamma-1}{2+(\gamma-1)M^2}\right]^{0\cdot5} \tag{2.16}$$

from (2.14) when $M = 1$, $q = a_*$,

$$\frac{a_0^2}{a_*^2} = \frac{T_0}{T_*} = \frac{\gamma+1}{2} \tag{2.17a}$$

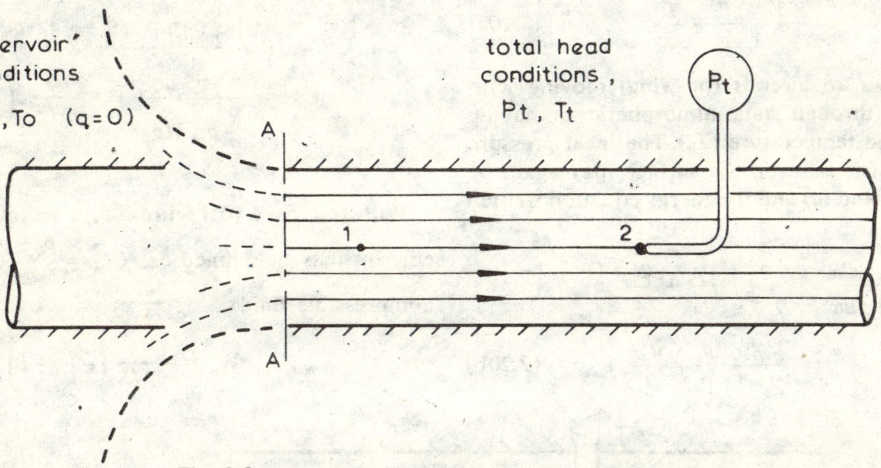

Fig. 2·3 reservoir conditions in steady flow

From the steady flow adiabatic energy equations (2.10),

$$\frac{\gamma}{\gamma-1}\frac{p_0}{\varrho_0} = \frac{\gamma}{\gamma-1}\frac{p}{\varrho}+\frac{q^2}{2} \tag{2.10a}$$

or

$$\frac{\gamma RT_0}{\gamma-1} = \frac{\gamma RT}{\gamma-1}+\frac{q^2}{2} \tag{2.10b}$$

$(a^2=\gamma RT)$

$$\therefore a_0^2 = a^2+\frac{\gamma-1}{2}q^2$$

or

$$q^2+\frac{2}{\gamma-1}a^2 = \frac{2}{\gamma-1}a_0^2 \tag{2.12a}$$

whence

$$q = \left(\frac{2}{\gamma-1}\right)^{\cdot5}(a_0^2-a^2)^{0\cdot5} \tag{2.12b}$$

when $a \to 0$, $T \to 0$, $p \to 0$, giving

(limit)

$$c = q_{\max} = \left(\frac{2}{\gamma-1}\right)^{0\cdot5} a_0 \tag{2.13}$$

alternatively, from (2.10b), $\frac{T_0}{T} = 1+\frac{\gamma-1}{2}\frac{q^2}{(\gamma RT)}$

adiabatic, giving $\frac{T_0}{T} = 1+\frac{\gamma-1}{2}M^2$ (2.14)

isentropic, $\frac{p_0}{p} = \left(\frac{T_0}{T}\right)^{\frac{\gamma}{\gamma-1}} = \left(1+\frac{\gamma-1}{2}M^2\right)^{\frac{\gamma}{\gamma-1}}$ (2.15a)

isentropic, $\frac{\varrho_0}{\varrho} = \left(\frac{T_0}{T}\right)^{\frac{1}{\gamma-1}} = \left(1+\frac{\gamma-1}{2}M^2\right)^{\frac{1}{\gamma-1}}$ (2.15b)

for air ($\gamma = 7/5$), $\frac{T_*}{T_0} = 0\cdot8333$, $\frac{p_*}{p_0} = 0\cdot5283$ (2.17b)

for other gases: ($\gamma = 1$), $\frac{p_*}{p_0} = 0\cdot6065$;

($\gamma = 1\cdot67$), $\frac{p_*}{p_0} = 0\cdot4867$

from (2.13) and (2.17a), $c = a_*\left(\frac{\gamma+1}{\gamma-1}\right)^{0\cdot5}$ (2.18a)

from (2.16), $\frac{q}{a_*} = M\left[\frac{\gamma+1}{2+(\gamma-1)M^2}\right]^{0\cdot5}$ (2.18b)

Combining the above reference conditions, the energy equation may be written thus,

$$\frac{\gamma}{\gamma-1}\frac{p_0}{\varrho_0} = \frac{\gamma}{\gamma-1}\frac{p}{\varrho}+q^2 = a_0^2 = \frac{a^2}{\gamma-1}+\frac{q^2}{2}$$

$$= \frac{\gamma+1}{2(\gamma-1)}a^2 = \frac{c^2}{2} = \frac{\gamma+1}{\gamma-1}\frac{a_*^2}{2} \tag{2.19}$$

2.4 PRESSURE COEFFICIENT

It is usually necessary to examine flow past a stationary model under laboratory conditions so that information may be related quickly and economically to the movement of the full-size body (aircraft or aircraft wing, etc.) through atmospheric air.

(a) Under laboratory conditions, the flow past a model wing section may be considered as shown in Figure 2.4. For subsonic flow ($q_\infty < a$) the presence of the model affects the streamline positions upstream of the model. The suffix (∞) is used to denote conditions in the duct sufficiently far upstream for the streamlines to be considered parallel and for flow properties to be considered uniform across the duct. At the model, local values of velocity (q), pressure (p) etc., vary along the surface of the model and the energy equation may be applied assuming steady adiabatic flow:

$$\frac{\gamma}{\gamma-1}\frac{p_0}{\varrho_0} = \frac{\gamma}{\gamma-1}\frac{p_\infty}{\varrho_\infty}+\frac{q_\infty^2}{2}$$

$$= \frac{\gamma}{\gamma-1}\frac{p}{\varrho}+\frac{q^2}{2} \tag{2.20a}$$

(b) Figure 2.5 shows an aircraft (or wing) moving with mean speed (q) through static atmospheric air having pressure (p_a) and temperature (T_a). The local pressure and velocity distribution on the surface may again be described by (p) and (q) and the energy equation written accordingly,

$$\frac{\gamma}{\gamma-1}\frac{p_0}{\varrho_0} = \frac{\gamma}{\gamma-1}\frac{p_a}{\varrho_a}+\frac{\bar{q}^2}{2}$$

$$= \frac{\gamma}{\gamma-1}\frac{p}{\varrho}+\frac{q^2}{2} \tag{2.20b}$$

$$C_p = \frac{p-p_\infty}{\frac{1}{2}\varrho_\infty q_\infty^2} \tag{2.21}$$

For an incompressible fluid, $\frac{1}{2}\varrho q^2$ ('dynamic pressure') is equal to the difference between the total head and static pressures, i.e., $p_0 = p+\frac{1}{2}\varrho q^2 = p_\infty+\frac{1}{2}\varrho q_\infty^2$

in (2.21)
$$C_p = 1-\frac{q^2}{q_\infty^2} \tag{2.22}$$

For a *compressible* fluid, however, $\frac{1}{2}\varrho q^2$ is no longer equal to the difference between the total head and static pressures:

$$\left(a^2 = \frac{\gamma p}{\rho}\right)\frac{1}{2}\rho q^2 = \frac{1}{2}\left(\frac{\rho}{\gamma p}\right)\gamma p q^2 = \frac{\gamma}{2}pM^2$$

$$\therefore C_p = \frac{p-p_\infty}{\frac{\gamma}{2}p_\infty M_\infty^2} = \frac{2}{\gamma M_\infty^2}\left(\frac{p}{p_\infty}-1\right) \tag{2.23}$$

With reference to Figure 2.5, a stagnation pressure coefficient may be defined as, $C_{p_0} = \dfrac{p_0-p}{\frac{1}{2}\varrho q^2}$ (= 1 for an incompressible fluid) (2.24a)

or
$$C_{p_0} = \frac{2}{\gamma M^2}\left(\frac{p_0}{p}-1\right) \tag{2.24b}$$

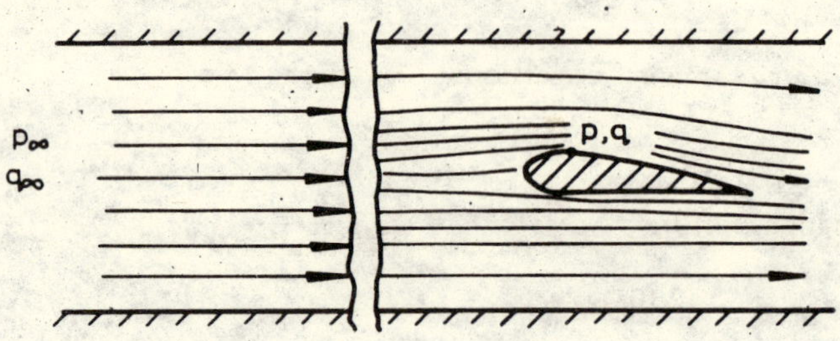

Fig. 2·4 flow past a model related to uniform upstream conditions

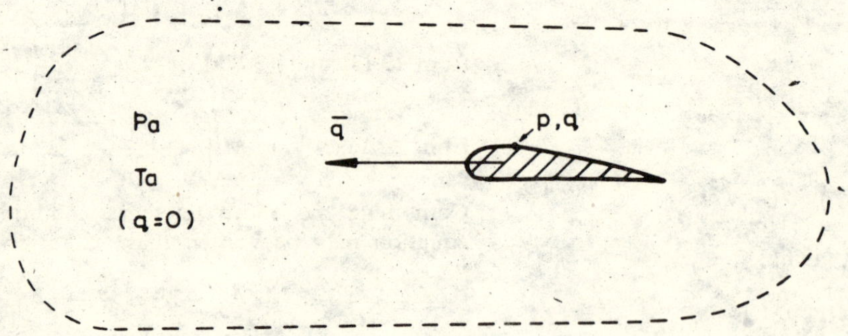

Fig. 2·5 aircraft wing moving through air at ambient conditions

(2.20a) and (2.20b) should be studied carefully for correct application and interpretation. *Static* ambient pressure and temperature conditions must not be confused with total head (stagnation) conditions.

The pressure coefficient, $C_p = \dfrac{\Delta p}{\frac{1}{2}\varrho q^2}$ is widely used in flow tests to relate the local pressures at different points on the model or body surfaces to some convenient reference condition. With reference to Figure 2.4,

from (2.15a), $\dfrac{p_0}{p} = (1+0{\cdot}2M^2)^{3\cdot5}$ for air ($\gamma = 7/5$)

binomial expansion, $\dfrac{p_0}{p} = 1+\dfrac{7}{10}M^2 +\dfrac{7}{40}M^4+\dfrac{7}{400}M^6+\dots$

in (2.24b)
$$C_{p_0} = 1+\frac{M^2}{4}+\frac{M^4}{40}+\frac{M^6}{1600}+\dots \tag{2.25}$$

comparing (2.24a) with (2.25) the error in assuming the flow to be incompressible is $\left(\frac{M^2}{4}+\frac{M^4}{40}+\frac{M^6}{1600}+\cdots\right)$

and values of C_{p_0} for M up to 1 are shown below:

$M =$	0	0·1	0·2	0·3	0·4	0·5
	0·6	0·7	0·8	0·9	1·0	
$C_{p_0} =$	1·000	1·003	1·010	1·023	1·041	1·064
	1·093	1·129	1·170	1·219	1·276	

Note: The pressure coefficient or Euler number corresponds to a ratio of pressure force to inertia force.

2.5 STEADY ISENTROPIC FLOW THROUGH A VARIABLE AREA DUCT

For a duct with gradual changes in area such that $\frac{\partial q}{\partial x} \gg \frac{\partial q}{\partial y}$ the flow may still be considered as one-dimensional and the basic fluid flow equations used to develop an expression for the speed of sound may be used, but with $dA \neq 0$. From Figure 2.6, (with dA considered to be small),

from (2.27) $\quad -a^2\,d\rho = \rho q\,dq$

or $$M^2\frac{dq}{q} = -\frac{d\rho}{\rho} \qquad (2.28)$$

from (2.26) $\therefore$ $$M^2\frac{dq}{q} = \frac{dq}{q}+\frac{dA}{A}$$

or $$(M^2-1)\frac{dq}{q} = \frac{dA}{A} \qquad (2.29)$$

When $M=1$, $\frac{dA}{A}=0$, and represents a minimum area condition, A_*.

When $M<1$, $\frac{dA}{dq}<0$, an increase (decrease) of velocity requires a decrease (increase) of area (c.f. incompressible flow).

When $M>1$, $\frac{dA}{dq}>0$, and increase (decrease) of velocity requires an increase (decrease) of area.

At supersonic speeds, the cross-sectional area of the duct must *increase* in the direction of motion to *increase* the

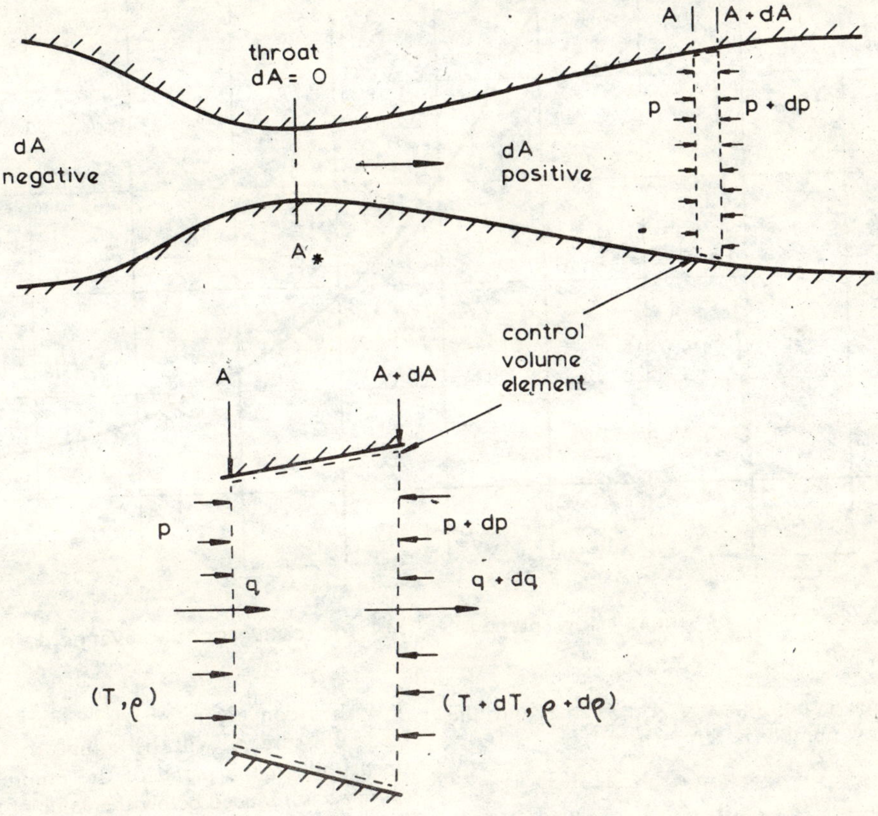

Fig. 2·6

continuity, $\quad d(\rho q A) = 0$

or $$\frac{d\rho}{\rho}+\frac{dq}{q}+\frac{dA}{A} = 0 \qquad (2.26)$$

(from (1.6)) *momentum,* $$-dp\,A = (A\rho q)\,dq \qquad (2.27)$$

isentropic, $$\frac{p}{\rho^\gamma} = k, \quad a^2 = \frac{dp}{d\rho}$$

velocity for steady mass flow; this is due to the more rapid *decrease* in density:

from (2.29), $$\frac{dq}{dA} \propto \frac{1}{(M^2-1)}$$

and from (2.28), $$\frac{d\rho}{dA} \propto \frac{-M^2}{(M^2-1)}$$

The mass flow at any section of the duct may be expressed as the mass flow per unit area,

$$\frac{m}{A} = \varrho q \quad \text{or} \quad \frac{m}{A}\left(\frac{T_0^{0\cdot5}}{p_0}\right) = \frac{p}{RT}\left(\frac{T_0^{0\cdot5}}{p_0}\right)\left(\frac{q}{a}\right)(\gamma RT)^{0\cdot5}$$

by using further substitutions (2.14), (2.15a), an expression for the mass flow number (MN) may be obtained, Figure 2.7,

$$MN = \frac{ma_0}{Ap_0} = \gamma M\left(1+\frac{\gamma-1}{2}M^2\right)^{\frac{\gamma+1}{2(\gamma-1)}} \quad (2.30)$$

of A, or $\frac{A}{A_*} > 1$, there are two values of M, one of which is subsonic $(M < 1)$ and the other supersonic $(M > 1)$. Table 1.2 presents solutions of (2.32) for air $(\gamma = 1{\cdot}4)$.

2.6 CONVERGENT NOZZLE

The convergent and convergent-divergent nozzles are examples of variable area ducts in which the steady adiabatic flow of a compressible fluid is governed by the equations established in the previous section.

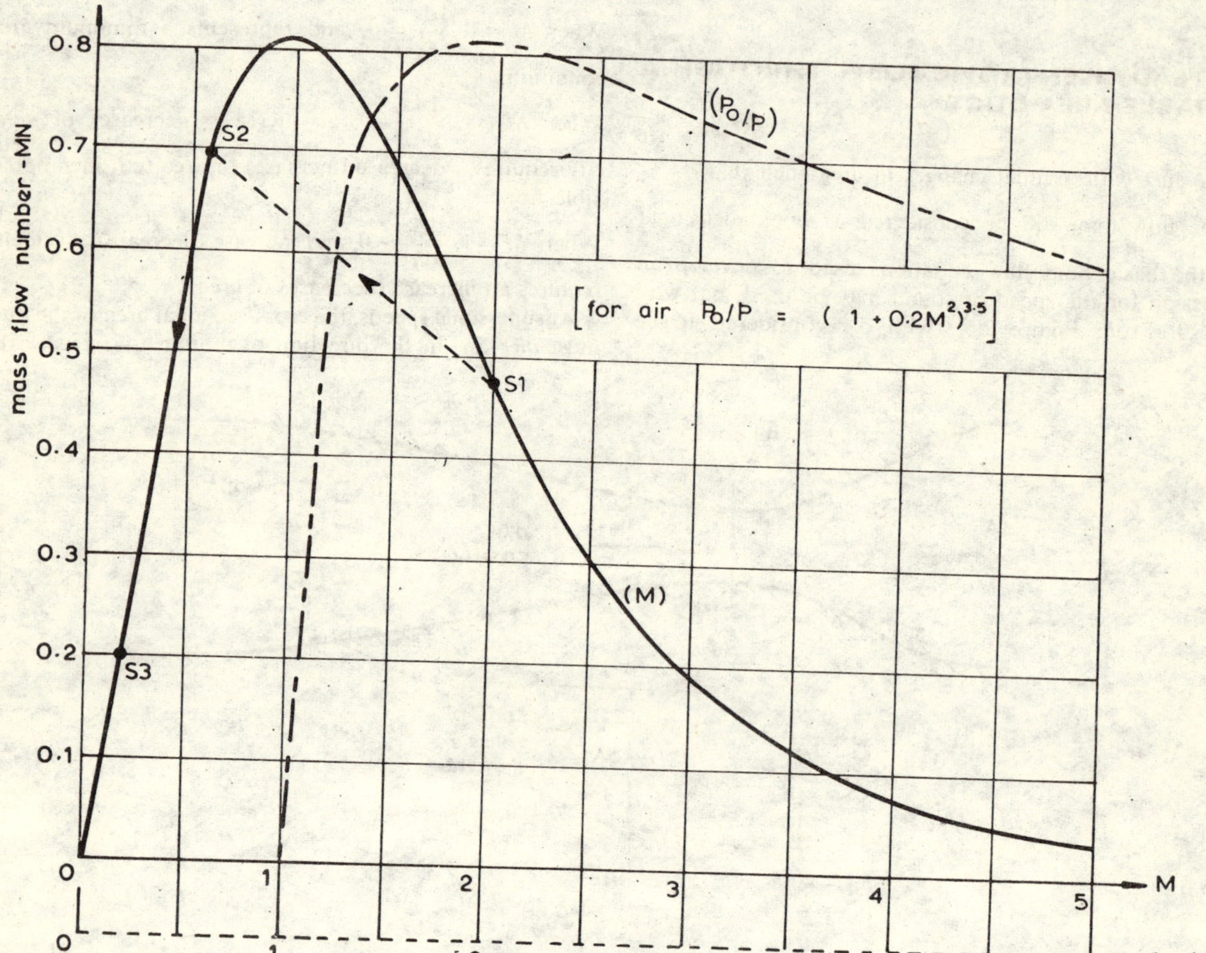

Fig. 2.7 variation of mass flow number - MN for convergent - divergent nozzle

For minimum area condition $A = A_*$, $M = 1$ and the maximum mass flow number is MN_* where,

$$MN_* = \frac{ma_0}{A_* p_0} = \gamma\left(\frac{2}{\gamma+1}\right)^{\frac{\gamma+1}{2(\gamma-1)}} \quad (2.31)$$

In general, for a given Mach number, the mass flow per unit area is a function of the total head conditions p_0 and $T_0(a_0)$.
From (2.30) and (2.31)

$$\frac{A}{A_*} = \frac{1}{M}\left[\frac{2}{\gamma+1}\left(1+\frac{\gamma-1}{2}M^2\right)\right]^{\frac{\gamma+1}{2(\gamma-1)}} \quad (2.32)$$

(See also EXAMPLE 2.7)

Inspection of this equation indicates that for any value

With a convergent nozzle, Figure 2.8, the ambient or back pressure p_b will be equal to the exit pressure p_e (measured in the plane of the minimum area or throat) until p_b is reduced below the value of the critical pressure, p_*. p_* is the pressure to which the flow expands from a 'reservoir', p_0 condition to produce a sonic condition $M = 1$.

Figure 2.8 also shows the variation of mass flow (and mass flow number) as $p_b(=p_e)$ is reduced from the '*no flow*' condition, $p_b = p_0$. The mass flow remains constant for further reduction of $p_b < p_*$. When $p_e = p_*$ the sonic condition $(M = 1)$ in the exit plane prevents the flow responding to the apparent increase in pressure drop across the nozzle due to reduction in p_b which has effect only outside the nozzle. There can be no further increase of mass flow and the flow is said to be choked. As before (2.17b) $p_*/p_0 = 0{\cdot}528$ for air $(\gamma = 1{\cdot}4)$.

When $p_e = p_*$ the emerging jet is parallel. For $p_b < p_*$, a Mach line pattern develops as the flow expands after leaving the nozzle (see Figure 2.10c). Such effects will be discussed in Section 4 dealing with Prandtl–Meyer processes.

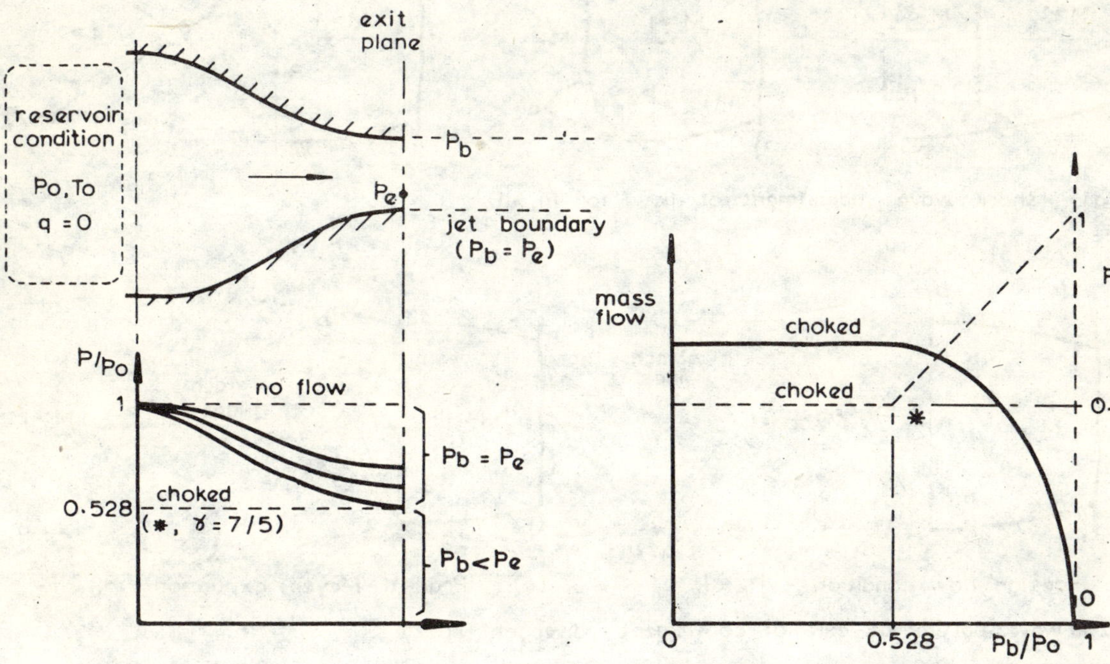

Fig. 2.8 some flow variables for the convergent nozzle

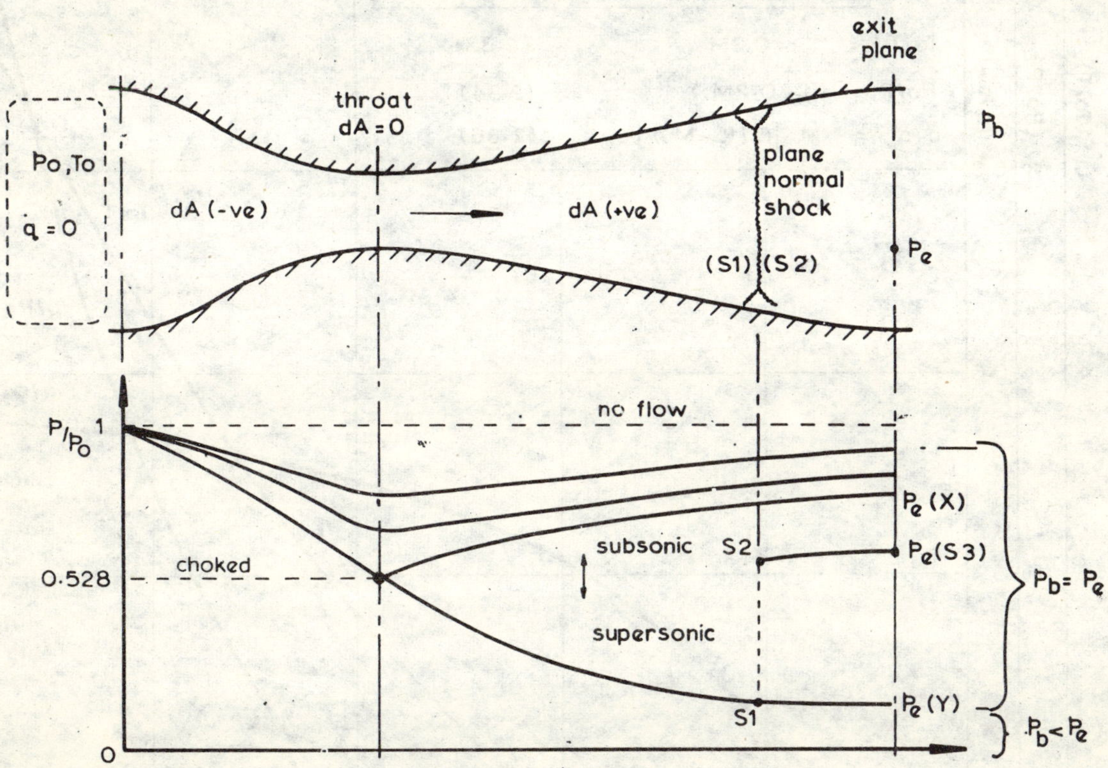

Fig. 2·9 pressure distribution for compressible flow through a convergent - divergent nozzle

2.7 CONVERGENT-DIVERGENT NOZZLE

Figure 2.9 shows a typical profile of a convergent-divergent nozzle designed to expand a compressible fluid to supersonic speeds.

Equation (2.29), $(M^2-1)\,dq/q = dA/A$, indicates that for $M > 1$ the flow speed will increase with increase of nozzle area after sonic conditions have been established at the throat. This increase in speed requires that the pressure decreases, (2.27), along the length of the nozzle and will therefore be determined by the back pressure (p_b) as with the operation of the convergent nozzle.

Figure 2.9 shows some different operating conditions of the convergent-divergent nozzle for different values of p_b. As p_b is reduced from the '*no flow*' condition $p_b = p_0$,

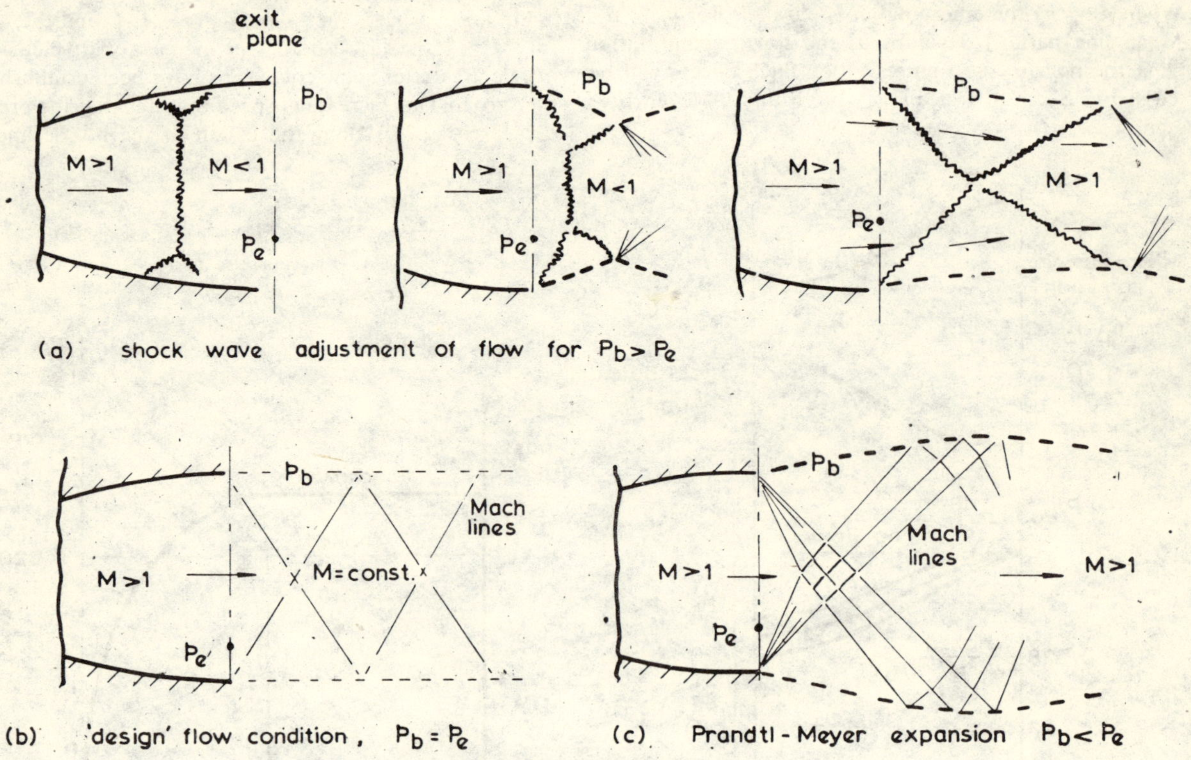

Fig. 2·10 wave pattern at exit of convergent - divergent nozzle

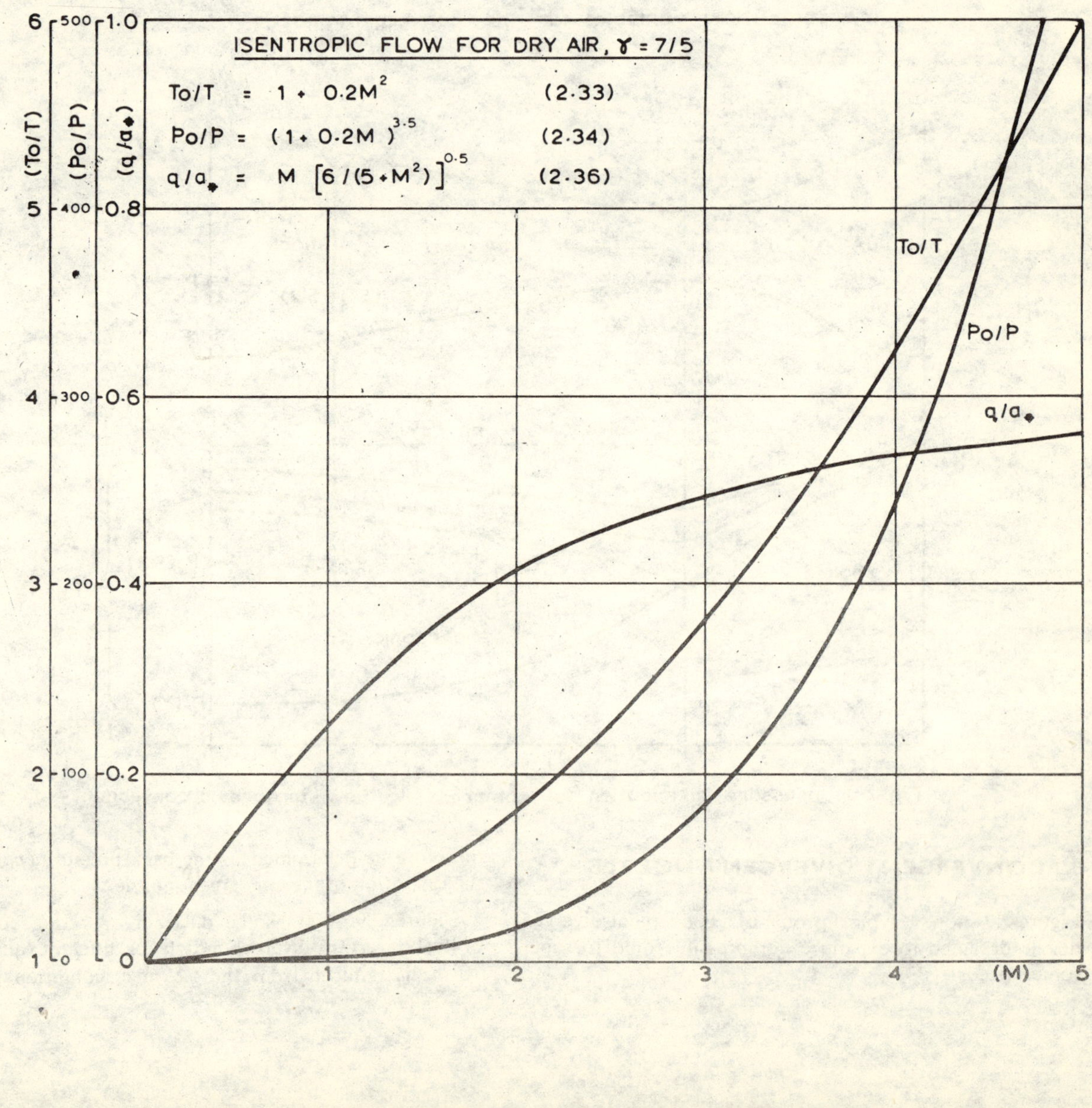

Fig. 2·11

$p_b = p_e$ until the critical condition $p_* \rightarrow M = 1$, is reached at the throat.

Until this condition is reached the flow accelerates towards the throat (in (2.29) for $M < 1$ if dA is $-ve$, dq is $+ve$) where the flow is still subsonic and the divergent portion then acts as a diffuser with increase of pressure and reduction of flow speed. At the critical throat condition, the nozzle is choked and there are two values for the exit pressure (p_e) corresponding to the two values of Mach number (2.15a) for a given area ratio A/A_*, (2.32).

The higher value of exit pressure $p_e(X)$ corresponds to subsonic isentropic flow in the divergent portion of the nozzle while the lower value $p_e(Y)$ corresponds to supersonic flow. As with the convergent nozzle, reduction of $p_b < p_e(X)$ has no effect on the mass flow due to the sonic condition divergence to a subsonic isentropic compression compatible with the exit pressure which is greater than the value $p_e(Y)$ required for full supersonic expansion (shown as the required 'design' flow condition in Figure 2.10b).

For an exit pressure $p_e(S3)$ the Mach number will increase from $M = 1$, at the throat, until a section is reached at which a sudden change, $S1 \rightarrow S2$, would permit an isentropic subsonic flow, $S2 \rightarrow S3$, for the subsequent divergence of the nozzle. Figure 2.7 shows the variation of mass flow number ($\dot{M}N$) with pressure ratio and with Mach number. $S1 \rightarrow S2$ is shown as a dashed line since the shock wave represents a discontinuous or non-equilibrium change in all thermodynamic and flow properties. The flow through the normal shock wave is adiabatic but no longer reversible and $p_{01} \neq p_{02}$.

Fig. 2·12

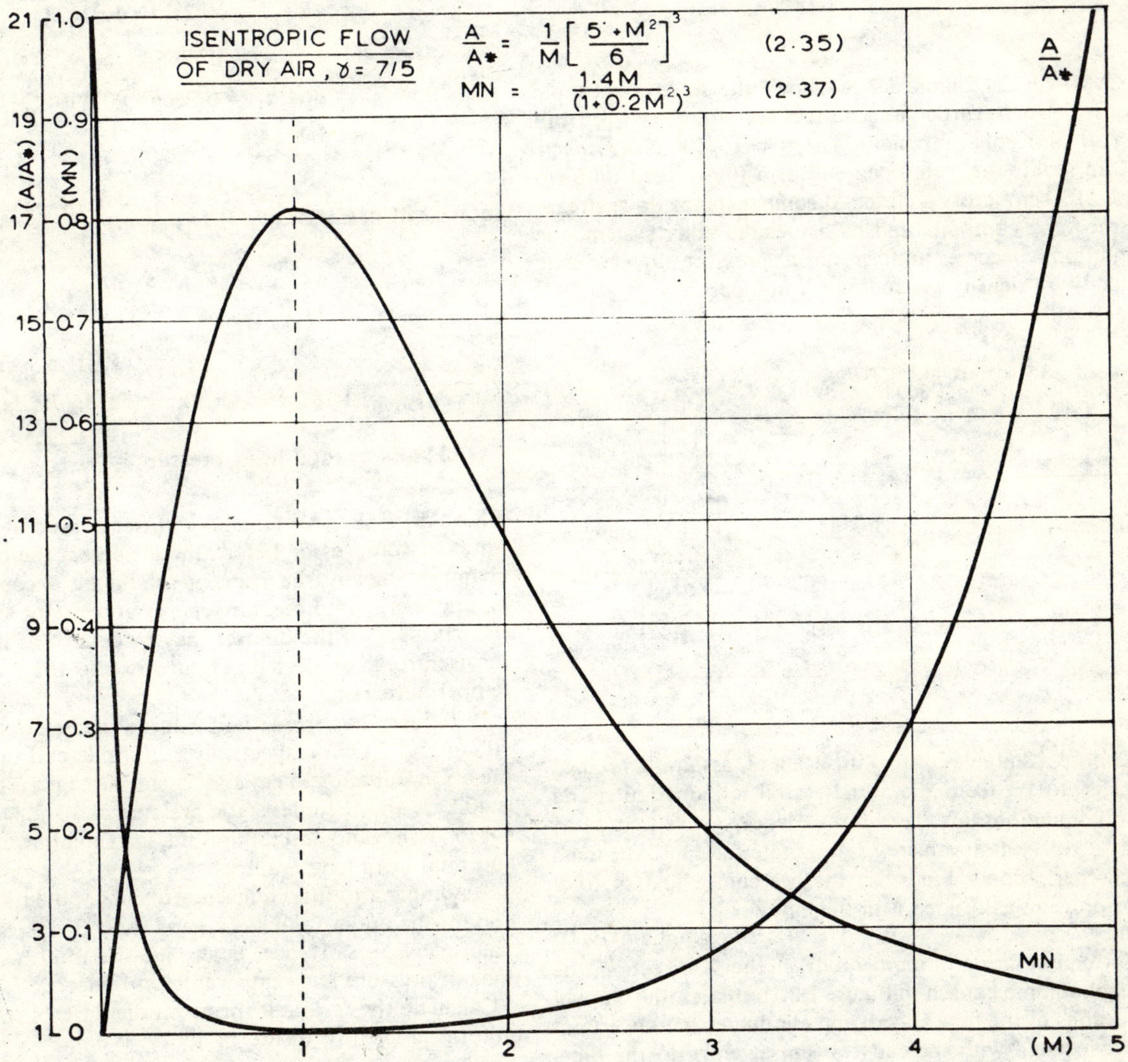

at the throat preventing an *effective* increase in pressure drop across the convergent portion. Reduction of $p_b < p_e(Y)$ for which $p_b \neq p_e$ creates an expansion outside the nozzle, Figure 2.10c.

For values of $p_b = p_e$ when $p_e(X) > p_e > p_e(Y)$ the flow is complicated by the occurrence of an irreversible pressure discontinuity or shock wave which is stationary. The shock wave is a self-adjusting mechanism of the flow for correction of an initially supersonic isentropic expansion in the

The shock wave is considered to lie in a plane normal to the flow and is analysed in Chapter 4. The position of the shock wave (inside or outside the nozzle, Figure 2.10a), depends on the area ratio, A_e/A_*, the total head pressure, p_0 and the back pressure, p_b. Real gas effects cause boundary layer growth with shock wave interaction giving rise to an oblique shock wave mechanism (analysed in Chapter 7) of flow deflection which is shown in the wave diagram in Figure 2.10a.

2.8 SUMMARY OF EQUATIONS

The following are the equations used in Table 1.2 for steady isentropic flow of dry air ($\gamma = 7/5$) and are plotted in Figures 2.11 and 2.12.

from (2.14), $$\frac{T_0}{T} = 1+0{\cdot}2M^2 \qquad (2.33)$$

from (2.15a), $$\frac{p_0}{p} = (1+0{\cdot}2M^2)^{3{\cdot}5} \qquad (2.34)$$

from (2.32), $$\frac{A}{A_*} = \frac{1}{M}\left[\frac{5+M^2}{6}\right]^3 \qquad (2.35)$$

from (2.18b), $$\frac{q}{a_*} = M\left[\frac{6}{5+M^2}\right]^{0{\cdot}5} \qquad (2.36)$$

from (2.30), $$MN = \frac{1{\cdot}4M}{(1+0{\cdot}2M^2)^3} \qquad (2.37)$$

EXAMPLE 2.1 Figure E2.1a shows the steady subsonic inflow of air (regarded as a perfect gas with $\gamma = 1{\cdot}4$) from rest at atmospheric conditions ($p_0 = 1\times10^5$ N/m², $a_0 = 340$ m/s) into the plain open end of a thin-walled uniform pipe of 90 cm² cross section. At some distance downstream of the vena contracta and its associated eddy system, the conditions become uniform at a pressure p_1, acoustic velocity a_1, density ϱ_1, and Mach number M_1.

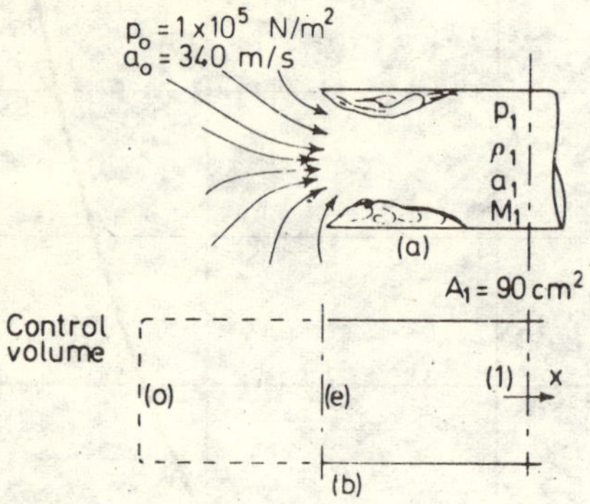

Fig. E2.1

Choose a suitable control surface and, assuming no heat transfer to or from the wall, establish momentum and energy equations for the inflow. Hence, given that $p_1 = 0{\cdot}74 \times 10^5$ N/m², evaluate M_1, a_1 and ρ_1 and find the mass inflow per second. Find also the percentage loss of total head pressure associated with the inflow.

(U. Birmingham 1960)

For conservation of momentum in the X-direction, the control volume chosen in Figure E2.1b relates the known conditions of the flow system and eliminates the uncertainties associated with the eddy system at entry to the pipe.

continuity, mass flow $= \varrho_1 A_1 q_1$

momentum in $= 0$, momentum out $= (\varrho_1 A_1 q_1)q_1$ ($\xrightarrow{+} x$)

forces causing change in momentum $= p_0A_1 - p_1A_1$ ($\xrightarrow{+} x$)

$\therefore p_0 - p_1 = \varrho_1 q_1^2$

or, $p_0/p_1 = 1+(\varrho_1 q_1^2)/p_1 = 1+\gamma M_1^2$ $\qquad (a^2 = \gamma p/\varrho)$

$\therefore M_1^2 = (p_0/p_1 - 1)/\gamma = (1/0{\cdot}74 - 1)/1{\cdot}4 = 0{\cdot}2510$

$\therefore \underline{M_1 = 0{\cdot}501}$

conservation of energy gives $c_pT_0 = c_pT_1 + q_1^2/2$ $\qquad$ (2.10c)

whence, since total head temperature is constant in an adiabatic flow system, $T_0 = T_{01}$,

$$\frac{T_0}{T_1} = 1+\frac{\gamma-1}{2}M^2 = 1+0{\cdot}2\times0{\cdot}251 \qquad (2.14)$$

$\therefore T_0/T_1 = 1{\cdot}0502$ (also from Table 1, for $M_1 = 0{\cdot}501$)

$a_0 = (\gamma RT_0)^{0{\cdot}5} = 20(T_0)^{0{\cdot}5}$ $\qquad$ (2.6)

$\therefore T_0 = (340/20)^2 = 289$ K

$\therefore T_1 = 289/1{\cdot}0502 = \underline{275{\cdot}2 \text{ K}}$

$a_1 = 20(T_1)^{0{\cdot}5}$ or $a_1/a_0 = (T_1/T_0)^{0{\cdot}5}$

$\therefore a_1 = 340/(1{\cdot}0502)^{0{\cdot}5} = \underline{331{\cdot}8 \text{ m/s}}$

$q_1 = 0{\cdot}501\times331{\cdot}8 = \underline{166{\cdot}2 \text{ m/s}}$

density, $$\varrho_1 = \frac{p_1}{RT_1} = \frac{0{\cdot}74\times10^5}{0{\cdot}2871\times10^3\times275{\cdot}2} =$$

$= 0{\cdot}9366$ kg/m³ $\qquad$ (Note 10^3!)

mass flow $= \varrho_1 A_1 q_1 = 0{\cdot}9366\times90\times10^{-4}\times166{\cdot}2$

$= \underline{1{\cdot}401 \text{ kg/s}}$

total head pressure at (1), $$p_{01} = p_1\left(\frac{\gamma-1}{2}M_1^2\right)^{\frac{\gamma}{\gamma-1}} \qquad (2.15a)$$

or, $p_{01}/p_1 = (T_0/T_1)^{3{\cdot}5} = 1{\cdot}187$ (also from Table 1, for $M_1 = 0{\cdot}501$)

$\therefore p_{01} = 1{\cdot}187\times0{\cdot}74\times10^5 = 0{\cdot}878\times10^5$ N/m²

$$\therefore \% \text{ loss of total head pressure} = \frac{1-0{\cdot}878}{1} = \underline{12{\cdot}2\%}$$

EXAMPLE 2.2 An annular diffuser is used at the exit of an axial compressor before the air enters the combustion chamber. Ideally the process should be a conversion of kinetic energy into pressure energy at constant entropy and the efficiency of the diffuser may be defined as the ratio of the isentropic static temperature rise to the actual static temperature rise.

A diffuser has inside and outside diameters at inlet of 40 cm and 50 cm respectively and inside and outside diameters at outlet of 35 cm and 55 cm. If the mass flow of air is 40 kg/s and the inlet static pressure and temperature are 4×10^5 N/m² and 450 K, find the inlet total head pressure and total head temperature.

Assuming a diffuser efficiency of 80%, obtain an expression for the outlet velocity (q_2) and show that q_2 is approximately 86·3 m/s. Hence, determine the total head conditions of pressure and temperature at outlet.

Calculate the resultant thrust on the diffuser assuming a uniform ambient pressure of 1×10^5 N/m².

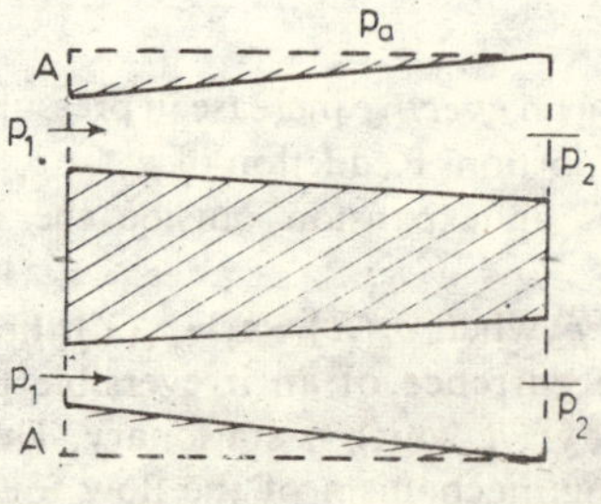

Fig. E2.2

Consider the control volume shown in Figure E2.2 for flow through the diffuser.

$$A_1 = \frac{\pi}{4}(D_1^2 - d_1^2) = \frac{\pi}{4}\times 10\times 90\times 10^{-4} = 0{\cdot}0707\ \text{m}^2$$

$$A_2 = 2A_1$$

continuity, mass flow, $M = \varrho_1 A_1 q_1 = \varrho_2 A_2 q_2$

$$\therefore \varrho_2/\varrho_1 = q_1/(2q_2) \qquad \text{(A)}$$

$(p = \varrho RT)$ at inlet, $\varrho_1 = (4\times 10^5)/(0{\cdot}2871\times 10^3\times 450)$

$$= 3{\cdot}096\ \text{kg/m}^3$$

$$q_1 = 40/(0{\cdot}0707\times 3{\cdot}096) = \underline{182{\cdot}7\ \text{m/s}}$$

incompressible, $p_{01} = p_1 + \varrho q_1^2/2 = 4{\cdot}517\times 10^5\ \text{N/m}^2$

N.B., $a_1 = 20(450)^{0{\cdot}5} = 424{\cdot}3$ m/s

$$\therefore M_1 = q_1/a_1 = 182{\cdot}7/424{\cdot}3 = 0{\cdot}43$$

$\therefore p_{01}/p_1 = 1{\cdot}1355$ and $T_{01}/T_1 = 1{\cdot}0370$ (Table 1)

alternatively, $p_{01}/p_1 = \left(1+\frac{\gamma-1}{2}M_1^2\right)^{\gamma/\gamma-1}$

$$= (1+0{\cdot}2\times 0{\cdot}43^2)^{3{\cdot}5}$$

whence, $p_{01} = 1{\cdot}1355\times 4\times 10^5 = \underline{4{\cdot}542\times 10^5\ \text{N/m}^2}$

(the error in assuming the flow to be incompressible increases as the Mach number increases)

$$T_{01} = 1{\cdot}0370\times 450 = \underline{466{\cdot}6\ \text{K}}$$

isentropic temperature ratio, $T_{2s}/T_1 = (\varrho_2/\varrho_1)^{\gamma-1}$

from (A), $$\frac{T_{2s}}{T_1} = \left(\frac{q_1}{2q_2}\right)^{0{\cdot}4} \qquad \text{(B)}$$

isentropic efficiency, $0{\cdot}8 = \dfrac{T_{2s}-T_1}{T_2-T_1}$

substituting for T_{2s} from (2) gives

$$T_2 = 562{\cdot}5\left(\frac{q_1}{2q_2}\right)^{0{\cdot}4} - 112{\cdot}5 \qquad \text{(C)}$$

energy, $T_{01} = T_{02} = T_2 + q_2^2/(2c_p)$

or, $q_2^2 = 2\times 1{\cdot}0049\times 10^3\left[466{\cdot}6 - 562{\cdot}5\left(\frac{q_1}{2q_2}\right)^{0{\cdot}4} + 112{\cdot}5\right]$

whence, by trial and error, $\underline{q_2 = 86{\cdot}3\ \text{m/s}}$ (approx.)

in (C), $T_2 = 562{\cdot}5\left(\frac{182{\cdot}7}{174}\right)^{0{\cdot}4} - 112{\cdot}5 = \underline{461{\cdot}1\ \text{K}}$

in (A), $\varrho_2 = 3{\cdot}096\times 182{\cdot}7/(2\times 86{\cdot}3) = 3{\cdot}277\ \text{kg/m}^3$

$(p = \varrho RT)$ at outlet, $p_2 = 3{\cdot}277\times 0{\cdot}2871\times 10^3\times 461{\cdot}1$

$$= 4{\cdot}338\times 10^5\ \text{N/m}^2$$

$T_{02}/T_2 = 1{\cdot}0119 \quad \therefore p_{02}/p_2 = (T_{02}/T_2)^{\gamma/\gamma-1} = (1{\cdot}0119)^{3{\cdot}5}$

$$= 1{\cdot}0423$$

(also, by interpolation in Table 1, whence $M_2 = 0{\cdot}244$)

$$p_{02} = 1{\cdot}0423\times 4{\cdot}338\times 10^5 = \underline{4{\cdot}522\times 10^5\ \text{N/m}^2}$$

Consider conservation of momentum for the control volume shown between stations (1) and (2), which includes the force, F_x, exerted by the solid boundaries of the diffuser on the fluid in the control volume (X-direction positive)

net change in momentum = pressure forces$+F_x$

$$m(q_2-q_1) = p_1A_1 + p_aA - p_2A_2 + F_x$$

$$\therefore 40(86{\cdot}3-182{\cdot}7) = 10^5(4\times 0{\cdot}0707 + 1\times\frac{\pi}{4}\times 5\times 105\times 10^{-4}$$
$$-4{\cdot}338\times 0{\cdot}1414)$$

$$\therefore F_x = (-0{\cdot}3856 - 2{\cdot}828 - 0{\cdot}412 + 6{\cdot}134)\times 10^4 = 25\,080\ \text{N}$$

$\therefore$ Net thrust *on* diffuser is $\underline{-25\,080\ \text{N} \leftarrow}$

EXAMPLE 2.3 A steady subsonic flow of air occurs in a uniform pipe of cross-sectional area A_1, the pressure, density, velocity and Mach number being respectively p_1, ϱ_1, q_1, M_1. A sudden enlargement occurs to a cross-sectional area $A_2 = KA_1$, after which the pipe is again uniform and is coaxial with the narrower portion.

If, at a station sufficiently far downstream of the enlargement for the flow to have become uniform across the pipe, the pressure, density, velocity and Mach numbers are p_2, ϱ_2, q_2 and M_2, show that $(\gamma = 7/5)$: –

$$p_1 - p_2 = q_2^2\varrho_2 - q_1^2\varrho_1/K$$

$$\frac{p_1}{p_2} = K\frac{(5+7M_2^2)}{(5K+7M_1^2)}$$

$$\frac{M_2^2(5+M_2^2)}{(5+7M_2^2)^2} = \frac{M_1^2(5+M_1^2)}{(5K+7M_1^2)^2}$$

The pressure on the transverse annular wall of the enlargement may be assumed equal to that in the narrower pipe and the flow is considered to be frictionless and adiabatic.

If $p_1 = 2{\cdot}7\times 10^5\ \text{N/m}^2$, $a_1 = 300$ m/s, $q_1 = 250$ m/s and $K = 2{\cdot}0$ show that:

$$401\,250 = 525q_2 + p_2$$
$$512\,500 = q_2^2 + p_2q_2/75$$

Determine q_2 and p_2 and hence ϱ_2, T_2. By comparing the temperature ratio (T_2/T_1) with that due to isentropic compression through the same pressure ratio, show that a gain in entropy occurs.

(U. Birmingham 1964)

Consider the control volume shown, Figure E 2.3

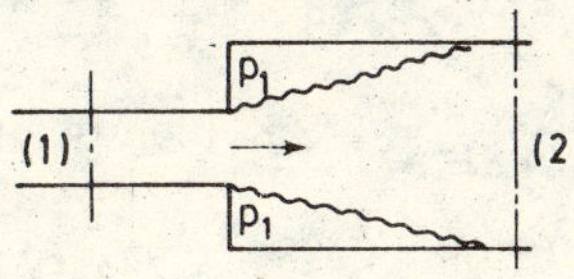

Fig. E2.3

continuity, mass flow, $m = \varrho_1 q_1 A_1 = \varrho_2 q_2 A_2$

$$\therefore \varrho_1 q_1 = K\varrho_2 q_2 \qquad \text{(A)}$$

momentum, $\varrho qA(q_2-q_1) = p_1A_1 - p_2A_2 + p_1(A_2-A_1)$

or, $\varrho_2 q_2^2 A_2 - \varrho_1 q_1^2 A_1 = (p_1-p_2)A_2$

whence, $$\underline{p_1 - p_2 = q_2^2\varrho_2 - q_1^2\varrho_1/K} \qquad \text{(B)}$$

energy, $q_1^2 + 5a_1^2 = q_2^2 + 5a_2^2$

or, $$(M_1^2+5)a_1^2 = (M_2^2+5)a_2^2 \qquad \text{(C)}$$

Now $a^2 = \gamma RT = \gamma p/\varrho$

$$\therefore \frac{a_2^2}{a_1^2} = \frac{T_2}{T_1} = \frac{p_2}{p_1}\frac{\varrho_1}{\varrho_2} = \frac{p_2}{p_1}K\frac{q_2}{q_1} \quad \text{from (A)}$$

or, $$\frac{M_1^2}{M_2^2} = K\frac{p_2}{p_1}\frac{q_1}{q_2} \quad \text{(D)}$$

from (B), $$\frac{p_1}{p_2} - 1 = \gamma M_2^2 - \frac{\varrho_2 q_1 q_2}{p_2} = \gamma M_2^2 - \frac{q_1}{q_2}\gamma M_2^2$$

from (D), $$\frac{p_1}{p_2} - 1 = \gamma M_2^2 - \frac{M_1^2}{M_2^2}\frac{1}{K}\frac{p_1}{p_2}\gamma M_2^2$$

$(\gamma = 7/5)$, or $$\frac{p_1}{p_2} = \frac{(1+\gamma M_2^2)}{(1+\gamma M_1^2/K)} = K\left[\frac{5+7M_2^2}{5K+7M_1^2}\right] \quad \text{(E)}$$

from (D) $$K\frac{p_2}{p_1}\frac{q_1}{q_2} = \frac{M_1^2}{M_2^2} \quad \text{(F)}$$

from (C) $$K\frac{p_2}{p_1}\frac{q_2}{q_1} = \frac{a_2^2}{a_1^2} = \frac{M_1^2+5}{M_2^2+5} \quad \text{(G)}$$

multiply (F)×(G), $$\left(\frac{p_2}{p_1}\right)^2 = \frac{1}{K^2}\frac{M_1^2}{M_2^2}\left[\frac{M_1^2+5}{M_2^2+5}\right]$$

equating with (E) gives,

$$\frac{M_2^2(5+M_2^2)^2}{(5+7M_2^2)^2} = \frac{M_1^2(5+M_1^2)}{(5K+7M_1^2)^2}$$

from (A), $$\varrho_1 q_1 = 2\varrho_2 q_2 \quad \text{(H)}$$

$(a^2 = \gamma p/\varrho)$, $$\therefore \varrho_1 = \frac{1.4\times2.7\times10^5}{9\times10^4} = 4.3 \text{ kg/m}^3$$

from (B), $$p_1 - p_2 = \varrho_1 q_1(q_2 - q_1)/2$$
$$2.7\times10^5 - p_2 = 4.2\times250(q_2-250)/2$$
$$\therefore 401\,250 = 525q_2 + p_2 \quad \text{(I)}$$

energy $$q_1^2+5a_1^2 = q_2^2+5a_2^2 = q_2^2+5\gamma p_2/\varrho_2$$

from (H), $$250^2+5\times300^2 = q_2^2+5\times1.4\times p_2\times2q_2/(4.2\times250)$$

or, $$512\,500 = q_2^2+p_2q_2/75 \quad \text{(J)}$$

from (H), $$6q_2^2-5350q_2+512\,500 = 0$$

whence, $q_2 = 109.2$ m/s and $p_2 = 3.439\times10^5$ N/m²

from (J), $$\varrho_2 = \frac{4.2\times250}{109.2\times2} = 4.81 \text{ kg/m}^3$$

$$\therefore T_2 = \frac{3.439\times10^5}{0.2871\times10^3\times4.81} = 249 \text{ K (non-isentropic)}$$

$(a^2 = \gamma RT)$, $T_1 = (300/20)^2 = 225$ K $\therefore T_2/T_1 = 1.1067$

isentropic, $$\left(\frac{T_2}{T_1}\right)_{is} = \left(\frac{p_2}{p_1}\right)^{\frac{\gamma-1}{\gamma}} = \left(\frac{3.439}{2.7}\right)^{1/3.5} = 1.0716$$

$\therefore \left(\frac{T_2}{T_1}\right)_{is} < \frac{T_2}{T_1}$ and there will be an increase in entropy (see 1.20a)

EXAMPLE 2.4 Write down the steady flow adiabatic energy equations for compressible and incompressible fluids in terms of static pressure (p), total head pressure (p_0), density (ϱ), velocity (q) and γ the ratio of specific heats. Assuming no change in potential energy and ideal fluids, show that, respectively:

$$q = \left[\frac{2\gamma}{\gamma-1}\frac{p}{\varrho}\left(\left[\frac{p_0}{p}\right]^{\frac{\gamma-1}{\gamma}}-1\right)\right]^{0.5} \quad \text{(A)}$$

$$q = \left[\frac{2}{\varrho}(p_0-p)\right]^{0.5} \quad \text{(B)}$$

For a compressible fluid, by writing $p_0/p = 1+x$, where x is a small quantity, use the Binomial Theorem to show that:

$$\left[\left(\frac{p_0}{p}\right)^{\frac{\gamma-1}{\gamma}}-1\right]^{0.5} = \left(\frac{\gamma-1}{\gamma}x\right)^{0.5}\left[1-\frac{1}{4}\frac{x}{\gamma}+\frac{(8\gamma+5)}{96}\left(\frac{x}{\gamma}\right)^2+\dots\right]$$

Use this result in (A) and compare with (B) to show that the *first order* error in neglecting compressibility is

$$+\frac{100}{4\gamma}\left(\frac{p_0-p}{p}\right)\%$$

For air ($\gamma = 1.4$) flowing at 256 m/s in a duct in which the static temperature is -17°C, calculate the % error in velocity measurement by Pitot-static tube (p_0 and p) if the fluid is assumed to be incompressible. What is the % error for one quarter of the Mach number?

(U. Leeds 1971)

energy, compressible, $$\frac{\gamma}{\gamma-1}\frac{p_0}{\varrho_0} = \frac{\gamma}{\gamma-1}\frac{p}{\varrho}+\frac{q^2}{2} \quad \text{(2.10a)}$$

$$\therefore q = \left[\frac{2\gamma}{\gamma-1}\frac{p}{\varrho}\left(\frac{p_0}{p}\frac{\varrho}{\varrho_0}-1\right)\right]^{0.5}$$

isentropic, $$q = \left[\frac{2\gamma}{\gamma-1}\left(\frac{p}{\varrho}\right)\left(\frac{p_0}{p}\right)^{\frac{\gamma-1}{\gamma}}-1\right]^{0.5} \quad \text{(A)}$$

energy, incompressible, $$p_0/\varrho = p/\varrho+q^2/2 \quad \text{(2.9)}$$

$$\therefore q = \left[\frac{2}{\varrho}(p_0-p)\right]^{0.5} \quad \text{(B)}$$

putting $p_0/p = 1+x$, $(p_0/p)^{\frac{\gamma-1}{\gamma}} = (1+x)^{\frac{\gamma-1}{\gamma}}$

which gives, for $x = (p_0-p)/p$ a small quantity, using the Binomial Theorem,

$$(p_0/p)^{\frac{\gamma-1}{\gamma}} = 1+\frac{\gamma-1}{\gamma}x+\frac{\gamma-1}{2\gamma}\left(-\frac{1}{\gamma}\right)x^2+ +\frac{\gamma-1}{6\gamma}\left(-\frac{1}{\gamma}\right)\left(-\frac{\gamma+1}{\gamma}\right)x^3+\dots$$

$$\therefore (p_0/p)^{\frac{\gamma-1}{\gamma}}-1 = \frac{\gamma-1}{\gamma}x\left[1-\left(\frac{1}{2}\frac{x}{\gamma}-\frac{\gamma+1}{6}\left(\frac{x}{\gamma}\right)^2+\dots\right)\right]$$

again, using the Binomial Theorem,

$$\left[(p_0/p)^{\frac{\gamma-1}{\gamma}}-1\right]^{0.5} = \left(\frac{\gamma-1}{\gamma}x\right)^{0.5}\left[1-\left(\frac{1}{2}\frac{x}{\gamma}-\frac{\gamma+1}{6}\left(\frac{x}{\gamma}\right)^2\right)\right]^{0.5}$$

$$= \left(\frac{\gamma-1}{\gamma}x\right)^{0.5}\left[1-\frac{1}{2}\left(\frac{1}{2}\frac{x}{\gamma}-\frac{\gamma+1}{6}\left(\frac{x^2}{2}\right)\right)\right.$$

$$\left.+\frac{1}{2}\cdot\frac{1}{2}\left(-\frac{1}{2}\right)\left(\frac{1}{2}\frac{x}{\gamma}-\frac{\gamma+1}{6}\left(\frac{x^2}{2}\right)\right)^2+\ldots\right]$$

(neglecting higher orders)

$$= \left(\frac{\gamma-1}{\gamma}x\right)^{0.5}\left[1-\frac{1}{4}\frac{x}{\gamma}+\frac{\gamma+1}{12}\left(\frac{x}{\gamma}\right)^2-\frac{1}{32}\left(\frac{x}{\gamma}\right)^2+\ldots\right]$$

$$\therefore \left[(p_0/p)^{\frac{\gamma-1}{\gamma}}-1\right]^{0.5}$$

$$= \left(\frac{\gamma-1}{\gamma}x\right)^{0.5}\left[1-\frac{1}{4}\frac{x}{\gamma}+\left(\frac{8\gamma+5}{96}\right)\left(\frac{x}{\gamma}\right)^2+\ldots\right]$$

$$\therefore \text{in (A)},\ q = \left[\frac{2}{\varrho}(p_0-p)\right]^{0.5}\left[1-\frac{1}{4\gamma}\left(\frac{p_0-p}{p}\right)\right.$$

$$\left.+\left(\frac{8\gamma+5}{96\gamma^2}\right)\left(\frac{p_0-p}{p}\right)^2+\ldots\right]$$

and, comparing with (B), the first order error is

$$1-\left[1-\frac{1}{4\gamma}\left(\frac{p_0-p}{p}\right)\right] = +\frac{100}{4\gamma}\left(\frac{p_0-p}{p}\right)\%$$

($a^2 = \gamma RT, \gamma = 7/5$,) $\therefore\ a = 20(256)^{0.5} = 320$ m/s

$\therefore$ Mach number, $M = 256/320 = 0.8$

from Table 1, $p_0/p = 1.5243$

$\therefore$ error is $\frac{100}{5.6}\times 0.5243 = 9.36\%$

for $M = 0.2$, Table 1 gives $p_0/p = 1.0283$

and the error is $\frac{100}{5.6}\times 0.0283 = 0.51\%$ (negligible)

EXAMPLE 2.5 An aircraft is flying at a height at which the pressure and temperature are 0.5×10^5 N/m^2 and -17°C. A window of area 0·1 m^2 is situated towards the nose of the aircraft. Calculate the force acting on the window if the aircraft interior is pressurised at 1×10^5 N/m^2 and determine the air speed and Mach number at the window when (1) $M = 0.2$ for the aircraft, $C_p = -0.8$ for the window, (2) $M = 0.7$ for the aircraft, $C_p = -0.5$ for the window.

(1) for $M = 0.2$, it may be assumed that the air flow is incompressible and the following relation used for the pressure coefficient, (U. Leeds 1972)

$$C_p = \frac{p-p_\infty}{\varrho_\infty q_\infty^2/2} = 1-q^2/q_\infty^2 \quad (2.21)\ (2.22)$$

($a^2 = \gamma RT$), $a_\infty = 20(T_\infty)^{0.5} = 20(256)^{0.5} = 320$ m/s

$\therefore\ q_\infty = 0.2\times320 = 64$ m/s

($p = \varrho RT$), $\varrho_\infty = 0.5\times10^5/(0.2871\times10^3\times256)$
$= 0.68$ kg/m^3

from (2.21), $p-p_\infty = -0.8\times0.5\times0.68\times64^2$
$= -1\,114$ N/m^2

$\therefore\ p = (0.5-0.011\,14)\times10^5 = 0.488\,86\times10^5$ N/m^2

$\therefore$ load exerted on window $= (1-0.488\,86)\times10^5\times0.1$
$= 5\,111.4$ N

from (2.22), $q/q_\infty = (1-C_p)^{0.5} = 1.8^{0.5}$ $\therefore\ q = 85.9$ m/s

$\therefore\ M(\text{window}) = q/a_\infty = 0.268$ (assuming no temperature or density change)

(2) for $M = 0.7$, compressibility effects are significant and the following equation applies,

$$(\gamma/2 = 0.7) \qquad C_p = \frac{p-p_\infty}{\varrho_\infty q_\infty^2/2} = \frac{p-p_\infty}{0.7p_\infty M_\infty^2} \quad (2.23)$$

$\therefore\ p-p_\infty = -0.5\times0.7\times0.5\times10^5\times(0.7)^2 = -8\,575$ N/m^2

$\therefore\ p = (0.5-0.085\,75)\times10^5 = 0.414\,25\times10^5$ N/m^2

$\therefore$ load exerted on the window $=$
$(1-0.414\,25)\ \times10^5\times0.1 = 5\,857.5$ N

$$p_0/p_\infty = \left(1+\frac{\gamma-1}{2}M_\infty^2\right)^{\gamma/\gamma-1} = (1+0.2\times0.49)^{3.5}$$

or, from Tables, for $M_\infty = 0.7$, $p_0/p_\infty = 1.3871$ (Table 1)

$\therefore\ p_0 = 1.3871\times0.5\times10^5 = 0.6936\times10^5$ N/m^2

(N.B. $p_0 \neq p_\infty$)

$\therefore\ p_0/p = 0.6936/0.4143 = 1.6742$

from Tables, by interpolation, M (window) $= 0.891$ (Table 1)

$T/T_\infty = (p/p_\infty)^{\gamma-1/\gamma}$ $\therefore\ T = 256(0.4143/0.5)^{1/3.5} = 242.6$ K

($a^2 = \gamma RT$), $\therefore\ a = 20(242.6)^{0.5} = 311.5$ m/s

$\therefore$ window air speed, $q = M\times a = 277.6$ m/s

EXAMPLE 2.6 If a body is placed in a stream of air ($\gamma = 1.4$) which, at infinite distance upstream, is uniform flow with free-stream conditions q_∞, p_∞, ϱ_∞, M_∞, etc., write the expression for the dimensionless pressure coefficient C_p.

(*a*) Show that the value of the pressure coefficient corresponding to the appearance of the critical velocity is given by

$$C_{p*} = \left(\left(\frac{2}{\gamma+1}\right)^{\frac{\gamma}{\gamma-1}}\left[1+\frac{\gamma-1}{2}M_\infty^2\right]^{\frac{\gamma}{\gamma-1}}-1\right)\Big/ M_\infty^2\,\frac{\gamma}{2}$$

(*b*) If the body is a Pitot-static tube measuring total head and static pressures, write the equations which you would use to determine the flow speed, in terms of these pressures, for (i) incompressible (ii) subsonic (iii) supersonic flow conditions. What other property measurements would be required in each case?

(*c*) An aircraft is flying with a velocity of 238 m/s in air at a temperature of 16°C. What is the maximum pressure coefficient which may be obtained on the wings without the speed anywhere becoming supersonic?

(U. Leeds 1970)

(a) Figure 2.4 identifies the local conditions p, ϱ, M etc. on the body, for which the following expression for C_p is appropriate:

$$C_p = \frac{p-p_\infty}{\frac{1}{2}\varrho_\infty q_\infty^2} = \frac{p/p_\infty-1}{M_\infty^2\gamma/2} \quad (2.23)$$

$$(M = 1, p = p_*),\ \therefore\ C_{p*} = \frac{(p_*/p_0)(p_0/p_\infty)-1}{M_\infty^2\gamma/2}$$

$$\text{for isentropic conditions, } p_0/p = \left(1+\frac{\gamma-1}{2}M^2\right)^{\frac{\gamma}{\gamma-1}} \quad (2.15a)$$

$$\therefore\ p_0/p_* = \left(\frac{\gamma+1}{2}\right)^{\frac{\gamma}{\gamma-1}}$$

$$\therefore\ C_{p*} = \left[\left(\frac{2}{\gamma+1}\right)^{\frac{\gamma}{\gamma-1}}\left(1+\frac{\gamma-1}{2}M_\infty^2\right)^{\frac{\gamma}{\gamma-1}}-1\right]\Big/ M_\infty^2\frac{\gamma}{2}$$

(b) (i) incompressible energy equation (assuming no change in potential energy) for an ideal fluid is

$$p_0 = p+\frac{\varrho q^2}{2}$$

$$\therefore\ q = [2(p_0-p)/\varrho]^{0\cdot5}$$

(ii) compressible energy equation

$$\frac{\gamma}{\gamma-1}\cdot\frac{p_0}{\varrho_0} = \frac{\gamma}{\gamma-1}\cdot\frac{p}{\varrho}+\frac{q^2}{2}$$

or

$$\frac{\gamma RT_0}{\gamma-1} = \frac{\gamma RT}{\gamma-1}+\frac{q^2}{2}$$

$$\therefore\ q = \left[\left(\frac{2}{\gamma-1}\right)\gamma RT_0\left(1-\frac{T}{T_0}\right)\right]^{0\cdot5}$$

$(a^2 = \gamma RT)$

isentropic flow ($M < 1$),

$$q = a_0\left[\frac{2}{\gamma-1}\left(1-(p/p_0)^{\frac{\gamma-1}{\gamma}}\right)\right]^{0\cdot5}$$

(T_0, or T, must be measured in addition to p and p_0)

(iii) supersonic ($M > 1$)—the supersonic Pitot-tube (see EXAMPLE 4.6)

$$\frac{p_{02}}{p_2} = \left[1+\frac{\gamma-1}{2}M_2^2\right]^{\frac{\gamma}{\gamma-1}}$$

and

$$M_2^2 = \frac{(\gamma-1)M_1^2+2}{2\gamma M_1^2-(\gamma-1)}$$

(to obtain q_1 from the above equations $T_1(a_1)$ is required; p_2 is difficult to obtain experimentally from a Pitot-static tube)

(c) at 16°C $a^2 = \gamma RT$ or $a = 20(289)^{0\cdot5} = 340$ m/s

$\therefore\ M_\infty = 238/340 = 0\cdot7$ $\therefore\ p_0/p_\infty = 1\cdot3871$ (Table 1)

$$\therefore\ C_{p*} = \frac{0\cdot5283\times1\cdot3871-1}{0\cdot7\times0\cdot49} = -0\cdot779$$

EXAMPLE 2.7 (a) Show that for the reversible adiabatic one-dimensional flow of a perfect gas in a convergent-divergent nozzle expanding fully from an inlet stagnation pressure (p_0) to an exit static pressure (p_2), the ratio of exit area A_2 to throat area A_* is given by:

$$\frac{A_2}{A_*} = \left[\left(\frac{2}{\gamma+1}\right)^{\frac{1}{\gamma-1}}\left(\frac{p_0}{p_2}\right)^{\frac{1}{\gamma}}\right]\Big/\left[\frac{\gamma+1}{\gamma-1}\left(1-\left[\frac{p_0}{p_2}\right]^{\frac{1-\gamma}{\gamma}}\right)\right]^{0\cdot5}$$

(b) The nozzle of a rocket motor is designed to expand the propellant gas from a combustion chamber stagnation pressure of 32×10^5 N/m² and stagnation temperature 3 300 K to a sea level static pressure of 1×10^5 N/m².

The nozzle throat area is 500 cm². Assuming that the combustion chamber conditions remain steady and that the expansion is reversible and adiabatic, determine:

(1) The nozzle exit area,

(2) The mass flow rate of propellant gas, and

(3) The thrust developed at an altitude where the atmosphere pressure is $0\cdot2\times10^5$ N/m². The propellant gas may be assumed to behave as a perfect gas having a mean molecular weight 25·8 and a mean $\gamma = 1\cdot25$. The universal gas constant is 8·3143 kw/kg-mole K.

(U. London 1966)

Figure E2.7 shows a typical convergent-divergent nozzle used to expand a gas from the reservoir or total head condition (p_0, T_0)

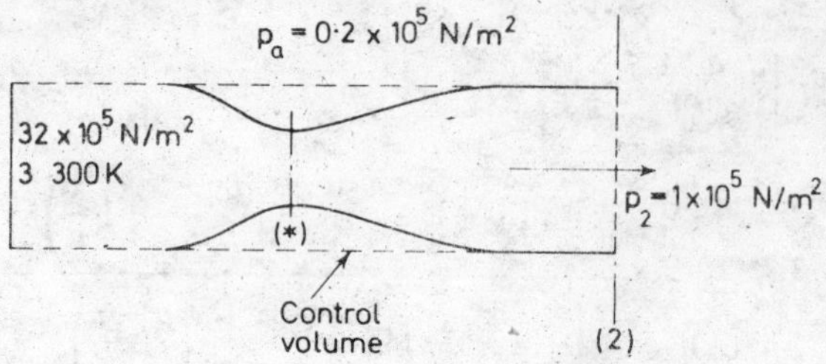

Fig. E2.7

continuity, mass flow, $m = \varrho_* q_* A_* = \varrho_2 q_2 A_2$ (A)

energy, $$\frac{\gamma}{\gamma-1}\frac{p_0}{\varrho_0} = \frac{\gamma}{\gamma-1}\frac{p_*}{\varrho_*}+\frac{q_*^2}{2} = \frac{\gamma}{\gamma-1}\frac{p_2}{\varrho_2}+\frac{q_2^2}{2}$$

$$q_2^2 = \frac{2a_0^2}{\gamma-1}\left[1-\left(\frac{p_2}{p_0}\right)^{\frac{\gamma-1}{\gamma}}\right] \quad \text{(B)}$$

also, *isentropic*, $$p_0/p = \left(1+\frac{\gamma-1}{2}M^2\right)^{\frac{\gamma}{\gamma-1}} \quad (2.15a)$$

$$(M = 1,\ q_* = a_*),\ p_0/p_* = \left(\frac{\gamma+1}{2}\right)^{\frac{\gamma}{\gamma-1}} \quad \text{(C)}$$

isentropic,

$$\therefore\ \frac{\varrho_2}{\varrho_*} = \left(\frac{p_2}{p_*}\right)^{\frac{1}{\gamma}} = \left(\frac{p_2}{p_0}\right)^{\frac{1}{\gamma}}\left(\frac{p_0}{p_*}\right)^{\frac{1}{\gamma}} = \left(\frac{p_2}{p_0}\right)^{\frac{1}{\gamma}}\left(\frac{\gamma+1}{2}\right)^{\frac{1}{\gamma-1}}$$

from (A) and (B), $\dfrac{A_2}{A_*} = \dfrac{\varrho_*}{\varrho_2}\dfrac{q_*}{q_2}$

$$= \left[\frac{q_*}{a_0}\left(\frac{\gamma-1}{2}\right)^{0\cdot5}\left(\frac{p_0}{p_2}\right)^{\frac{1}{\gamma}}\left(\frac{2}{\gamma+1}\right)^{\frac{1}{\gamma-1}}\right]\Big/\left[1-\left(\frac{p_2}{p_0}\right)^{\frac{\gamma-1}{\gamma}}\right]^{0\cdot5}$$

$$(q_* = a_*),\ \therefore\ \frac{a_*}{a_0} = \frac{q_*}{a_0} = \left(\frac{T_*}{T_0}\right)^{0\cdot5} = \left(\frac{2}{\gamma+1}\right)^{0\cdot5} \quad \text{(D)}$$

$$\therefore\ \frac{A_2}{A_*} = \left[\left(\frac{2}{\gamma+1}\right)^{\frac{1}{\gamma-1}}\left(\frac{p_0}{p_2}\right)^{\frac{1}{\gamma}}\right]\Big/\left[\frac{\gamma+1}{\gamma-1}\left(1-\left(\frac{p_0}{p_2}\right)^{\frac{1-\gamma}{\gamma}}\right)\right]^{0\cdot5} \quad \text{(E)}$$

Note:—This expression can also be derived using (2.31). i.e.

$$MN_* = \frac{ma_0}{A_* p_0} = \gamma\left(\frac{2}{\gamma+1}\right)^{\frac{\gamma+1}{2(\gamma-1)}}$$

($\gamma = 5/4$), for substitution in (E), $(p_0/p_2)^{1/\gamma} = (32)^{4/5} = 16$

and, $$\left(\frac{p_0}{p_2}\right)^{\frac{1-\gamma}{\gamma}} = (32)^{-1/5} = 0\cdot5$$

$$\therefore\ \frac{A_2}{A_*} = \frac{(8/9)^4\times16}{(9\times0\cdot5)^{0\cdot5}} = 4\cdot708 \quad \therefore\ A_2 = 2\ 354\ \text{cm}^2$$

mass flow $= \varrho_* q_* A_*$

from (D), $\dfrac{T_0}{T_*} = \dfrac{\gamma+1}{2} = 1{\cdot}125$

$\therefore T_* = 3\,300/1{\cdot}125 = 2\,933$ K

gas constant, $R = R_0/M = 8{\cdot}3143/25{\cdot}8 = 0{\cdot}3223$ kJ/kg K

$(a^2 = \gamma RT)$, $a_* = q_* = (1{\cdot}25\times 0{\cdot}3223\times 10^3\times 2\,933)^{0{\cdot}5}$

$= 1\,087$ m/s

from (C), $\dfrac{p_0}{p_*} = \left(\dfrac{\gamma+1}{2}\right)^{\frac{\gamma}{\gamma-1}} = (1{\cdot}125)^5 = 1{\cdot}802$

$\therefore p_* = (32/1{\cdot}802)\times 10^5 = \underline{17{\cdot}76\times 10^5 \text{ N/m}^2}$

$(p = \varrho RT)$, $\therefore \varrho_* = (17{\cdot}76\times 10^5)/(0{\cdot}3223\times 10^3\times 2\,933)$

$= \underline{1{\cdot}879 \text{ kg/m}^3}$

$\therefore$ mass flow $= 1{\cdot}879\times 1087\times 500\times 10^{-4} = \underline{102{\cdot}1 \text{ kg/s}}$

$\left(\text{alternatively } MN_* = \dfrac{m a_0}{A_* P_0} = \gamma\left(\dfrac{2}{\gamma+1}\right)^{\frac{\gamma+1}{2(\gamma-1)}} \text{ where } \gamma = \dfrac{5}{4}, \text{ giving } m = 102{\cdot}1 \text{ kg/s}\right)$

Consider the control volume shown for conservation of momentum with F_x, the force exerted *on* the fluid *by* the rocket, in the positive X-direction.
momentum into control volume is zero
momentum out of control volume = pressure forces acting on fluid inside control volume + F_x

$$\therefore mq_2 = -(p_2-p_a)A_2+F_x \qquad \text{(F)}$$

$\dfrac{\varrho_2}{\varrho_*} = \left(\dfrac{p_2}{p_*}\right)^{1/\gamma}$ $\therefore \varrho_2 = 1{\cdot}879\times(1/17{\cdot}76)^{0{\cdot}8} = 0{\cdot}188$ kg/m^3

from (A), $q_2 = 102{\cdot}1/(0{\cdot}188\times 2\,354\times 10^{-4}) = 2\,307$ m/s

in (F), $102{\cdot}1\times 2\,307 = [-(1-0{\cdot}2)2\,354]10^5\times 10^{-4}+F_x$

$\therefore F_x = 2{\cdot}543\times 10^5$ N

$\therefore$ the thrust *on* the rocket is $-2{\cdot}543\times 10^5$ N $\leftarrow$

Note: for $p_a < p_2$, the nozzle exit pressure of a supersonic flow is not affected and the control volume need only contain the entire rocket and is of arbitrary size.

EXAMPLE 2.8 Figure E2.8a shows a convergent—divergent nozzle comprising two sections flanged together at F, the throat and exit areas being 10 cm^2 and 14·95 cm^2 respectively. Air from a large reservoir at 10×10^5 N/m^2 and 17°C flows steadily through the nozzle before discharging to atmosphere as shown. Regarding air as a perfect gas ($\gamma = 1{\cdot}4$) show that the pressure in the exit plane is higher than atmospheric, and determine its value. Hence find the net thrust exerted by the air on the detachable section of the nozzle.

Use may be made of the following table for the function $y = (1-x^{2/7})^{0{\cdot}5}x^{5/7}$

x	0·158	0·160	0·162	0·164
y	0·17134	0·17244	0·17353	0·17460

(U. Birmingham 1960)

continuity, mass flow, $m = \varrho_* A_* q_* = \varrho_2 q_2 A_2$

$(\gamma = 7/5)$, $q_* = a_* = 20(T_*)^{0{\cdot}5}$ and $T_0/T_* = (\gamma+1)/2 = 1{\cdot}2$ (2.14)

$\therefore T_* = 290/1{\cdot}2 = 241{\cdot}7$ K.

$\therefore q_* = 20(241{\cdot}7)^{0{\cdot}5} = 310{\cdot}9$ m/s

similarly, $p_0/p_* = ((\gamma+1)/2)^{\gamma/\gamma-1} = 1{\cdot}2^{3{\cdot}5} = 1{\cdot}893$ (2.15a)

$p_* = 10^6/1{\cdot}893 = 5{\cdot}283\times 10^5$ N/m^2

$\varrho_* = p_*/RT_* = (5{\cdot}283\times 10^5)/(0{\cdot}2871\times 10^3\times 241{\cdot}7)$

$= 7{\cdot}6133$ kg/m^3

$m = 7{\cdot}6133\times 0{\cdot}001\times 310{\cdot}9 = \underline{2{\cdot}367 \text{ kg/s}}.$

energy, $\dfrac{\gamma}{\gamma-1}\dfrac{p_0}{\varrho_0} = \dfrac{q_2^2}{2}+\dfrac{\gamma}{\gamma-1}\dfrac{p_2}{\varrho_2}$ (2.10a)

$\therefore q_2^2 = \dfrac{2\gamma}{\gamma-1}\dfrac{p_0}{\varrho_0}\left(1-\dfrac{p_2}{p_0}\dfrac{\varrho_0}{\varrho_2}\right)$

$(\gamma = 1{\cdot}4)$, $q_2^2 = 7RT_0(1-(p_2/p_0)^{2/7})$

$(x = p_2/p_0)$, $q_2 = (7\times 0{\cdot}2871\times 10^3\times 290)^{0{\cdot}5}(1-x^{2/7})^{0{\cdot}5}$ (A)

$q_2 = m/(\varrho_2 A_2)$

$\varrho_2 = \varrho_0(p_2/p_0)^{1/\gamma} = \varrho_0 x^{5/7}$

$\varrho_0 = p_0/RT_0 = 10^6/(0{\cdot}2871\times 10^3\times 290) = 12{\cdot}01$ kg/m^3

substituting for q_2 in (A), $m/(\varrho_2 A_2) = 763{\cdot}4(1-x^{2/7})^{0{\cdot}5}x^{5/7}$

or, $y = 2{\cdot}367/(12{\cdot}01\times 0{\cdot}001495\times 763{\cdot}4) = (1-x^{2/7})^{0{\cdot}5}x^{5/7}$

$y = 0{\cdot}1727$, and interpolation in the table above gives $x = 0{\cdot}1605$

$\therefore \underline{p_2 = 1{\cdot}605\times 10^5 \text{ N/m}^2}$

$\varrho_2 = \varrho_0(p_2/p_0)^{5/7} = 3{\cdot}251$ kg/m^3

$\therefore q_2 = 2{\cdot}367/(3{\cdot}251\times 0{\cdot}001495) = 487$ m/s

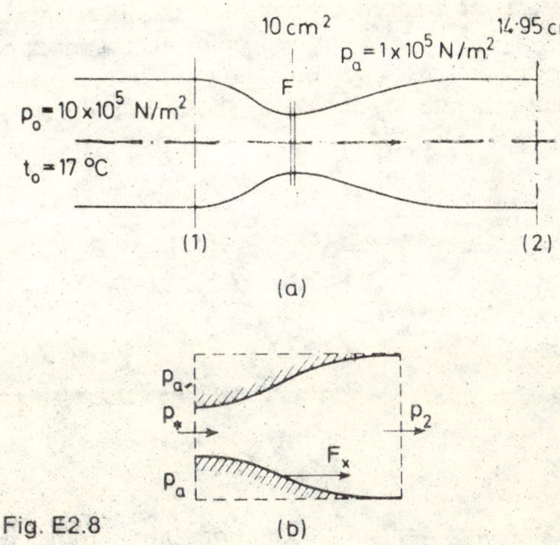

Fig. E2.8

conservation of momentum is considered for the positive X-direction and the control volume is chosen as shown in Figure 2.8b to include the unknown force (F_x) exerted by the solid boundary of the nozzle on the fluid.

momentum in $= m\times q_*$, momentum out $= m\times q_2$

forces causing momentum change $= p_a(A_2-A_*) + p_*A_*-p_2A_2$

$\therefore 2{\cdot}367(487-310{\cdot}9) = 10^5\times 10^{-4}[1(14{\cdot}95-10) + 5{\cdot}283\times 10-1{\cdot}605\times 14{\cdot}95]+F_x$

$\therefore F_x = 78{\cdot}9$ N $\rightarrow$

Since F_x is the force exerted *on* the fluid, the fluid exerts a thrust of $-78{\cdot}9$ N($\leftarrow$) on the detachable divergent section of the nozzle.

3 PRANDTL-MEYER FLOW

3.1 MACH LINES AND SOME LINEARISED RELATIONS IN TWO-DIMENSIONAL FLOW

In Section 2.1 the concept of a Mach line was introduced with respect to a point disturbance moving through a stationary gas at a speed greater than the local speed of sound ($M > 1$, $q > a$). For fixed point disturbances on a flat smooth surface, supersonic flow of a compressible fluid would generate Mach lines as shown in Figure 3.1.

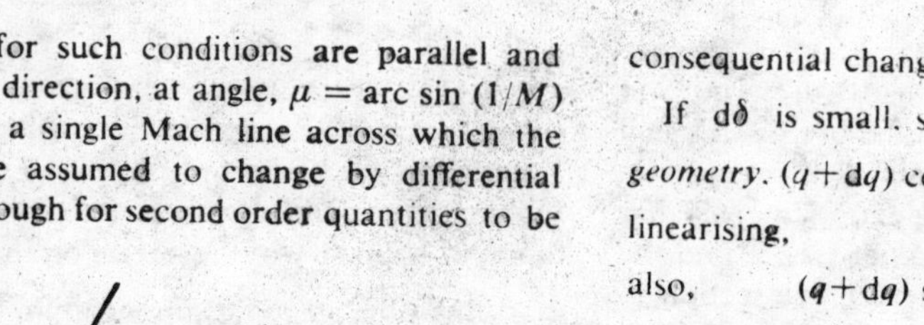

Fig. 3.1 Mach lines

The Mach lines for such conditions are parallel and inclined, to the flow direction, at angle, $\mu = \text{arc sin}\,(1/M)$.

Figure 3.2 shows a single Mach line across which the flow parameters are assumed to change by differential increments small enough for second order quantities to be negligible. The resulting first order relationships established between parameters across the Mach line, thus represent linearised equations for steady, reversible adiabatic flow. The Mach line is considered as a characteristic of the flow.

In general, the component of velocity at right angles to a Mach line is always the acoustic speed and may be considered to change from (a) to ($a + da$). The velocity component (m) parallel to the Mach line does not change. Thus, for a change (da) in one component of the flow velocity (q) the

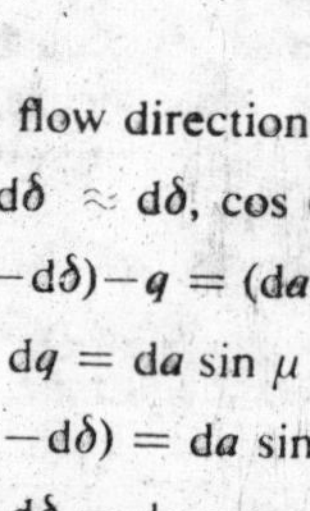

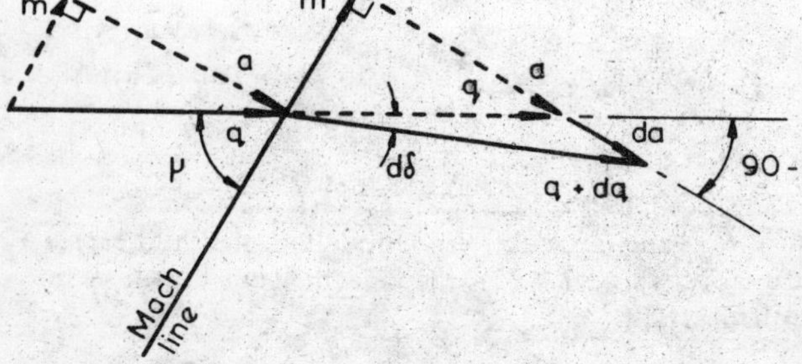

Fig. 3.2 Mach line kinematic changes

consequential change in flow direction is shown as ($-d\delta$).

If $d\delta$ is small, $\sin d\delta \approx d\delta$, $\cos d\delta \approx 1$ (3.1)

geometry, $(q+dq)\cos(-d\delta) - q = (da)\cos(90-\mu)$

linearising, $$dq = da \sin\mu \qquad (3.2)$$

also, $$(q+dq)\sin(-d\delta) = da\sin(90-\mu)$$

linearising, $$-q\, d\delta = da\cos\mu$$

from (3.2), $$d\delta = -\frac{dq}{q}\frac{1}{\tan\mu} \qquad (3.3a)$$

from Figure 3.1, $$d\delta = -\frac{dq}{q} \pm (M^2-1)^{0.5} \qquad (3.3b)$$

(Clearly M must be greater than or equal to unity)

momentum, $\varrho q[(q+dq)-q)] = -dp$

$$\therefore dp = \varrho q\, dq$$

$(a^2 = \gamma p/\varrho)$, $$\therefore \frac{dp}{p} = -\gamma M^2\frac{dq}{q} \qquad (3.4)$$

combining (3.3b) and (3.4)

$$\frac{dp}{p} = \frac{\gamma M^2}{\pm(M^2-1)^{0.5}}\,d\delta = -\gamma M^2 \frac{dq}{q} \tag{3.5}$$

energy,

$$\frac{q^2}{2} + \frac{a^2}{\gamma-1} = \text{constant} \tag{2.19}$$

differentiating,

$$dq = -\frac{2}{(\gamma-1)M}\,da \tag{3.6}$$

$$M = \frac{q}{a}$$

differentiating,

$$\frac{dM}{M} = \frac{dq}{q}\left(1 - M\frac{da}{dq}\right)$$

from (3.6),

$$\frac{dM}{M} = \frac{dq}{q}\left(1 + \frac{\gamma-1}{2}M^2\right) \tag{3.7}$$

From (3.5), four different types of Mach lines may be identified as shown in Figures 3.3a, b, c, d. The Mach lines are either characteristics of Type I, making a negative acute angle with the flow direction or characteristics of Type II making a positive acute angle with the flow direction.

From (3.3a), for a characteristic of Type I (μ $-ve$)$d\delta \propto dq/q$.

Figure 3.3a shows ($d\delta$) negative with $dq < 0$ and, from (3.4), $dp > 0$; the process is one of compression and deceleration.

Figure 3.3b shows ($d\delta$) positive with $dq > 0$, $dp < 0$; expansion and acceleration

From (3.3a), for a characteristic of Type II ($\mu + ve$) $d\delta \propto -dq/q$

Figure 3.3c shows ($d\delta$) negative with $dq > 0$, $dp < 0$; expansion and acceleration (as shown in Figure 3.2).

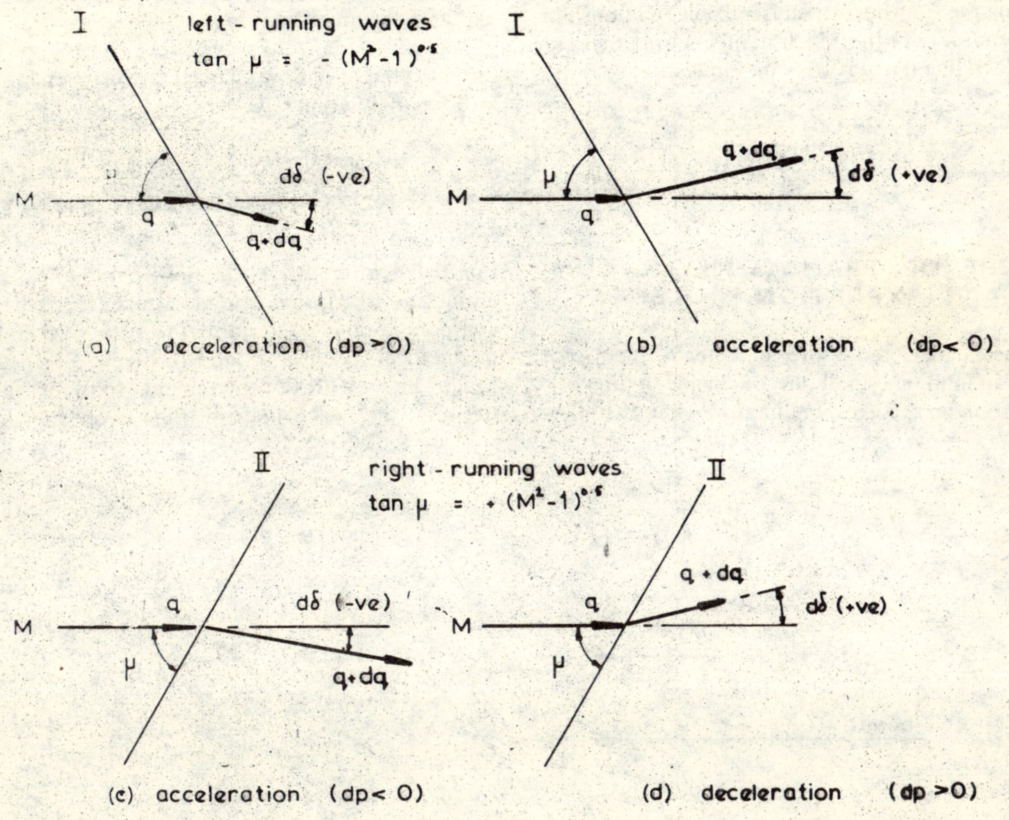

Fig. 3.3 Mach line variations

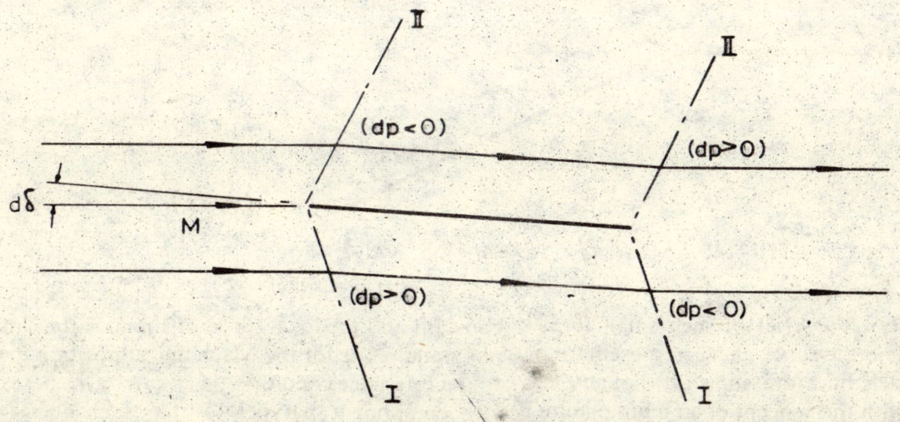

Fig 3.4 supersonic flow past a thin flat plate

Figure 3.3d shows ($d\delta$) positive with $dq < 0$, $dp > 0$; compression and deceleration.

Type I waves are also referred to as left-running waves and Type II waves as right-running waves—both types are illustrated in Figure 3.4 for superonic flow past a flat plate inclined at a small angle ($d\delta$) to the flow.

With reference to Figure 3.1, equations (3.5) and (3.7) indicate that, for steady flow, with no change in velocity ($dq = 0$) there is no change in pressure or in Mach number, hence the Mach lines are parallel.

A linearised expression for the pressure coefficient may also be obtained from (2.23) for incremental changes in flow parameters,

$$C_p = \frac{2(p-p_\infty)}{\gamma p_\infty M_\infty^2} = \frac{dp}{p}\left(\frac{2}{\gamma M^2}\right) \quad \text{for } M = M_\infty$$

from (3.5)
$$C_p = \frac{2d\delta}{\pm(M^2-1)^{0.5}} = -2\frac{dq}{q} \tag{3.8}$$

Note: By using the fully developed isentropic relationships, if M_1 and M_2 are the Mach numbers on either side of a Mach line it can readily be shown that

$$C_p = \frac{2}{\gamma M_1^2}\left[\left(\frac{2+(\gamma-1)M_1^2}{2+(\gamma-1)M_2^2}\right)^{\frac{\gamma}{\gamma-1}} - 1\right] \tag{3.9}$$

3.2 CONCEPT OF PRANDTL-MEYER COMPRESSION AND EXPANSION PROCESSES

The disturbance, considered above, in a steady supersonic flow could be an infinitely small turn or change in direction of a smooth boundary, corresponding to a change in flow direction ($d\delta$). In Figure 3.5 a series of such small turns would form part of a smooth curve and a representative number of different Mach lines are shown although the flow field is continuous with the concept of an infinite number of Mach lines.

For Type II waves (right-running) with $d\delta$ negative, $dq > 0$, $dM > 0$, the Mach angle μ will increase from $\mu_1(M_1)$ to $\mu_2(M_2)$. Intermediate Mach lines will therefore diverge outwards and form an expansion 'fan' between the uniform flow conditions $M_1 > 1$ and $M_2 > M_1$, before and after the turn respectively.

Analysis of the process is simplified by assuming that the divergent Mach lines radiate from the same point (Figure 3.5). In isentropic (irrotational) flow, any flow boundary may be replaced by a streamline and the Prandtl–Meyer expansion can be considered for flow round a smooth curve or round a sharp corner. The flow area diverges to accommodate the expansion (EXAMPLE 3·3).

Alternatively, for supersonic flow approaching a smooth concave corner with $d\delta$ positive (Figure 3.6), the Type II (right-running) waves produce a compression process $dq < 0$, $dM < 0$. If the upper boundary ABC is further away than the point at which the Mach lines coalesce, then the pressure increments accumulate to produce a substantial and abrupt change in the flow conditions. This convergence of Mach waves produces an oblique shock wave across which the flow changes are irreversible and which will be analysed in Chapter 8.

3.3 PRANDTL–MEYER FLOW ANALYSIS

For the centred wave the flow parameters are constant along any Mach line (radial ordinate of the fan) and the component of velocity normal to such a Mach line is equal to the local speed of sound (a). Figure 3.7 shows two such Mach lines with an increment ($d\theta$) of displacement. Between the Mach lines all fluid properties are assumed constant and the local speed of sound is the same at positions r and $r+dr$ for the Mach line at position θ with respect to the reference condition $M = 1$, $a = a_*$, $\theta = 0$. The velocity component parallel to the Mach line is m and is a function of θ.

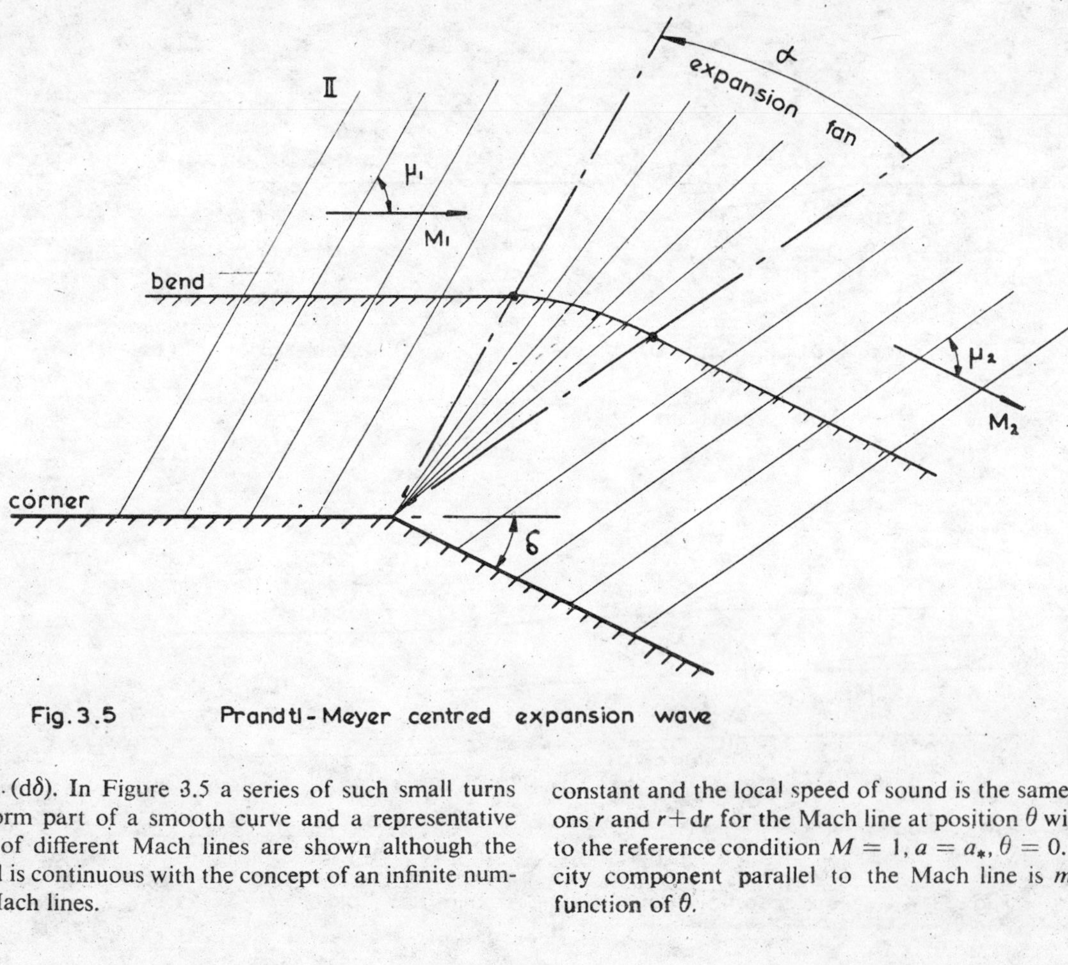

Fig. 3.5 Prandtl-Meyer centred expansion wave

For isentropic flow, the fluid flow is irrotational and the circulation will be zero for any closed contour in the fluid such as ABCD, thus:

$$\left(m+\frac{\partial m}{\partial \theta}\,d\theta\right)dr-m\,dr-a(r+dr)\,d\theta-ar\,d\theta=0$$

whence,
$$a=\frac{\partial m}{\partial \theta} \tag{3.10}$$

from the energy relations (2.19),

$$\frac{q^2}{2}+\frac{a^2}{\gamma-1}=\frac{c^2}{2}$$

or,
$$\frac{m^2+a^2}{2}+\frac{a^2}{\gamma-1}=\frac{c^2}{2} \tag{3.11}$$

from (3.10), $a=\dfrac{\partial m}{\partial \theta}=\left[\dfrac{\gamma-1}{\gamma+1}\,(c^2-m^2)\right]^{0\cdot5}$

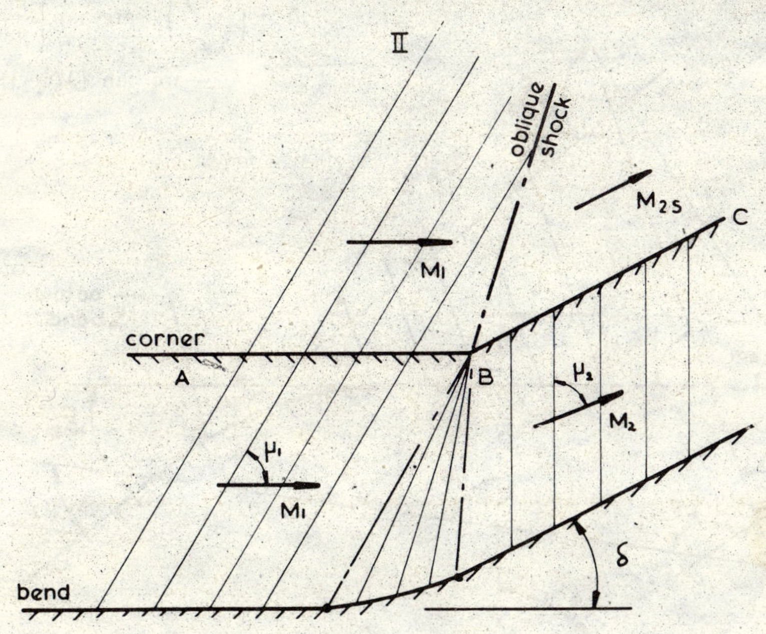

Fig. 3.6 Prandtl - Meyer compression wave

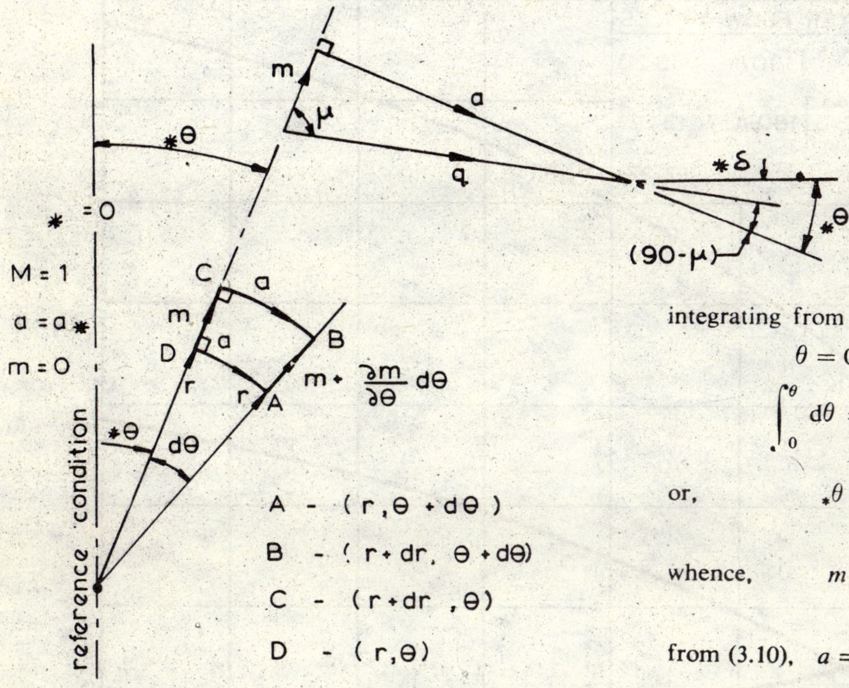

Fig. 3.7 Prandtl - Meyer analysis

integrating from the reference condition,

$$\theta=0,\ M=1,\qquad a=a_*,\ m=0,$$

$$\int_0^{\theta} d\theta=\left(\frac{\gamma+1}{\gamma-1}\right)^{0\cdot5}\int_0^{m}\frac{1}{(c^2-m^2)^{0\cdot5}}\,dm$$

or,
$$_*\theta=\left(\frac{\gamma+1}{\gamma-1}\right)^{0\cdot5}\text{arc sin}\left(\frac{m}{c}\right)$$

whence,
$$m=c\sin\left[\left(\frac{\gamma-1}{\gamma+1}\right)^{0\cdot5}{}_*\theta\right] \tag{3.12}$$

from (3.10),
$$a=\left(\frac{\gamma-1}{\gamma+1}\right)^{0\cdot5}c.\cos\left[\left(\frac{\gamma-1}{\gamma+1}\right)^{0\cdot5}{}_*\theta\right] \tag{3.13}$$

$$=a_*\cos\left[\left(\frac{\gamma-1}{\gamma+1}\right)^{0\cdot5}{}_*\theta\right] \tag{3.13a}$$

$$M=\left[\frac{m^2+a^2}{a^2}\right]^{0\cdot5}$$

from (3.12) and (3.13),

$$M = \left[1 + \frac{\gamma+1}{\gamma-1}\tan^2\left[\left(\frac{\gamma-1}{\gamma+1}\,{}_*\theta\right)^{0\cdot5}\right]\right] \qquad (3.14)$$

rearranging gives,

$${}_*\theta = \left(\frac{\gamma+1}{\gamma-1}\right)^{0\cdot5} \text{arc tan}\left[\frac{\gamma-1}{\gamma+1}(M^2-1)\right]^{0\cdot5} \qquad (3.15)$$

where ${}_*\theta$ is the angle the Mach line makes with the reference condition $M = 1$, ${}_*\theta = 0$, with the Mach line perpendicular to the flow direction.

Figure 3.7 shows the geometrical relationship between ${}_*\theta$, μ and ${}_*\delta$ the angle by which the flow direction has changed from tne reference condition.

$${}_*\delta = {}_*\theta - \left(\frac{\pi}{2} - \mu\right) \qquad (3.16)$$

or,

$${}_*\delta = \left(\frac{\gamma+1}{\gamma-1}\right)^{0\cdot5} \text{arc tan}\left[\frac{\gamma-1}{\gamma+1}(M^2-1)\right]^{0\cdot5} - \text{arc tan}\,(M^2-1)^{0\cdot5} \qquad (3.17)$$

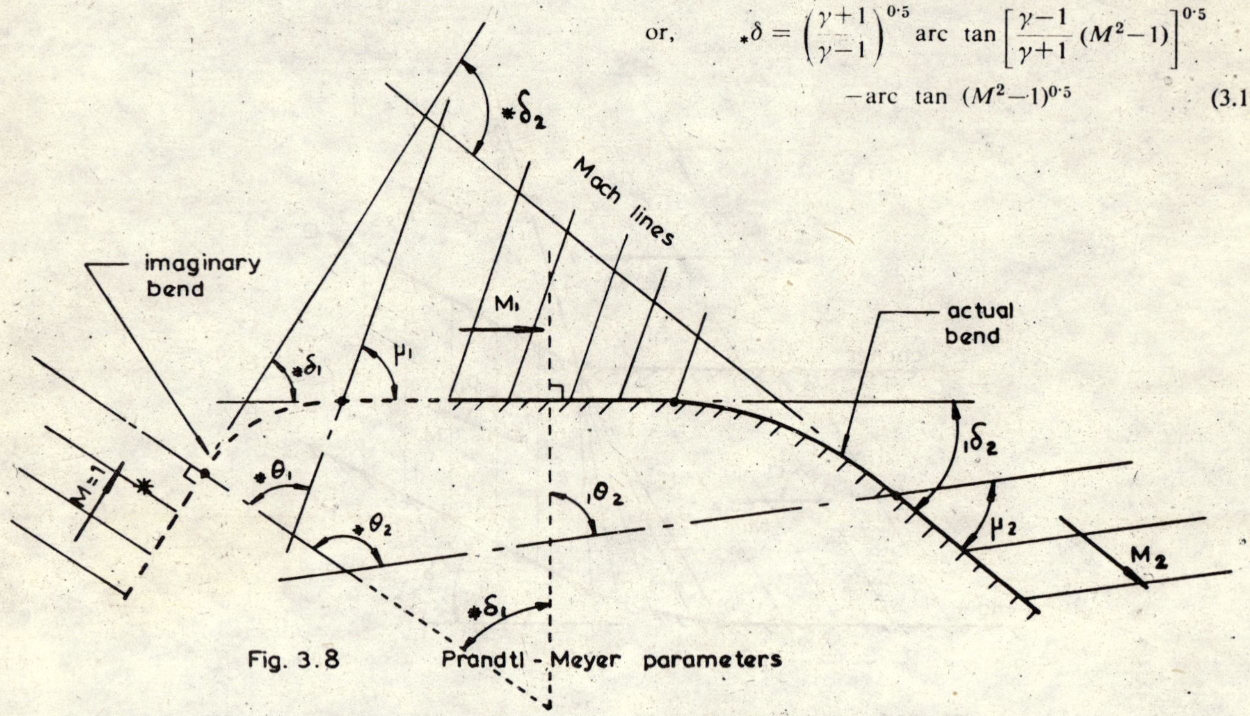

Fig. 3.8 Prandtl - Meyer parameters

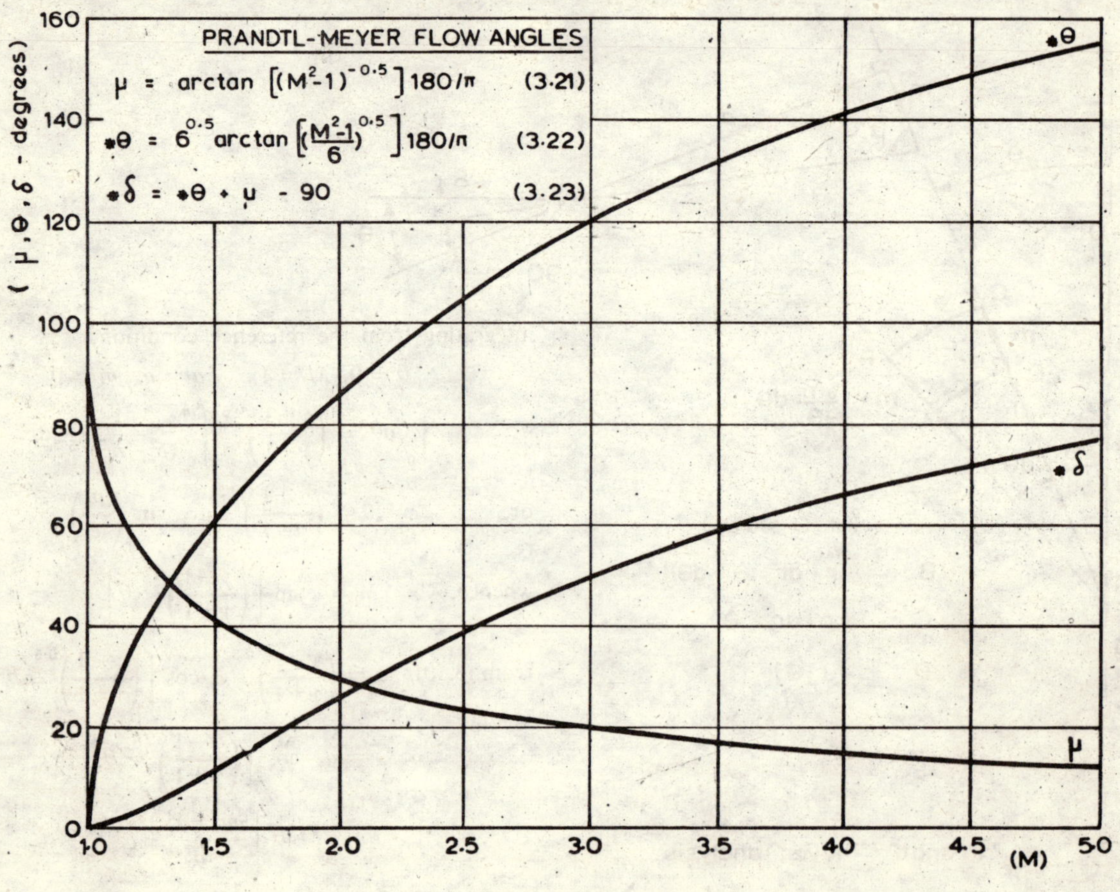

Fig. 3.9

For Prandtl–Meyer Flow round any corner or bend from a uniform flow condition (1) M_1, q_1, to a uniform flow condition (2) M_2, q_2, the flow process is analysed by considering (Figure 3.8) the *actual* corner or bend to be preceded by an *imaginary* corner or bend which (for an expansion) accelerates the flow from the sonic reference condition to the actual initial condition (1). From Figure 3.8 it may be seen that,

$$ {}_1\delta_2 = {}_*\delta_2 - {}_*\delta_1 \qquad (3.18)$$

and,

$$ {}_1\theta_2 = {}_*\theta_2 - {}_*\theta_1 \qquad (3.19)$$

The maximum turn or change in direction of the flow of a compressible fluid will be obtained if the flow is expanded from an initial sonic condition to the limit velocity, $c = q_{max} = [2a_0^2/(\gamma-1)]^{0\cdot5}$ when $a = 0$, i.e., a theoretical expansion from $M = 1$ to $M = \infty$.

From (3.17), $$\delta_{max} = {}_*\delta_\infty = \left[\left(\frac{\gamma+1}{\gamma-1}\right)^{0\cdot5}\frac{\pi}{2}\right] - \frac{\pi}{2} \qquad (3.20)$$

$$= 130{\cdot}4° \text{ for } \gamma = 1{\cdot}4$$

3.4 SUMMARY OF EQUATIONS

The following are the equations used in Table 1 for steady isentropic flow of dry air ($\gamma = 7/5$) — Prandtl–Meyer flow ($M \geqslant 1$), and are plotted in Figure 3.9.

from Figure 3.3, $$\mu = \text{arc tan}\,\frac{1}{(M^2-1)^{0\cdot5}}\times\frac{180}{\pi} \text{ degrees} \qquad (3.21)$$

from (3.15), $${}_*\theta = 6^{0\cdot5}\,\text{arc tan}\left(\frac{M^2-1}{6}\right)^{0\cdot5}\times\frac{180}{\pi} \text{ degrees} \qquad (3.22)$$

from (3.16), $${}_*\delta = {}_*\theta + \mu - 90 \qquad (3.23)$$

EXAMPLE 3.1 For dry air ($\gamma = 7/5$) flowing at a uniform supersonic speed of Mach number M, which is diverted through a small angle θ, show, using linearised theory, that the pressure change is given by,

$$C_p = \frac{p-p_\infty}{(\varrho_\infty q_\infty^2/2)} = \frac{2\theta}{\pm(M-1)^{0\cdot5}}$$

where the suffix (∞) refers to uniform upstream conditions.

If $M = \sqrt{5}$ and $\varrho_\infty q_\infty^2/2 = 2{\cdot}8\times10^5$ N/m² determine the lift and drag per metre span, on the airfoil shown in Figure E3.1a, in terms of the incident angle α.

(U. Leeds 1962)

from (3.4) $$\frac{dp}{p} = -\gamma M^2\,\frac{dq}{q}$$

from (3.3b) $$d\delta = \theta = \pm(M^2-1)^{0\cdot5}\left(-\frac{dq}{q}\right)$$

combining, $$\frac{dp}{p} = \frac{p-p_\infty}{p_\infty} = +\theta\,\frac{\gamma M^2}{\pm(M^2-1)^{0\cdot5}}$$

$(a^2 = \gamma p/\varrho)$, $$C_p = \frac{p-p_\infty}{\varrho_\infty q_\infty^2/2} = \frac{p-p_\infty}{p_\infty\gamma M_\infty^2/2}$$

$$= \frac{2\theta}{\pm(M^2-1)^{0\cdot5}} \quad \text{where } M = M_\infty$$

Note: the ($\pm$) is included to emphasise the distinction which must be made for $\mu = \tan^{-1}(M^2-1)^{0\cdot5}$ having positive values for Type II right-running waves and negative values for Type I left-running waves.

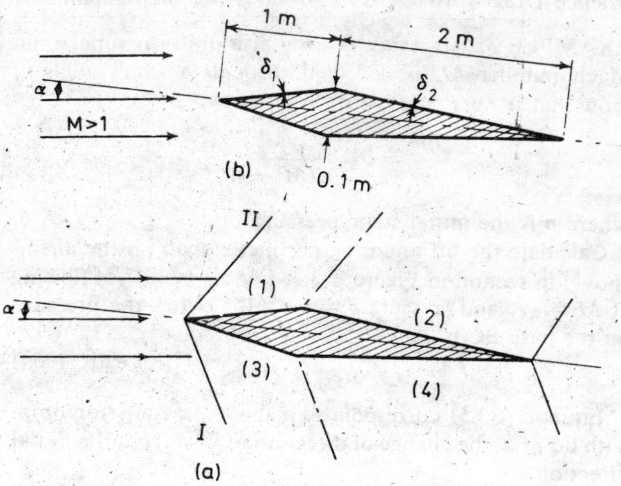

Fig. E3.1

Figure E3.1b shows the wave types for the airfoil considered

from above, $$C_p = \frac{p-p_\infty}{\varrho_\infty q_\infty^2/2} = \frac{2\theta}{\pm(M^2-1)^{0\cdot5}}$$

$$\therefore p - p_\infty = 2{\cdot}8\times10^5\times(\pm\theta)$$

$$\therefore p = p_\infty \pm 2{\cdot}8\times10^5\,\theta = p_\infty \pm k\theta$$

for all small angles (e.g. δ) it will be assumed that $\cos\delta = 1$, $\tan\delta = \sin\delta = \delta$; for the airfoil it can then be shown that $\delta_1 = 2\delta_2 = 2(0{\cdot}05) = 0{\cdot}1$

if θ is the angle by which the flow direction in a given region is above the original (∞) direction then (α positive),

$$\theta_1 = \delta_1 - \alpha = 2\delta_2 - \alpha \text{ (Type II) } p_1 = p_\infty + k\theta_1$$

$$\theta_2 = \theta_1 - \delta_1 - \delta_2 = -\delta_2 - \alpha \text{ (Type II) } p_2 = p_\infty + k\theta_2$$

$$\theta_3 = -(\delta_1 + \alpha) = -2\delta_2 - \alpha \text{ (Type I) } p_3 = p_\infty - k\theta_3$$

$$\theta_4 = \theta_3 + \delta_1 + \delta_2 = \delta_2 - \alpha \text{ (Type I) } p_4 = p_\infty - k\theta_4$$

The total lift force (L) and drag force (D) exerted on the airfoil per metre span due to the pressure distributions on the airfoil surfaces are obtained by resolving the pressure forces (which are normal to their surfaces) in directions respectively normal to and parallel to the direction of motion.

$$\text{Lift, } L = \left[p_4\left(\frac{2\times1}{\cos\delta_2}\right)\right]\cos\theta_4 + \left[p_3\left(\frac{1\times1}{\cos\delta_1}\right)\right]\cos\theta_3 - \left[p_2\left(\frac{2\times1}{\cos\delta_2}\right)\right]\cos\theta_2 - \left[p_1\left(\frac{1\times1}{\cos\delta_1}\right)\right]\cos\theta_1$$

$$\text{(small angles)} = 2p_4 + p_3 - 2p_2 - p_1$$

$$\therefore L = 2[p_\infty - k(\delta_2-\alpha)] + 1[p_\infty - k(-2\delta_2-\alpha)] - 2[p_\infty + k(-\delta_2-\alpha)] - 1[p_\infty \times k(2\delta_2-\alpha)]$$

or $L = 6k\alpha$ whence Lift $= 16{\cdot}8\times10^5\alpha$ N per metre span

$$\text{Drag, } D = \left[p_1\left(\frac{1\times1}{\cos\delta_1}\right)\right]\sin\theta_1 + \left[p_2\left(\frac{2\times1}{\cos\delta_2}\right)\right]\sin\theta_2 - \left[p_3\left(\frac{1\times1}{\cos\delta_1}\right)\right]\sin\theta_3 - \left[p_4\left(\frac{2\times1}{\cos\delta_2}\right)\right]\sin\theta_4$$

(N.B. signs for drag force consideration are implicit in ϑ)

$$\therefore D = p_1\theta_1 + 2p_2\theta_2 - p_3\theta_3 - 2p_4\theta_4$$

giving $D = 6k\alpha^2 + 12k\delta_2^2$

whence Drag $= 16{\cdot}8\times10^5\alpha^2 + 8\,400$ N per metre span

EXAMPLE 3.2 If a gas, flowing at a uniform supersonic Mach number M, is deflected through a small angle θ, show that the pressure rise dp is given by,

$$\frac{dp}{p} = \theta \cdot \frac{\gamma M^2}{\pm(M^2-1)^{0{\cdot}5}}$$

where p is the initial static pressure.

Calculate the lift and drag per metre span on the airfoil shown in section in Figure E 3.2. For air ($\gamma = 7/5$) flowing at $M = \sqrt{5}$ and pressure $p = 0{\cdot}9\times10^5$ N/m², the pressure on the base BC is $0{\cdot}3p$.

(U. Leeds 1964)

Equation (3.5) corresponds to the expression for dp/p, with $d\delta = \theta$, the change of direction of flow from the initial direction.

In Figure E 3.2 the airfoil is shown, with waves indicated, to which linearised theory will be applied.

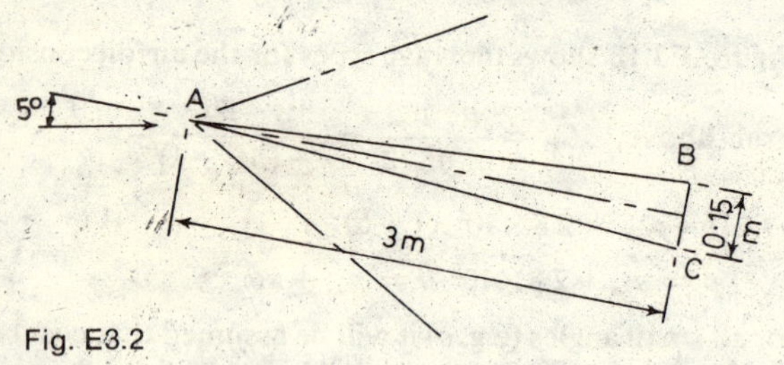

Fig. E3.2

angle, $BAC = 2\alpha$ where $\alpha = \tan^{-1}(0{\cdot}15/3) = 0{\cdot}05$ rad

angle of incidence $5° = 0{\cdot}08727$ rad

$\therefore\ \theta_{AB} = -0{\cdot}0873 + 0{\cdot}05 = -0{\cdot}0373$ rad (Type II—μ positive)

$$\frac{dp_{AB}}{p} = \theta_{AB}\frac{\gamma M^2}{+(M^2-1)^{0{\cdot}5}}$$

$$\therefore p_{AB} - p = -0{\cdot}0373\left(\frac{1{\cdot}4\times5\times0{\cdot}9\times10^5}{2}\right)$$

$$= -0{\cdot}1175\times10^5\ \text{N/m}^2$$

$\theta_{AC} = -0{\cdot}0873 - 0{\cdot}05 = -0{\cdot}1373$ rad (type I—μ negative)

$$\frac{dp_{AC}}{p} = \theta_{AC}\frac{\gamma M^2}{-(M^2-1)^{0{\cdot}5}}$$

$$\therefore p_{AC} - p = 0{\cdot}1373\times3{\cdot}15\times10^5 = +0{\cdot}4325\times10^5\ \text{N/m}^2$$

Lift, $L = (p_{AC}\times\text{area}_{AC})\cos\theta_{AC} - (p_{AB}\times\text{area}_{AB})\cos\theta_{AB} + (p_{BC}\times0{\cdot}15\times1)\sin5°$

(per metre span),

$$L = 1{\cdot}3325\times10^5\times\frac{(3\times1)}{\cos(0{\cdot}05)}\times\cos(0{\cdot}1732)$$

$$-0{\cdot}7825\times10^5\times\frac{(3\times1)}{\cos(0{\cdot}05)}\cos(0{\cdot}0372)$$

$$+0{\cdot}3\times0{\cdot}9\times10^5\times(0{\cdot}15\times1)\sin(0{\cdot}0872)$$

(for small angles (e.g. α) $\cos\alpha = 1$, $\tan\alpha = \sin\alpha = \alpha$)

$\therefore$ Lift $= 159\,200$ N per metre span

Drag, $D = (p_{AC}\times\text{area}_{AC})\sin\theta_{AC} + (p_{AB}\times\text{area}_{AB})\sin\theta_{AB} - (p_{BC}\times0{\cdot}15\times1)\cos5°$

$$= (1{\cdot}3325\times10^5\times3\times0{\cdot}1381) - (0{\cdot}7825\times10^5\times3\times0{\cdot}0373)$$

$$-(0{\cdot}3\times0{\cdot}9\times10^5\times0{\cdot}15\times0{\cdot}09963)$$

$\therefore$ Drag $= 42\,440$ N per metre span

EXAMPLE 3.3 In Figure E 3.3, the flow along a wall ABC is respresented by a streamline DEF which passes through an expansion fan centred at B. The wall bends through $17{\cdot}1°$ and the incident flow conditions are $t_1 = 16°$C and $u_1 = 479{\cdot}4$ m/s.

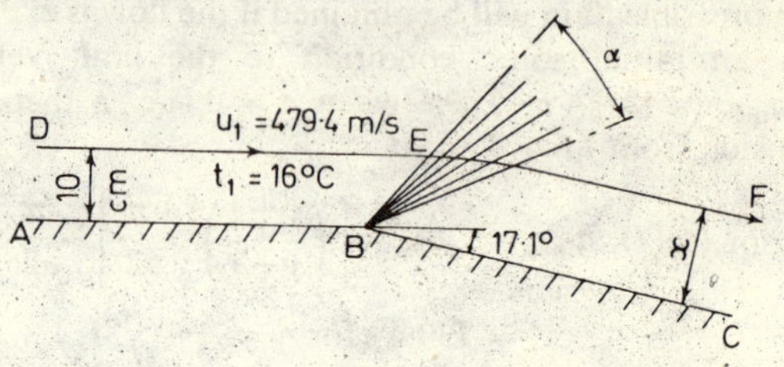

Fig. E3.3

With reference to the undisturbed stream along AB, determine the pressure coefficient over BC and calculate the flow velocity along BC, the distance x cm and the included angle, α, of the expansion fan.

speed of sound, $a = (\gamma RT)^{0{\cdot}5}$

$\therefore a_1 = 20(289)^{0{\cdot}5} = 340$ m/s

$\therefore M_1 = u_1/a_1 = 479{\cdot}4/340 = 1{\cdot}41$

from Table 1, ${}_*\delta_1 = 9{\cdot}28°$ and $p_{01}/p_1 = 3{\cdot}2275$

$\therefore {}_*\delta_2 = {}_*\delta_1 + {}_1\delta_2 = 9{\cdot}28 + 17{\cdot}1 = 26{\cdot}38°$

from Table 1, $M_2 = 2$ and $p_{02}/p_2 = 7{\cdot}8244$

$$\text{pressure coefficient,}\quad C_p = \frac{2}{\gamma M_1^2}\left(\frac{p_2}{p_1} - 1\right) \qquad (2.23)$$

and, $\dfrac{p_2}{p_1} = \dfrac{p_{01}}{p_1}\times\dfrac{p_2}{p_{02}}$ since, for isentropic flow, $p_{01} = p_{02}$

$\therefore p_2/p_1 = 3{\cdot}2275/7{\cdot}8244 = 0{\cdot}4125$

similarly, $T_2/T_1 = (T_0/T_1)/(T_0/T_2)$
$= 1{\cdot}3976/1{\cdot}80 = 0{\cdot}7764$

$$\therefore C_p = -\frac{0{\cdot}5875}{0{\cdot}7\times1{\cdot}41^2} = -0{\cdot}422$$

continuity, $m = \varrho_1u_1A_1 = \varrho_2u_2A_2$

$\therefore$ for unit width, $\dfrac{x}{10} = \dfrac{u_1}{u_2}\times\dfrac{\varrho_1}{\varrho_2}$

where $\varrho_1/\varrho_2 = (p_1/p_2)\times(T_2/T_1) = 0{\cdot}7764/0{\cdot}4125 = 1{\cdot}8822$
and, from Table 1, $u_1/u_2 = (u_1/a_*)\times(a_*/u_2) = 1{\cdot}3065/1{\cdot}633 = 0{\cdot}8001$

$\therefore x = 10\times0{\cdot}8001\times1{\cdot}8822 = 15{\cdot}06$ cm

$u_2 = 479{\cdot}4/0{\cdot}8001 = 599{\cdot}2$ m/s

from Table 1, $\mu_1 = 45{\cdot}17°$ and $\mu_2 = 30°$

$\therefore \alpha = 45{\cdot}17 - (30 - 17{\cdot}1) = 32{\cdot}27°$

alternatively, $\alpha = {}_*\theta_2 - {}_*\theta_1 = 86{\cdot}38 - 54{\cdot}1 = 32{\cdot}28°$

—the difference due to approximation to two figures after the decimal.

Note: Although the streamline distance has increased, the velocity does not decrease as it would for incompressible flow. The more significant *decrease* in density associated with the Prandtl–Meyer expansion results in an *increase* in velocity.

EXAMPLE 3.4 Figure E 3.4 shows a steady two-dimensional isentropic air flow along a solid boundary which bends and then becomes straight again. If the initial conditions are pressure, $p_1 = 0{\cdot}74 \times 10^5$ N/m², flow speed, $q_1 = 339{\cdot}2$ m/s and local acoustic speed, $a_1 = 320$ m/s and, after the bend, the pressure $p_2 = 0{\cdot}5 \times 10^5$ N/m² determine:

(a) The direction of a Mach line after the bend
(b) the direction of the flow after the bend
(c) the components of the flow velocity q_2, parallel and normal to the Mach line direction.

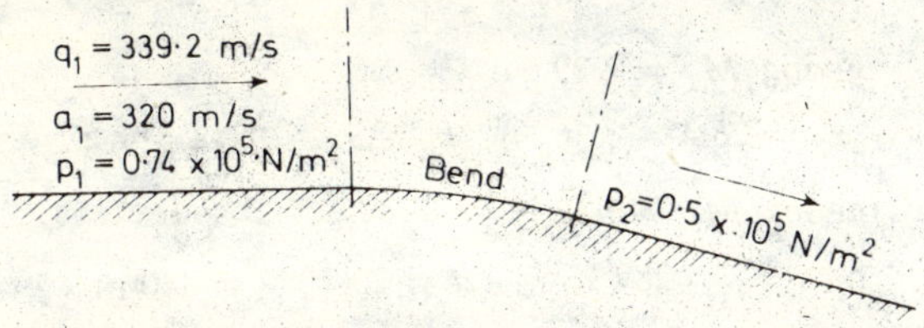

Fig. E3.4

Reference should be made to Figure 3.8 which shows such a bend with a complete presentation of the appropriate Prandtl–Meyer angles. The Mach lines shown represent Type II waves (Figure 3.3c) and the imaginary bend is assumed to accelerate (and expand) the flow from a sonic condition to the given initial condition (1).

$$M_1 = q_1/a_1 = 339{\cdot}2/320 = 1{\cdot}06$$

from Table 1, ${}_*\delta_1 = 0{\cdot}64°$, ${}_*\theta_1 = 20{\cdot}01°$, $\mu_1 = 70{\cdot}63°$ and $p_0/p_1 = 2{\cdot}033$

$$\frac{p_0}{p_2} = \frac{p_0}{p_1} \times \frac{p_1}{p_2} = 2{\cdot}033 \times 0{\cdot}74/0{\cdot}5 = 3{\cdot}009$$

from Table 1, $M_2 = 1{\cdot}36$ (say)

$$\therefore {}_*\delta_2 = 7{\cdot}84°,\ {}_*\theta_2 = 50{\cdot}51°,\ \mu_2 = 47{\cdot}33°$$

(a) ${}_1\theta_2 = {}_*\theta_2 - {}_*\delta_1$ (3.19)

$$\therefore {}_1\theta_2 = 50{\cdot}51 - 0{\cdot}64 = 49{\cdot}87°$$

which is the angle the Mach line at (2) makes with a direction perpendicular to the original flow.

(b) the angle through which the flow has turned is

$${}_1\delta_2 = {}_*\delta_2 - {}_*\delta_1 = 7{\cdot}84 - 0{\cdot}64 = 7{\cdot}2°$$

(c) from Table 1, $q_1/a_* = 1{\cdot}0492$, $q_2/a_* = 1{\cdot}2729$

$$\therefore q_2 = \frac{q_2}{a_*} \times \frac{a_*}{q_1} \times q_1 = \frac{1{\cdot}2729}{1{\cdot}0492} \times 339{\cdot}2 = 411{\cdot}5 \text{ m/s}$$

velocity component normal to Mach line at (2) is a_2

where $a_2 = q_2/M_2 = 411{\cdot}5/1{\cdot}36 = 302{\cdot}6$ m/s

velocity component parallel to Mach line at (2) is m_2

where $m_2 = (q_2^2 - a_2^2)^{0.5} = 278{\cdot}9$ m/s.

(or $m_2 = q_2 \cos \mu_2$)

EXAMPLE 3.5 Figure E 3.5a shows steady two-dimensional isentropic air-flow over a wall which is initially straight (*AB*) and thereafter parabolic, obeying the equation, $y = x^2/20$ with (*B*) the origin. The initial flow (along *AB*) is sonic and at a uniform pressure; on turning upwards to follow *BCD* it becomes supersonic. Points (*E*) and (*F*) are on the *X*-axis, the Mach number (M_E), pressure (p_E) and gas speed (q_E) at (*E*) being 1·63, $0{\cdot}351 \times 10^5$ N/m² and 326 m/s and the pressure (p_F) at *F* being $0{\cdot}302 \times 10^5$ N/m².

Neglecting the influence of other walls and of boundary layers, use the Prandtl–Meyer data to find the positions of *E* and *F*, the directions of gas flow at these points and the local acoustic velocity at *F*.

(U. Birmingham 1961)

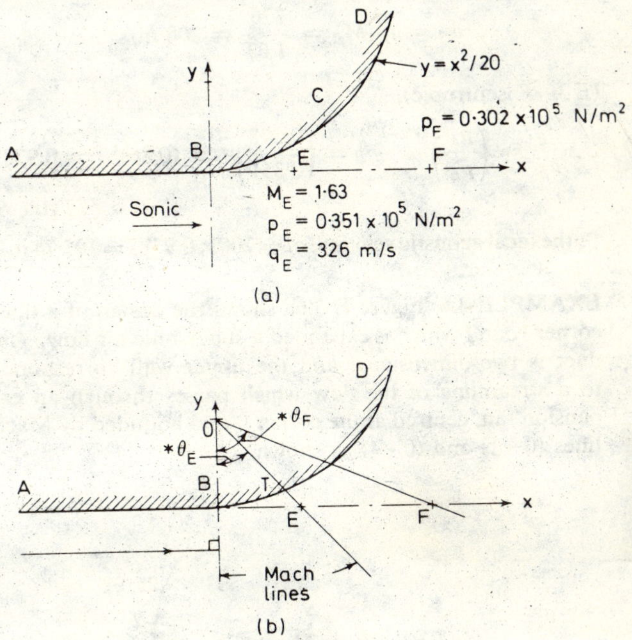

Fig. E3.5

Figure E 3.5b indicates the centred (*O*) Prandtl–Meyer expansion with two Mach lines passing through (*E*) and (*F*) and making angles (${}_*\theta_E$) and (${}_*\theta_F$) with the line (*OB*) at right angles to the original flow direction (by definition).

The line tangent to the parabolic boundary at (*T*), the intersection of the parabola by (*OE*), represents the flow direction (since the parabolic boundary is a streamline) for conditions which are uniform between (*T*) and (*E*). Prandtl–Meyer data for $M_E = 1{\cdot}63$ gives ${}_*\delta_E = 15{\cdot}75°$ (Table 1) tangency condition for $y = x^2/20$ is $dy/dx = x/10$ at *T*,

$$\therefore dy/dx = x/10 = \tan(15{\cdot}75°)$$

$$\therefore x_T = 2{\cdot}82$$

Line *OE* has the equation $y = -m_E x + C$ where *C* is an intercept on the *y*-axis, i.e. $C = OB$ and $m_E = \tan(90 - {}_*\theta_E)$

$$= \tan(90 - 67{\cdot}90°) = 0{\cdot}4061 \quad \text{(Table 1)}$$

for (*T*), $y_T = -m_E x_T + C = x_T^2/20$

$$\therefore -0{\cdot}4061 \times 2{\cdot}82 + C = 0{\cdot}3977$$

giving $C = OB = 1{\cdot}543$

for intersection of line *OE* with the horizontal axis, $x = 0$

$$y_E = 0 = -m_E x_E + C$$

∴ the position of *E* is $x_E = C/m_E = 1{\cdot}543/0{\cdot}4061 = 3.8$

for $M_E = 1{\cdot}63$ $\quad p_0/p_E = 4{\cdot}4442$ (Table 1)

at *F*, $p_0/p_F = p_0/p_E \times p_E/p_F$

$$\therefore p_0/p_F = 4{\cdot}4442 \times 0{\cdot}351/0{\cdot}302 = 5{\cdot}165$$

from Table 1, $M_F = 1{\cdot}73$ $_*\theta_F = 73{\cdot}38°$ and $_*\delta_F = 18{\cdot}69°$

$_*\delta_F$ is the direction of the gas flow and the line OF has the equation $y = -m_F x + C$ where C is the same intercept as above because of the assumption of a centred wave.

$$\therefore m_F = \tan(90 - {_*\theta_F}) = \tan 16{\cdot}62 = 0{\cdot}2985$$

for intersection of line OF with the horizontal axis, $x = 0$

$$y_F = 0 = -m_F x_F + C$$

$\therefore$ the position of F is $x_F = C/m_F = 1{\cdot}543/0{\cdot}2985 = 5{\cdot}17$

at E the local acoustic velocity

$$a_E = q_E/M_E = \frac{326}{1{\cdot}63} = 200 \text{ m/s}$$

($E \rightarrow F$ isentropic),

$$\therefore \frac{a_F}{a_E} = \left(\frac{p_F}{p_E}\right)^{(\gamma-1)/2\gamma} = \left(\frac{0{\cdot}302}{0{\cdot}351}\right)^{1/7} = (0{\cdot}86)^{1/7} = 0{\cdot}9787$$

(Table 6)

$\therefore$ the local acoustic velocity, $a_F = 200 \times 0{\cdot}9787 = 195{\cdot}74$ m/s

EXAMPLE 3.6 Figure E 3.6a shows the design of a duct corner round which is expanded a supersonic air flow. The duct is two-dimensional and the upper wall corresponds to a streamline of the flow which passes through an expansion fan, centred at the corner 0 and bounded by Mach lines (0—1) and (0—4) as shown.

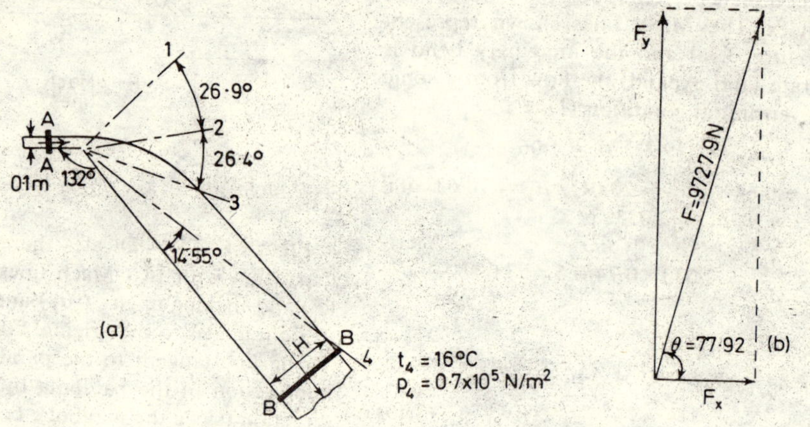

Fig. E3.6

The duct corner is flanged at A and B at which section the uniform conditions of the flow are pressure, $p_4 = 0{\cdot}7 \times 10^5$ N/m² and temperature, $t_4 = 17$°C.

Determine,

(1) the included angle 1—0—4 of the expansion fan,

(2) the dimension H metres of the duct if the width is constant,

(3) the Mach number and the local speed of sound at the Mach line positions (0—2) and (0—3),

(4) the mass flow through the duct (per unit width),

(5) the magnitude and direction of the force acting on the flanged section (per unit width).

(U. Leeds 1968)

(1) Mach angle $\mu_4 = 14{\cdot}55°$ $\therefore M_4 = 3{\cdot}98$ (Table 1) (the Mach lines represent Type II waves—Figure 3.3c)

also, $_*\delta_4 = 65{\cdot}52°$ $\therefore {_*\delta_1} = {_*\delta_4} - {_1\delta_4}$

$\therefore {_*\delta_1} = 65{\cdot}52 - 48 = 17{\cdot}52°$ giving $M_1 = 1{\cdot}69$ (Table 1)

then Mach angle $\mu_1 = 36{\cdot}28°$

and $_1\alpha_4 = \mu_1 + {_1\delta_4} - \mu_4 = 36{\cdot}28 + 48 - 14{\cdot}55 = 69{\cdot}73°$

(2) continuity per 1 cm width,

$$\varrho_1 q_1 \times 0{\cdot}1 = \varrho_4 q_4 \times H$$

or, $$H = \frac{q_1}{a_*} \times \frac{a_*}{q_4} \times \frac{\varrho_1}{\varrho_0} \times \frac{\varrho_0}{\varrho_4} \times 0{\cdot}1$$

$(p = \varrho RT)$, $\therefore \dfrac{\varrho}{\varrho_0} = \dfrac{p}{p_0} \times \dfrac{T_0}{T}$ values obtained from Table 1

then $$H = \frac{1{\cdot}4769}{2{\cdot}1355} \times \frac{1{\cdot}5712}{4{\cdot}8622} \times \frac{147.835}{4.1681} \times 0{\cdot}1 = 0{\cdot}793 \text{ m}$$

(3) from the Prandtl–Meyer expansion fan and using data from Table 1

$_1\alpha_2 = 26{\cdot}9°$ $\therefore {_*\theta_2} = {_*\theta_1} + {_1\alpha_2} = 71{\cdot}24 + 26{\cdot}9 = 98{\cdot}14°$

giving $M_2 = 2{\cdot}29$

$_2\alpha_3 = 26{\cdot}4°$ $\therefore {_*\theta_3} = {_*\theta_2} + {_2\alpha_3} = 98{\cdot}14 + 26{\cdot}4 = 124{\cdot}54°$

giving $M_3 = 3{\cdot}17$

now $T_4 = 289$ K $\therefore a = 20(T)^{0{\cdot}5} = 340$ m/s

$$\therefore q_4 = 3{\cdot}98 \times 340 = 1353{\cdot}2 \text{ m/s}$$

for $M_4 = 3{\cdot}98$, $q_4/a_* = 2{\cdot}1355$ (Table 1)

$$\therefore a_* = 1353{\cdot}2/2{\cdot}1355 = 633{\cdot}7 \text{ m/s}$$

from (3.13a) with $\gamma = 7/5$, $a = a_* \cos({_*\theta}/\sqrt{6})$

$$\therefore a_2 = 633{\cdot}7 \cos(98{\cdot}14/\sqrt{6}) = 485{\cdot}9 \text{ m/s}$$

and $a_3 = 633{\cdot}7 \cos(124{\cdot}54/\sqrt{6}) = 400{\cdot}9$ m/s

(4) mass flow, $m = \varrho_4 q_4 H$ per cm width

$$\therefore m = 0{\cdot}8437 \times 1353{\cdot}2 \times 0{\cdot}793 \times 10^{-2} = 9{\cdot}05 \text{ kg/s}$$

(5) for consideration of momentum changes in the X and Y directions the control volume is conveniently drawn at the flanges,

consider momentum change in the X-direction ($\xrightarrow{+}$)

$$m(q_4 \cos 48° - q_1) = p_1 A_1 - (p_4 A_4) \cos 48° + F_x \qquad \text{(A)}$$

where F_x is the force exerted, in the X-direction, by the transition casing on the fluid in the control volume.

from Table 1, $q_1/a_* = 1{\cdot}4769$ $\therefore q_1 = 1{\cdot}4769 \times 633{\cdot}7 = 935{\cdot}9$ m/s also, $p_0/p_1 = 4{\cdot}8622$, $p_0/p_4 = 147{\cdot}835$

$\therefore p_1 = 147{\cdot}835 \times 0{\cdot}7 \times 10^5 / 4{\cdot}8622 = \underline{21{\cdot}28 \times 10^5 \text{ N/m}^2}$

in (A), $9{\cdot}05(1353{\cdot}2 \times 0{\cdot}66913 - 935{\cdot}9) = (21{\cdot}28 \times 0{\cdot}1 \times 10^{-2} - 0{\cdot}7 \times 0{\cdot}793 \times 10^{-2} \times 0{\cdot}6691)10^5 + F_x$

$\therefore F_x = -2032$ N

$\therefore$ force exerted *on* casing $= \underline{+2032 \text{ N}} (\rightarrow)$

consider momentum change in Y-direction $(+\downarrow)$

$$m(q_4 \sin 48°) = -(p_4 A_4) \sin 48° + F_y$$

$9{\cdot}05 \times 1353{\cdot}2 \times 0{\cdot}7431 = -0{\cdot}7 \times 0{\cdot}793 \times 10^{-2} \times 0{\cdot}7431 + F_y$

$\therefore F_y = 9513{\cdot}3$ N

$\therefore$ force exerted *on* casing $= \underline{-9\,513{\cdot}3 \text{ N}} (\uparrow)$

(Figure 3.6b) resultant force $F = (F_x^2 + F_y^2)^{0.5} = \underline{9727{\cdot}9 \text{ N}}$

at an angle $\theta = \tan^{-1}(F_y/F_x) = \underline{77{\cdot}92°}$

4 THE PLANE NORMAL SHOCK WAVE

4.1 CONCEPT AND ANALYSIS

The plane normal shock wave is essentially an irreversible discontinuity in a flow field, at which there is a sudden change in the characteristics of the flow. A stationary shock wave may occur in a duct or nozzle and Figure 4.1 shows

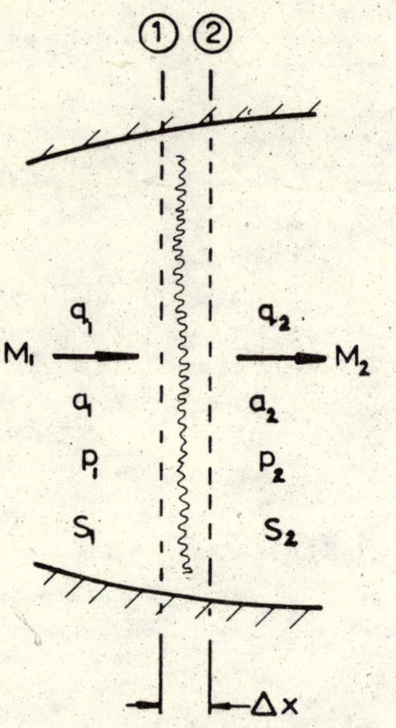

Fig. 4.1 control volume for plane normal shock wave

a control volume for which the change in flow characteristics across a shock wave may be determined. Since the thickness (Δx) of the shock wave is of the order of 10^{-2} mm the effects of area change (ΔA) in the duct or nozzle may be neglected.

For non-isentropic flow through the stationary shock wave the equations of continuity, momentum and energy are used for the analysis.

continuity, mass flow, $m = \varrho q A$ and $A_1 = A_2$

$$\therefore \varrho_1 q_1 = \varrho_2 q_2 = \varrho q \tag{4.1}$$

energy,
$$q_1^2 - q_2^2 = \frac{2\gamma}{\gamma-1}\left(\frac{p_2}{\varrho_2} - \frac{p_1}{\varrho_1}\right) \tag{4.2}$$

momentum,
$$\varrho q A(q_2 - q_1) = (p_1 - p_2)A \tag{4.3}$$

or,
$$q_1(q_1 - q_2) = (p_2 - p_1)\frac{1}{\varrho_1}$$

and,
$$q_2(q_1 - q_2) = (p_2 - p_1)\frac{1}{\varrho_2}$$

adding,
$$q_1^2 - q_2^2 = (p_2 - p_1)\left(\frac{1}{\varrho_1} + \frac{1}{\varrho_2}\right) \tag{4.4}$$

By equating (4.2) with (4.4) and with some manipulation using the state relationship

$$\frac{\varrho_2}{\varrho_1} = \frac{p_2 T_1}{p_1 T_2}, \text{ it can be shown that } \frac{T_2}{T_1} = \frac{\dfrac{\gamma+1}{\gamma-1} + \dfrac{p_2}{p_1}}{\dfrac{\gamma+1}{\gamma-1} + \dfrac{p_1}{p_2}} \tag{4.5}$$

and
$$\frac{\varrho_2}{\varrho_1} = \frac{\dfrac{\gamma+1}{\gamma-1}\dfrac{p_2}{p_1} + 1}{\dfrac{\gamma+1}{\gamma-1} + \dfrac{p_2}{p_1}} = \frac{q_1}{q_2} \tag{4.6}$$

Equations (4.5) and (4.6) are known as the Rankine–Hugoniot equations and Figure 4.2 shows a graph of T_2/T_1 plotted against p_2/p_1 for (4.5) and for the isentropic relationship $T_2/T_1 = (p_2/p_1)^{(\gamma-1)/\gamma}$ for dry air, $\gamma = 7/5$. The significance of the dashed portion of the shock line may be examined by considering the change of entropy across the shock,

$$s_2 - s_1 = c_p \log_e \frac{T_2}{T_1} - R \log_e \frac{p_2}{p_1} \tag{1.20a}$$

For the dashed line, $0 < p_2/p_1 < 1$ and, from Figure 4.2, for any given value of p_2/p_1 $(T_2/T_1)_{shock} < (T_2/T_1)_{isentropic}$. In (1.20a) this would indicate that there is a decrease of entropy for an adiabatic irreversible process. Since this contravenes the Second Law of Thermodynamics a shock

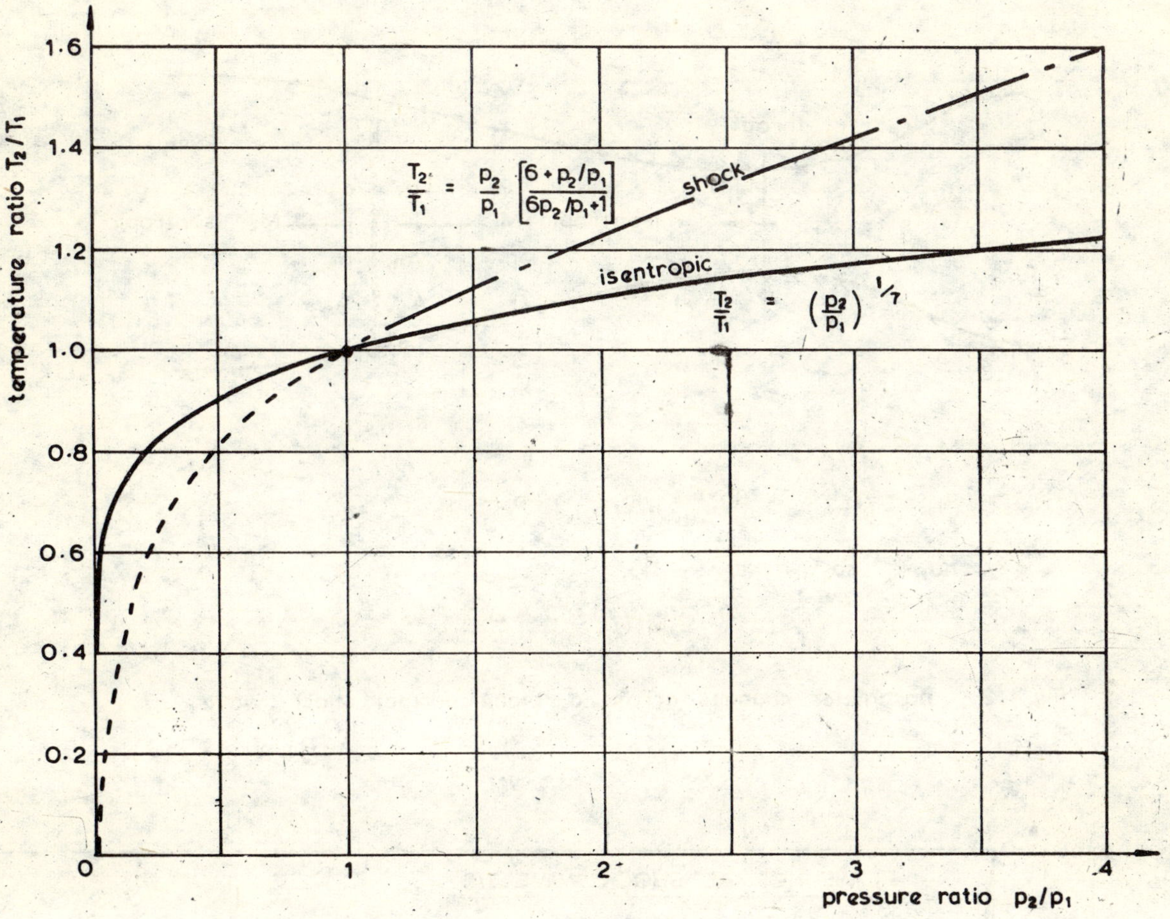

Fig. 4.2 isentropic and normal shock relationships for air ($\gamma = 7/5$)

can only exist when $(T_2/T_1)_{shock} > (T_2/T_1)_{isentropic}$ which means that a shock wave must be a compressive mechanism, $p_2/p_1 > 1$.

It is convenient to express the characteristic changes across a shock as ratios which are functions of the incident Mach number (M_1) and the range of values is so presented in Table 2. The following approach may be used to establish the relationship $p_2/p_1 = f(M)$.

From (4.1) and (4.3) and using $a^2 = \gamma p/\varrho$,

$$\frac{p_2}{p_1} - 1 = \gamma M_1^2\left(1 - \frac{q_2}{q_1}\right) \tag{4.7}$$

using (4.6) in (4.7) with some manipulation, the required result is obtained in the form,

$$\frac{p_2}{p_1} = 1 + \frac{2\gamma}{\gamma+1}(M_1^2 - 1) \tag{4.8a}$$

alternatively,

$$\frac{p_1}{p_2} = 1 + \frac{2\gamma}{\gamma+1}(M_2^2 - 1) \tag{4.8b}$$

In (4.8a) if $p_2/p_1 > 1$ for a plane normal shock wave to exist then $M_1 > 1$ and a shock can only occur in supersonic flow. In (4.8b) $M_2 < 1$ and the flow downstream of a stationary shock is always subsonic.

The Rankine–Hugoniot relationships (4.5) and (4.6) used with (4.8a) give the following equations:

$$\frac{T_2}{T_1} = \frac{[2\gamma M_1^2 - (\gamma-1)][(\gamma-1)M_1^2 + 2]}{(\gamma+1)^2 M_1^2} \tag{4.9}$$

$$\frac{\varrho_1}{\varrho_2} = \frac{(\gamma-1)M_1^2 + 2}{(\gamma+1)M_1^2} = \frac{q_2}{q_1} \tag{4.10}$$

with (4.8b),

$$M_2 = \left[\frac{(\gamma-1)M_1^2 + 2}{2\gamma M_1^2 - (\gamma-1)}\right]^{0.5} \tag{4.11}$$

from (1.20b), $s_2 - s_1 = c_p \log_e \dfrac{\varrho_1}{\varrho_2} + c_v \log_e \dfrac{p_2}{p_1}$

or,

$$\frac{s_2 - s_1}{R} = \frac{\gamma}{\gamma-1}\log_e\left[\frac{(\gamma-1)M_1^2+2}{(\gamma+1)M_1^2}\right] + \frac{1}{\gamma-1}\log_e\left[\frac{2\gamma M_1^2 - (\gamma-1)}{\gamma+1}\right] \tag{4.12}$$

On either side of the shock wave (e.g. in the divergent portion of the nozzle in Figure 4.3) the flow process is isentropic and the total head pressures and temperatures are constant but only the total head temperature is constant *across* the shock. Across the adiabatic irreversible shock wave the flow field cannot be considered irrotational and the total head pressure is not constant, $p_{01} > p_{02}$.

$$\frac{p_{02}}{p_{01}} = \frac{p_{02}}{p_2} \times \frac{p_1}{p_{01}} \times \frac{p_2}{p_1}$$

from (2.15) and (4.8a),

$$\frac{p_{02}}{p_{01}} = \left[\frac{(\gamma+1)M_1^2}{(\gamma-1)M_1^2+2}\right]^{\frac{\gamma}{\gamma-1}}\left[\frac{\gamma+1}{2\gamma M_1^2 - (\gamma-1)}\right]^{\frac{1}{\gamma-1}} \tag{4.13}$$

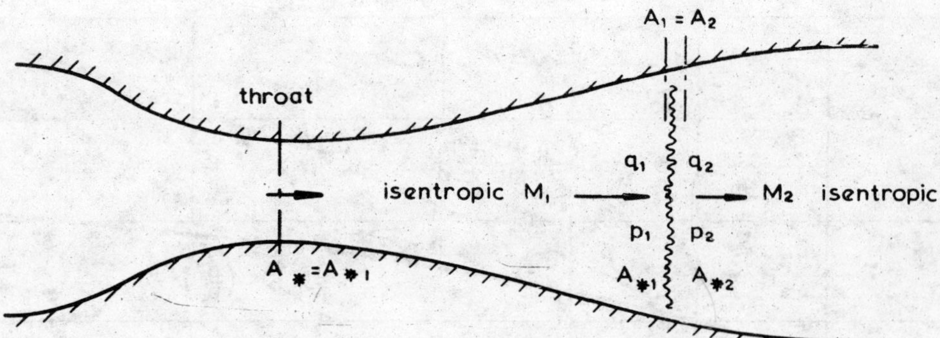

$M_1 > M_2$

$q_1 > q_2$

$p_{o_1} > p_{o_2}$

$T_{o_1} = T_{o_2}$

$T_1 < T_2$

$p_1 < p_2$

$s_1 < s_2$

$A_{*1} < A_{*2}$

Fig. 4. 3 parameter changes across a plane normal shock wave

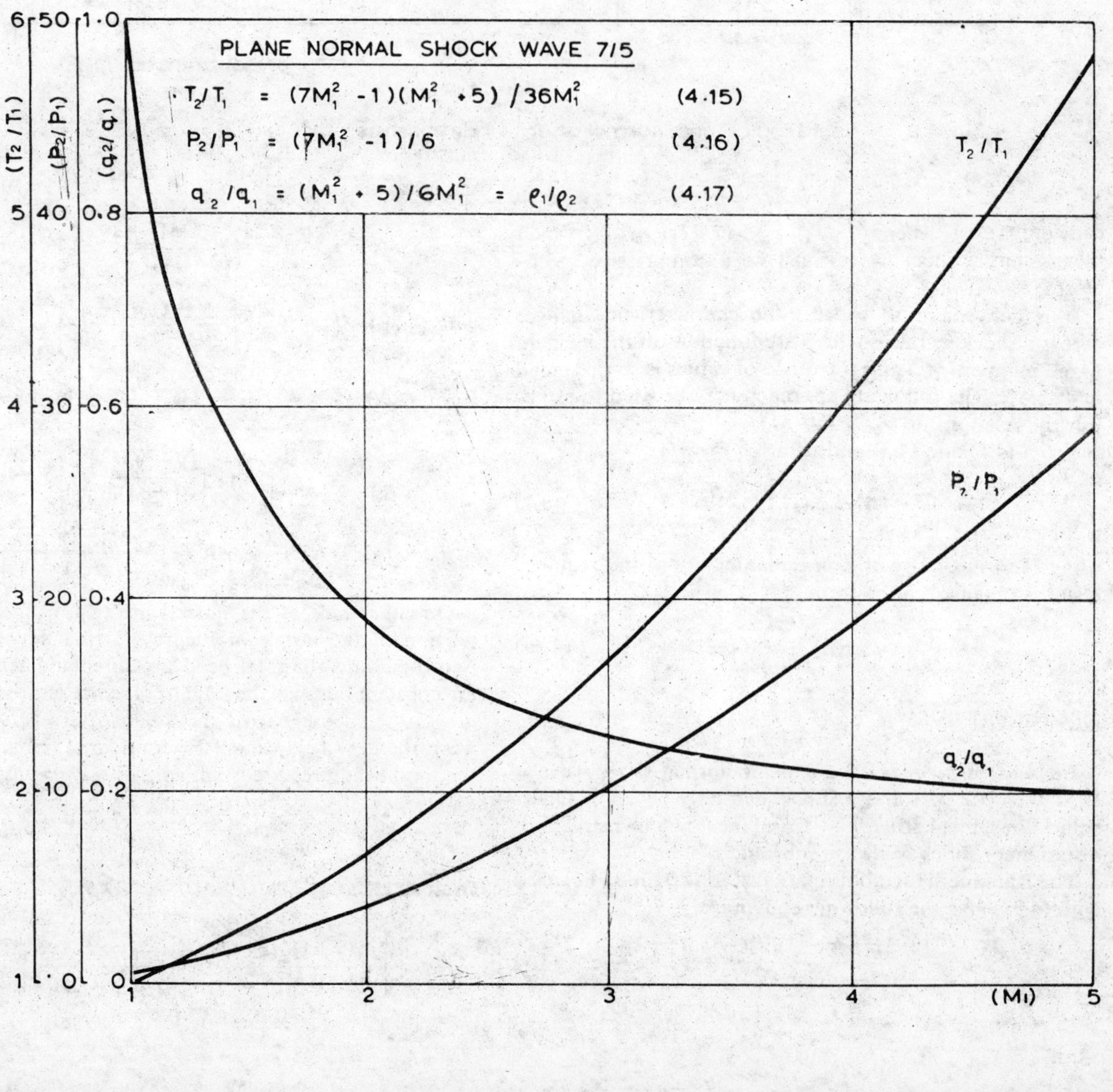

4.4

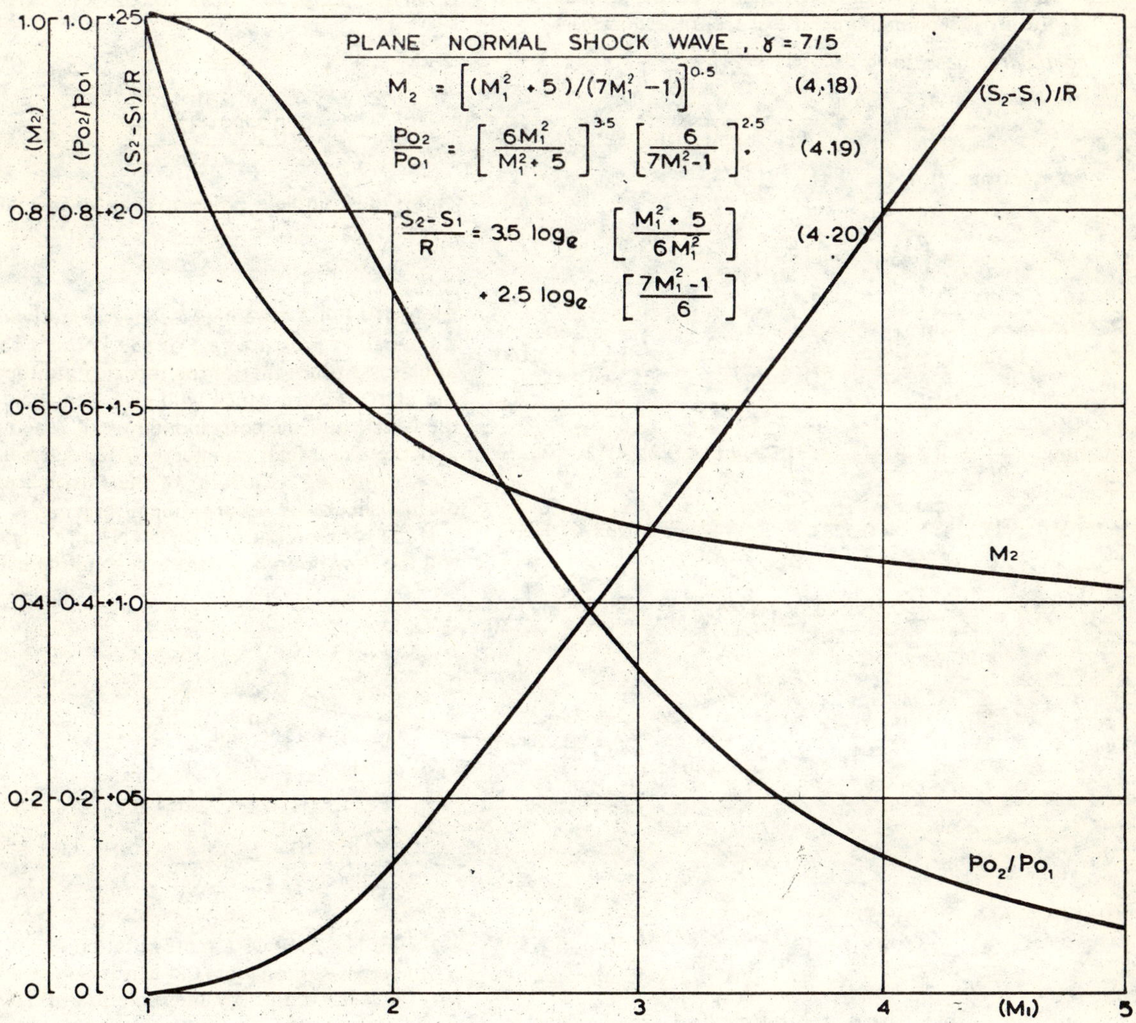

Fig. 4·5

from (2.31), $p_{02}/p_{01} = A_{*1}/A_{*2}$ (4.13a)

The increase in entropy across the shock is also given by (2.11).

$$\frac{s_2-s_1}{R} = \log_e\left(\frac{p_{01}}{p_{02}}\right) \log_e\left(\frac{A_{*2}}{A_{*1}}\right) \quad (4.14)$$

In the divergent section of a nozzle, Figure 4.3, because the shock wave is so narrow $A_1 = A_2$ but $M_1 > M_2$. However, A_1/A_* gives, from (2.32) a value $M_1 > 1$ and a value $M < 1$ which is not equal to M_2. M_2 corresponds to a different value of A/A_*. Since $A_1 = A_2$ the discrimination between the two sides of the shock wave must be $A_{*1} \neq A_{*2}$. A_{*1} corresponds to the real nozzle throat upstream of the shock and A_{*2} corresponds to an imaginary nozzle throat for the isentropic flow downstream of the shock.

A summary of parameter changes is shown in Figure 4.3.

4.2 SUMMARY OF EQUATIONS

The following are the equations used in Table 2 for steady adiabatic flow of dry air ($\gamma = 7/5$) through a plane normal shock wave and are plotted in Figures 4.4 and 4.5.

From (4.9), $$\frac{T_2}{T_1} = \frac{(7M_1^2-1)(M_1^2+5)}{36M_1^2} \quad (4.15)$$

from (4.8a), $$\frac{p_2}{p_1} = \frac{7M_1^2-1}{6} \quad (4.16)$$

from (4.10), $$\frac{q_2}{q_1} = \frac{M_1^2+5}{6M_1^2} = \frac{\varrho_1}{\varrho_2} \quad (4.17)$$

from (4.11), $$M_2 = \left[\frac{M_1^2+5}{7M_1^2-1}\right]^{0.5} \quad (4.18)$$

from (4.13), $$\frac{p_{02}}{p_{01}} = \left[\frac{6M_1^2}{M_1^2+5}\right]^{3.5}\left[\frac{6}{7M_1^2-1}\right]^{2.5} \quad (4.19)$$

from (4.12), $$\frac{s_2-s_1}{R} = 3.5\log_e\left[\frac{M_1^2+5}{6M_1^2}\right] + 2.5\log_e\left[\frac{7M_1^2-1}{6}\right] \quad (4.20)$$

EXAMPLE 4.1 Air ($\gamma = 7/5$) enters a convergent-divergent nozzle with a stagnation pressure of 12×10^5 N/m². A shock wave is observed in the divergence at a section where the Mach number, $M = 2$. Without using the compressible flow tables, find the ratio of the cross-sectional area at this section, to that at the throat of the nozzle.

If the cross-sectional area at the exist section is 4 times that at the throat, use the tables to find the Mach number and static pressure at exit from the nozzle (assuming the shock wave to be plane and normal in an ideal fluid).

Figure E 4.1 shows a representation of the problem with A_3 the cross-sectional area at exit from the nozzle.

continuity, mass flow $= \varrho_1 q_1 A_1 = \varrho_* q_* A_*$

$$\therefore \frac{A_1}{A_*} = \frac{q_*}{q_1} \times \frac{\varrho_*}{\varrho_1} = \frac{M_*}{M_1} \times \frac{a_*}{a_1} \frac{\varrho_*}{\varrho_1}$$

for isentropic flow (*) → (1),

$$\frac{a_*}{a_1} = \left(\frac{T_*}{T_1}\right)^{0.5} = \left(\frac{p_*}{p_1}\right)^{\frac{\gamma-1}{2\gamma}} = \left(\frac{\varrho_*}{\varrho_1}\right)^{\frac{\gamma-1}{2}}$$

$$(\gamma = 7/5), \quad \therefore \frac{A_1}{A_{*1}} = \frac{M_*}{M_1}\left(\frac{p_*}{p_1}\right)^{\frac{\gamma+1}{2\gamma}} = \frac{1}{2}\left(\frac{p_*}{p_1}\right)^{6/7} \quad \text{(A)}$$

$$\text{now,} \quad \frac{p_{01}}{p_1} = \left(1+\frac{\gamma-1}{2}M_1^2\right)^{\frac{\gamma}{\gamma-1}} = (1+0{\cdot}2\times 4)^{3\cdot 5} = 7{\cdot}8244$$

$$(M_* = 1), \quad \frac{p_*}{p_{01}} = 0{\cdot}5283 \qquad \therefore \frac{p_*}{p_1} = 0{\cdot}5283\times 7{\cdot}8244$$

$$\text{in } (A), \quad \frac{A_1}{A_{*1}} = 0{\cdot}5(0{\cdot}5283\times 7{\cdot}8244)^{6/7} = \underline{1{\cdot}6875}$$

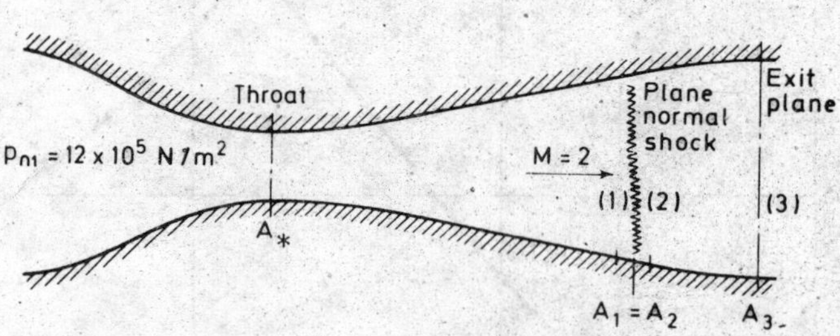

Fig. E4.1

[alternatively, (2.32) → (2.35),

$$\frac{A_1}{A_{*1}} = \frac{1}{M_1}\left(\frac{5+M_1^2}{6}\right)^3 = 1{\cdot}5\tfrac{3}{8} = 1{\cdot}6875]$$

For the second part of the question $A_{*1} \neq A_{*2}$ due to change in total head pressure across the plane normal shock wave.

from Table 2, $M_1 = 2$, $M_2 = 0{\cdot}5774$, $p_{02}/p_{01} = 0{\cdot}7209$

from Table 1, for $M_2 = 0{\cdot}5774$, A_2/A_{*2} is obtained by interpolation,

M	(diff.)	A/A_*	(diff.)
0·57		1·2263	
$M_2 = 0{\cdot}5774$	(0·0074)	$A_2/A_{*2} = 1{\cdot}2165$	
0·58	(0·0100)	1·2130	(0·0133)

(1·2165 = 1·2263 − 0·0133 × 74/100)

$A_2 = A_1 \quad \therefore A_{*2}/A_{*1} = 1{\cdot}6875/1{\cdot}2165$

$A_3/A_{*1} = 4 \quad \therefore A_3/A_{*2} = 4\times 1{\cdot}2165/1{\cdot}6875 = \underline{2{\cdot}8836}$

[alternatively, $A_{*2}/A_{*1} = p_{01}/p_{02}$ (4.13a)]

from Table 1, for $A_3/A_{*2} = 2{\cdot}8836$, M_3 and p_{03}/p_3 are obtained by interpolation

A/A_*	(diff.)	M (diff.)
2·9635		0·20
$A_3/A_* = 2{\cdot}8836$	(0·0799)	$M_3 = 0{\cdot}206$
2·8293	(0·1342)	0·21

(0·206 = 0·20 + 0·01 × 0·799/1·342)

Similarly, $p_{03}/p_3 = 1{\cdot}0283+0{\cdot}0029\times 0{\cdot}799/1{\cdot}342 = 1{\cdot}03$

$$p_3 = \frac{p_3}{p_{02}} \times \frac{p_{02}}{p_{01}} \times p_{01}$$

$$\therefore p_3 = \frac{0{\cdot}7209\times 12\times 10^5}{1{\cdot}03} = 8{\cdot}4\times 10^5 \text{ N/m}^2$$

Note: the complete process is shown in Figure 2.7 $(S_1 \to S_2 \to S_3)$

EXAMPLE 4.2 Air enters a convergent-divergent nozzle at a stagnation pressure (p_0) of 689 kN/m². The nozzle area at exit is 3 times that of the throat. Find the two values of the exit pressure which willl give isentropic flow through the nozzle and the corresponding exit Mach numbers.

If the exit Mach number is 0·32, determine the nozzle area, expressed as a ratio of the throat area, at which a normal shock will occur in the divergence.

If the back pressure is 30·05 kN/m², through what angle will the flow boundary deflect beyond the exit?

(U. Leeds 1969)

Figure E 4.2 shows a representation of the problem with A_3 the cross-sectional area at exit from the nozzle.

$\therefore A_3/A_* = 3{\cdot}0$ and, by interpolation in Table 1, the subsonic and supersonic isentropic flow solutions of (2.35) may be obtained.

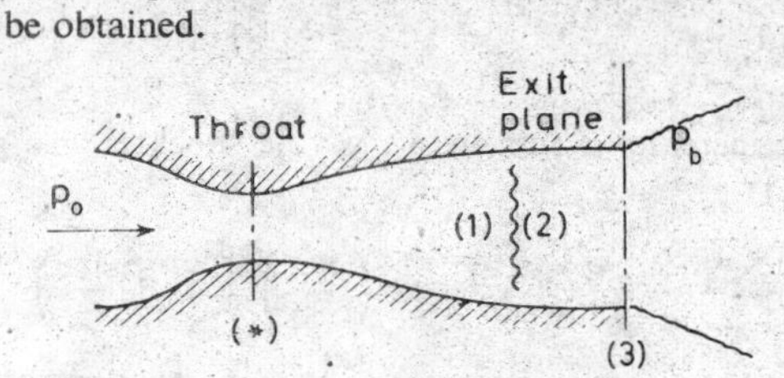

Fig. E4·2

subsonic A/A_*	(diff.)	M (diff.)
3·1123		0·19
$A_3/A_* = 3{\cdot}0000$	(0·1123)	$M_3 = 0{\cdot}1975$
2·9635	(0·1488)	0·20

(0·1975 = 0·19 + 0·01 × 1123/1488)

Similarly, $p_{03}/p_3 = 1{\cdot}0255+0{\cdot}0028\times 1123/1488 = 1{\cdot}0276$

$(p_{01} = p_{03})$, $\therefore p_3 = 689/1{\cdot}0266 = \underline{670{\cdot}5 \text{ kN/m}^2}$

supersonic A/A_*	(diff.)	M (diff.)
2·9791		2·63
$A_3/A_* = 3{\cdot}0000$	(0·0209)	$M_3 = 2{\cdot}6374$
3·0073	(0·0282)	2·64

(2·6374 = 2·63 + 0·01 × 209/282)

Similarly, $p_{03}/p_3 = 20{\cdot}901+0{\cdot}326\times209/282 = 21{\cdot}143$

$\therefore\ p_3 = 689/21{\cdot}143 = 32{\cdot}59$ kN/m^2

normal shock if $M_3 = 0{\cdot}32$, $A_3/A_{*2} = 1{\cdot}9219$

the change in total head pressure across a plane normal shock wave is directly related to the change in A_* ($A_{*1} \neq A_{*2}$) and may be obtained by inspection of equation (2.31) as,

$$p_{02}/p_{01} = A_{*1}/A_{*2} \qquad (4.13a)$$

since $A_3/A_{*1} = 3$, $p_{02}/p_{01} = 1{\cdot}9219/3 = 0{\cdot}6406$

the following interpolation links Table 1 with Table 2

Table 2

	p_{02}/p_{01}	(diff.)	M
	0·6419		2·17
$p_{02}/p_{01} =$	0·6406	(0·0013)	M_1
	0·6373	(0·0046)	2·18

Table 1

M		A/A_*	(diff.)	
2·17		1·9525		
M_1	$A_1/A_{*1} =$	1·9574	=	1·9525+0·0173×13/46
2·18		1·9698	(0.0173)	

at nozzle exit $p_0/p_3 = 21{\cdot}143$ and $p_3 = 32{\cdot}59$ kN/m^2

from Table 1, ${}_*\delta_3 = 42{\cdot}09+0{\cdot}22\times209/282 = 42{\cdot}253°$

for a back pressure $p_b = 30{\cdot}05$ kN/m^2 ($p_b < p_3$) the flow is isentropic throughout and, $p_0/p_b = 689/30{\cdot}05 = 22{\cdot}927$

from Table 1, ${}_*\delta_b = 43{\cdot}50°$

$\therefore$ flow boundary deflection $= 43{\cdot}40-42.253 = 1{\cdot}147°$

EXAMPLE 4.3 An intense plane normal shock occurs in a steady supersonic air stream, the temperature rise being such that allowance must be made for variations in specific heat. Assuming the air to obey the gas law $p = \varrho RT$ and its specific heat (c_p) at constant pressure to follow the law $c_p = A+BT$ where T is the absolute temperature and A and B constants, establish an equation relating the static pressure ratio (p_2/p_1) and the static temperatures T_2 and T_1, suffixes 1 and 2 referring respectively to upstream and downstream conditions.

If, in the above, $A = 0{\cdot}912$ kJ/kg K, $B = 17{\cdot}56\times10^{-5}$ kJ/kg K^2 and $R = 0{\cdot}2871$ kJ/kg K, determine the temperature rise if the static pressure ratio for the shock is 10 and $T_1 = 367$ K.

(U. Birmingham 1964).

For a plane normal shock wave (Figure 4.1) the equations of continuity, momentum and energy may be applied neglecting area change, thus,

continuity $\quad A\varrho_1 q_1 = A\varrho_2 q_2 = A\varrho q \qquad$ (A)

momentum, $\quad A\rho q(q_2-q_1) = A(p_1-p_2)$

or, $\quad \rho_1 q_1(q_1-q_2) = p_2-p_1$

and, $\quad \varrho_2 q_2(q_1-q_2) = p_2-p_1$

adding, $\quad q_1^2-q_2^2 = (p_2-p_1)\left(\dfrac{1}{\varrho_2}+\dfrac{1}{\varrho_1}\right) \qquad$ (B)

energy, $\quad h_0 = h_1+q_1^2/2 = h_2+q_2^2/2$

or, $\quad h_2-h_1 = (q_1^2-q_2^2)/2$

$$\therefore \int_{T_1}^{T_2} c_p\,dT = (q_1^2-q_2^2)/2 \qquad (C)$$

now, $$\int_{T_1}^{T_2} c_p dT = \int_{T_1}^{T_2}(A+BT)dT = A(T_2-T_1)+B(T_2^2-T_1^2)/2$$

substituting in (C) and equating with (B).

$$(p = \varrho RT),\ A(T_2-T_1)+B(T_2^2-T_1^2)/2 = (p_2-p_1)(T_1/p_1+T_2/p_2)R/2$$

whence, $$\frac{BT_1}{R}\left(\frac{T_2}{T_1}\right)^2+\left(\frac{2A}{R}+\frac{p_1}{p_2}-1\right)\frac{T_2}{T_1}+\left(1-\frac{p_2}{p_1}-\frac{2A}{R}-\frac{BT_1}{R}\right) = 0$$

substituting the data given produces a quadratic in (T_2/T_1) as follows,

$$\frac{17{\cdot}56\times10^{-5}\times367}{0{\cdot}2871}\left(\frac{T_2}{T_1}\right)^2+\left(\frac{2\times0{\cdot}912}{0{\cdot}2871}+\frac{1}{10}-1\right)\frac{T_2}{T_1}+\left(1-10-\frac{2\times0{\cdot}912}{0{\cdot}2871}-\frac{17{\cdot}56\times10^{-5}\times367}{0{\cdot}2871}\right) = 0$$

giving, $\quad (T_2/T_1)^2+24{\cdot}29(T_2/T_1)-69{\cdot}39 = 0$

whence, $T_2/T_1 = 2{\cdot}582$ $\therefore$ temperature rise $= 580{\cdot}6$ K

from Table 2, ($\gamma = 7/5$), $T_2/T_1 = 2{\cdot}6229$ (by interpolation)

$\therefore$ temperature rise $= 595{\cdot}6$ K (constant specific heats)

EXAMPLE 4.4 The flow of air through the frictionless duct shown in Figure E 4.4a is adiabatic and the outlet conditions at A_4 are $t_4 = 16$°C, $p_4 = 30$ in Hg abs., with a Pitot-tube measurement of 14.4 in Hg. The area $A_2 = 1685$ mm^2 and flow is sonic at $A_1 = 500$ mm^2. A plane normal shock occurs at $A_s = 1235{\cdot}3$ mm^2.

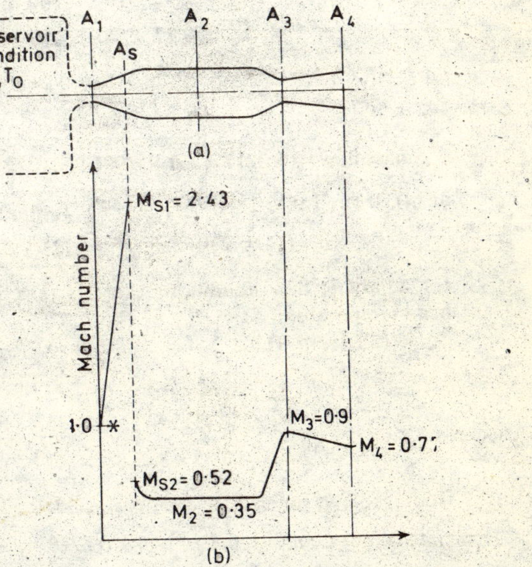

Fig. E4.4

If the Mach number at throat A_3 is not to exceed 0·9, calculate the minimum throat area, (A_3), and plot the variation of Mach number along the length of the duct.

Determine the mass flow of air through the duct, the reservoir conditions (p_0) and (T_0) and the flow speed at A_2. What are the percentage changes in total head pressure and temperature? (1 in Hg = 3379 N/m^2).

(U. Leeds 1971.)

At section (4), static pressure $p_4 = 30$ in Hg abs total head pressure, $p_{04} = 30+14{\cdot}4 = 44{\cdot}4$ in Hg abs.

$$\therefore\ p_{04}/p_4 = 1{\cdot}48$$

from Table 1, $M_4 = 0{\cdot}77$ and $A_4/A_{*2} = 1{\cdot}0519$

The normal shock must be located in supersonic incident flow which can only occur between A_1 and A_3; due to the consequent change in total head pressure $A_{*1} \neq A_{*2}$ and conditions on each side of the shock are referred to by suffixes (sl) and s2).

$$\frac{A_s}{A_{*1}} = \frac{1235{\cdot}3}{500} = 2{\cdot}4706 \quad \therefore\ M_{s1} = 2{\cdot}43 \text{ (from Table 1)}$$

and, from Table 2, $M_{s2} = 0{\cdot}52$ also $\dfrac{A_2}{A_{*2}} = 1{\cdot}3034$ (Table 1)

for $M_3 = 0{\cdot}9$, $\dfrac{A_3}{A_{*2}} = 1{\cdot}0089$ (Table 1)

$$\therefore\ \frac{A_3}{A_s} = \frac{1{\cdot}0089}{1{\cdot}3034} \quad \therefore\ A_3 = 0{\cdot}774\times 1235{\cdot}3 = 956{\cdot}2\ \text{mm}^2$$

similarly, $\dfrac{A_4}{A_s} = \dfrac{1{\cdot}0519}{1{\cdot}3034}$

$$\therefore\ A_4 = 0{\cdot}807\times 1235{\cdot}3 = 996{\cdot}9\ \text{mm}^2$$

$$\frac{A_2}{A_{*2}} = \frac{A_2}{A_4}\times\frac{A_4}{A_{*2}} = \frac{1685}{996{\cdot}9}\times 1{\cdot}0519 = 1{\cdot}778$$

from Table 1, $M_2 = 0{\cdot}35$

(the Mach number distribution is shown plotted in Figure E 4.4b)

$$\text{mass flow} = \varrho_4 q_4 A_4$$

$(a^2 = \gamma RT)$, $\quad \therefore\ a_4 = 20(289)^{0{\cdot}5} = 340$ m/s

$$\therefore\ q_4 = 0{\cdot}77\times 340 = 261{\cdot}8 \text{ m/s}$$

$(p = \varrho RT)$, $\quad \varrho_4 = (30\times 3379)/(0{\cdot}2871\times 10^3\times 289)$

$$= 1{\cdot}2217\ \text{kg/m}^3$$

$\therefore$ mass flow $= 1{\cdot}2217\times 261{\cdot}8\times 996{\cdot}9\times 10^{-6} = 0{\cdot}3188$ kg/s

at (4), from Table 1, for $M_4 = 0{\cdot}77$, $T_{04}/T_4 = 1{\cdot}1186$

$$\therefore\ T_{04} = 289\times 1{\cdot}1186 = 323{\cdot}3\ \text{K}$$

for adiabatic flow conditions there is no change in total head temperature, $T_0 = T_{04} = T_{02}$

$\therefore$ at (2), from Table 1, for $M_2 = 0{\cdot}35$, $T_{02}/T_2 = 1{\cdot}0245$

$$\therefore\ T_2 = 323.3/1{\cdot}0245 = 315.6 \text{ k}$$

$$\therefore\ a_2 = 20(315{\cdot}6)^{0{\cdot}5} = 355{\cdot}3 \text{ m/s}$$

$$\therefore\ q_2 = 0{\cdot}35\times 355.3 = 124{\cdot}3 \text{ m/s}$$

from Table 2, for $M_{s1} = 2{\cdot}43$, $p_{02}/p_{01} = 0{\cdot}5276 = p_{04}/p_0$

$\therefore$ total head condition, $p_0 = 44{\cdot}4/0{\cdot}5276 = 84{\cdot}16$ in Hg $=$

$$= 2{\cdot}844\times 10^5\ \text{N/m}^2$$

% change in total head pressure

$$= \frac{p_0 - p_{04}}{p_0} = 1 - p_{04}/p_0 = 57{\cdot}24\%$$

EXAMPLE 4.5 Show that the energy equation for one-dimensional, *steady*, adiabatic flow of a perfect gas may be written in the form $\dfrac{a^2}{\gamma-1} + \dfrac{q^2}{2} = \dfrac{1}{2}\dfrac{\gamma+1}{\gamma-1}a_*^2$ where (q) is the velocity of the fluid, (a) the local speed of sound in the fluid, and the asterisk relates to reference conditions in the flow where the Mach number $M = 1$. Hence show that if a perfect gas passes through a normal shock, $q_1 q_2 = a_*^2$ where (q_1) and (q_2) are the velocities of the gas before and after the shock, relative to the shock.

Just after the passage of a shock wave moving into stationary air in a shock tube, the velocity and temperature of the air are 670 m/s and 729 K respectively. Determine the speed of the wave relative to the stationary fluid.

(U. London 1968)

For one-dimensional, steady, adiabatic flow of a perfect gas, the total head enthalpy is a constant,

$$\therefore\ h_0 = h + q^2/2$$

giving,
$$c_p T_0 = c_p T + q^2/2$$

or,
$$\frac{\gamma R T_0}{\gamma-1} = \frac{\gamma R T}{\gamma-1} + q^2/2$$

$$(p = \varrho RT),\quad \frac{\gamma}{\gamma-1}\frac{p_0}{\varrho_0} = \frac{\gamma}{\gamma-1}\frac{p}{\varrho} + \frac{q^2}{2}$$

$$(a^2 = \gamma p/\varrho),\quad \frac{a_0^2}{\gamma-1} = \frac{a^2}{\gamma-1} + \frac{q^2}{2}$$

$$(M = 1,\ q_* = a_*),\quad \frac{a_0^2}{\gamma-1} = \frac{a^2}{\gamma-1} + \frac{q^2}{2} = \frac{1}{2}\frac{\gamma+1}{\gamma-1}a_*^2 \quad \text{(A)}$$

for the stationary normal shock wave using the above relation,

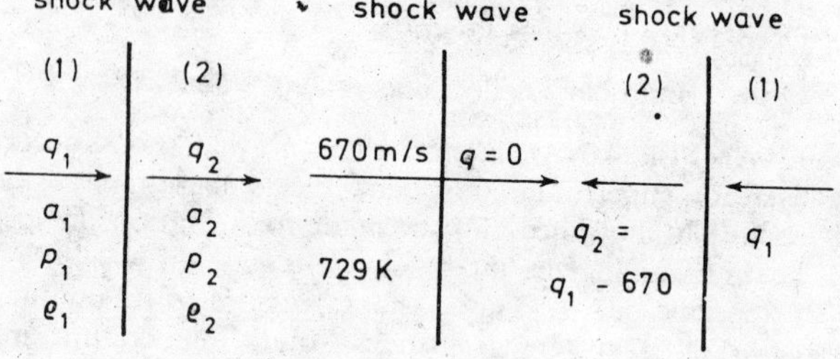

Fig. E4.5

$$q_1 + \frac{2\gamma}{\gamma-1}\frac{p_1}{\varrho_1 q_1} = \frac{\gamma+1}{\gamma-1}\frac{a_*^2}{q_1^2} \quad \text{(B)}$$

and
$$q_2 + \frac{2\gamma}{\gamma-1}\frac{p_2}{\varrho_2 q_2} = \frac{\gamma+1}{\gamma-1}\frac{a_*^2}{q_2} \quad \text{(C)}$$

(B) minus (C),

$$q_1 - q_2 + \frac{2\gamma}{\gamma-1}\left(\frac{p_1}{\varrho_1 q_1} - \frac{p_2}{\varrho_2 q_2}\right) = \frac{\gamma+1}{\gamma-1}a_*^2\left(\frac{q_2-q_1}{q_1 q_2}\right) \quad \text{(D)}$$

continuity, mass flow, $\quad m = \varrho_1 q_1 A_1 = \varrho_2 q_2 A_2 = \varrho q A$

$$(A_1 = A_2),\quad \therefore\ \varrho_1 q_1 = \varrho_2 q_2 = \varrho q \quad (4.1)$$

momentum $\quad \varrho q A(q_2 - q_1) = (p_1 - p_2)A$

using (4.1), $\quad q_2 - q_1 = \dfrac{p_1}{\varrho_1 q_1} - \dfrac{p_2}{\varrho_2 q_2}$

substituting in (D), $\quad -1 + \dfrac{2\gamma}{\gamma-1} = \dfrac{\gamma+1}{\gamma-1}\dfrac{a_*^2}{q_1 q_2}$

$$\text{or,}\quad q_1 q_2 = a_*^2 \quad \text{(E)}$$

which is the Prandtl relation (7.9c) for a plane normal shock wave.

The above equations for fluid flow may not be applied to the shock wave moving through a stationary fluid. Figure E 4.5a; if the wave is brought to rest by applying an equal and opposite velocity (q_1) to the fluid then Figure E 4.5b is appropriate to the application of (A) and (E) from (A), $(\gamma = 7/5)\ \dfrac{a_2^2}{0{\cdot}4} + \dfrac{q_2^2}{2} = 3a_*^2$

$(a^2 = \gamma RT)$, $\quad a_2 = 20(729)^{0{\cdot}5} = 540$ m/s

$$\therefore a_*^2 = 243\,000 + q_2^2/6 \qquad \text{(F)}$$

from (E), $q_2 = q_1 - 670 \quad \therefore (q_2 + 670)q_2 = a_*^2$

from (F), $\therefore q_2^2 - q_2^2/6 + 670\,q_2 - 243\,000 = 0$

or $\qquad q_2^2 + 742\,q_2 - 291\,600 = 0$

whence $\qquad q_2 = 284{\cdot}2$ m/s

the speed of the wave relative to the stationary fluid

$$= q_1 = 954{\cdot}2 \text{ m/s}$$

EXAMPLE 4.6. A Pitot-tube in a perfect gas is aligned so that its opening faces directly upstream in a supersonic flow of Mach number (M). Find an expression for the pressure (P) in the pitot-tube in terms of (M), the ratio of specific heats (γ), and the static pressure (p) of the undisturbed stream. You may assume that the density ratio (ϱ_1/ϱ) and the pressure ratio (p_1/p) across a normal shock-wave are connected by the relation,

$$\frac{\varrho_1}{\varrho} = \frac{1+\alpha(p_1/p)}{\alpha+(p_1/p)}$$

where $\alpha = \dfrac{\gamma+1}{\gamma-1}$ and suffix 1 refers to conditions immediately behind the shock-wave.

(U. London 1966)

Figure E 4.6 shows a diagrammatic representation of the opening of a Pitot-tube facing a supersonic flow which uses the mechanism of the shock wave to come to rest at

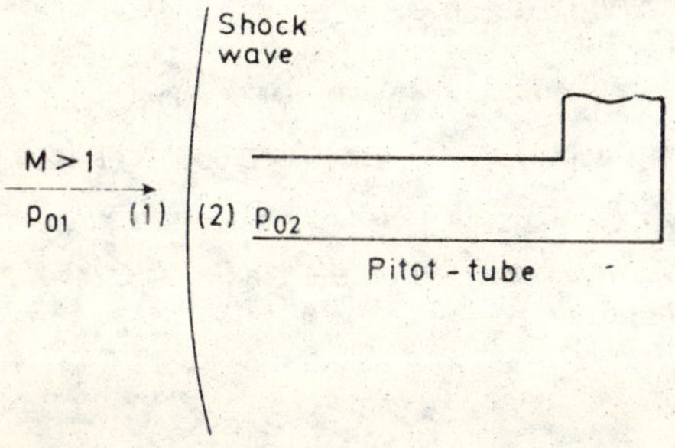

Fig. E4.6

the entry to the Pitot-tube which therefore registers $P = p_{02}$ the total head pressure behind the shock-wave. For the relatively small dimensions of the pitot-tube the shock wave is assumed to be a plane normal shock wave immediately in front of the Pitot-tube opening.

For the required relation

$$\frac{P}{p} = \frac{p_{02}}{p_1} = \frac{p_{02}}{p_{01}} \times \frac{p_{01}}{p_1}$$

since $$\frac{p_0}{p} = \left(1 + \frac{\gamma-1}{2}M^2\right)^{\frac{\gamma}{(\gamma-1)}}$$

then $$\frac{p_{02}}{p_1} = \left(1 + \frac{\gamma-1}{2}M_2^2\right)^{\frac{\gamma}{(\gamma-1)}}\left(\frac{p_2}{p_1}\right) \qquad \text{(A)}$$

It is therefore necessary to find expressions for (M_2) and (p_2/p_1) in terms of the upstream Mach number (M_1) as follows:

continuity across the shock, mass flow $m = \varrho qA$ and $A_1 = A_2$

$$\therefore \varrho_1 q_1 = \varrho_2 q_2 \qquad (4.1)$$

momentum $\qquad \varrho qA(q_2 - q_1) = (p_1 - p_2)A$

$$\therefore \frac{p_2}{p_1} - 1 = \frac{\varrho_1 q_1^2}{p_1}\left(1 - \frac{q_2}{q_1}\right)$$

$$\left(a^2 = \frac{\gamma p}{\varrho}\right), \qquad \therefore \frac{p_2}{p_1} - 1 = \gamma M_1^2\left(1 - \frac{q_2}{q_1}\right) \qquad (4.7)$$

from (4.1) and the relation for the density ratio given in the question

$$\frac{q_2}{q_1} = \frac{\alpha + p_2/p_1}{1 + \alpha p_2/p_1}$$

$\therefore$ in (4.7), $$\frac{p_2}{p_1} - 1 = \gamma M_1^2\left(\frac{1 + \alpha p_2/p_1 - \alpha - p_2/p_1}{1 + \alpha p_2/p_1}\right)$$

or, $$1 + \alpha p_2/p_1 = \gamma M_1^2(\alpha - 1)$$

$$\alpha = \frac{\gamma+1}{\gamma-1}, \qquad \therefore p_2/p_1 = \frac{2\gamma M_1^2 - (\gamma-1)}{\gamma+1}$$

or, $$p_2/p_1 = \frac{2\gamma}{\gamma+1}(M_1^2 - 1) + 1 \qquad \text{(4.8a)}$$

similarly, $$p_1/p_2 = \frac{2\gamma}{\gamma+1}(M_2^2 - 1) + 1 \qquad \text{(4.8b)}$$

$$\therefore M_2^2 = \left(\frac{p_1}{p_2} + \frac{\gamma-1}{\gamma+1}\right)\frac{\gamma+1}{2\gamma}$$

$$= \left(\frac{\gamma+1}{2\gamma M_1^2 - (\gamma-1)} + \frac{\gamma-1}{\gamma+1}\right)\left(\frac{\gamma+1}{2\gamma}\right)$$

$$M_2^2 = \frac{(\gamma-1)M_1^2 + 2}{2\gamma M_1^2 - (\gamma-1)} \qquad (4.11)$$

$\therefore$ substituting for M_2^2 and p_2/p_1 in (A)

$p_{02}/p_1 =$

$$= \left[1 + \frac{\gamma-1}{2}\left[\frac{2 + (\gamma-1)M_1^2}{2\gamma M_1^2 - (\gamma-1)}\right]\right]^{\frac{\gamma}{\gamma-1}}\left[\frac{2\gamma M_1^2 - (\gamma-1)}{\gamma+1}\right]$$

$$= \left[\frac{M_1^2(\gamma+1)^2}{2(2\gamma M_1^2 - (\gamma-1))}\right]^{\frac{\gamma}{\gamma-1}}\left[\frac{2\gamma M_1^2 - (\gamma-1)}{\gamma+1}\right]$$

$$\therefore \frac{P}{p} = p_{02}/p_1 = \left[\frac{\gamma+1}{2}M_1^2\right]^{\frac{\gamma}{\gamma-1}}\left[\frac{\gamma+1}{2\gamma M_1^2 - (\gamma-1)}\right]^{\frac{1}{\gamma-1}}$$

which is known as the Rayleigh Pitot-tube formula.

EXAMPLE 4.7 Figure E4.7 shows the design of a type of stationary supersonic compressor cascade for which it may be assumed that $M_1 = 2{\cdot}0$, $p_1 = 1 \times 10^5$ N/m^2 and that a normal shock occurs at $M_2 = 1{\cdot}39$ after the flow has passed through a centred Prandtl-Meyer wave $A'CD$. The air enters at an angle of 60° to the axial Y-direction in which the flow leaves.

Determine the Mach number and pressure at exit (BB') and the magnitude of the thrust per cm depth, in the X-direction on the pair of blades shown.

(U. Leeds 1969)

The Prandtl-Meyer compression $A'CD$ consists of Type I Mach waves (Figure 3.4a) and compresses the flow isentropically thus diminishing the loss of total head pressure if only shock wave processes were used to increase pressure. Since the flow at condition (3) is subsonic (3)–(4) is a diffuser process—again considered isentropic.

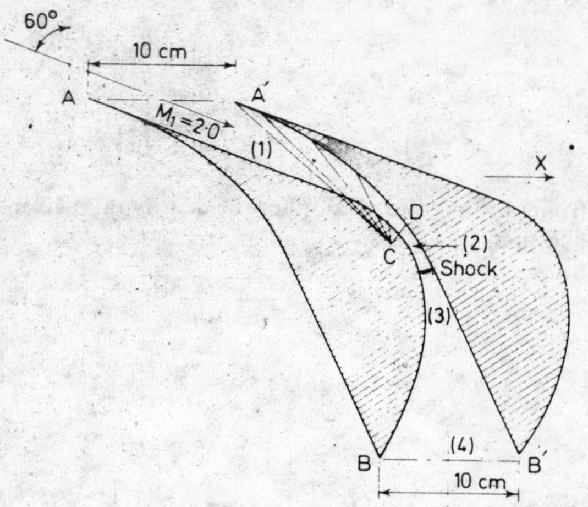

Fig. E4.7

from Table 1, for $M_1 = 2{\cdot}0$, ${}_*\delta_1 = 26{\cdot}38°$, $\mu_1 = 30°$

$$A_1/A_{*1} = 1{\cdot}6330, \quad p_{01}/p_1 = 7{\cdot}8244$$

for $M_2 = 1{\cdot}39$, ${}_*\delta_2 = 8{\cdot}70°$, $\mu_2 = 46{\cdot}01°$, $A_2/A_{*1} = 1.2932$

$${}_1\delta_2 = {}_*\delta_1 - {}_*\delta_2 = 26{\cdot}38 - 8{\cdot}70 = 17{\cdot}68°$$

angle $A'CD = \mu_2 + {}_1\delta_2 - \mu_1 = 46{\cdot}01 + 17{\cdot}68 - 30 = 33{\cdot}69°$

from Table 2, for $M_2 = 1{\cdot}39$, $M_3 = 0{\cdot}744$,

$$p_{03}/p_{01} = 0{\cdot}9607$$

from Table 1, for $M_3 = 0{\cdot}744$, $A_3/A_{*3} = 1{\cdot}0658$
(by interpolation)

If a dashed line is drawn through A' normal to the incident flow direction to intersect the second blade in A'' then $A'A''$ represents area $A_1 = AA' \sin 30°$ (per unit depth).

or, $A_1 = 0{\cdot}1 \times 1 \times 10^{-2} \times 0{\cdot}5$ (per cm depth)

$$= 0{\cdot}0005 \text{ m}^2$$

$\therefore A_4 = 2A_1$, $A_3 = A_2$ but $A_{*3} \neq A_{*2}$,

$$\frac{A_4}{A_{*4}} = \frac{A_4}{A_{*3}} = \frac{A_4}{A_3} \times \frac{A_3}{A_{*3}} = \frac{2A_1}{A_2} \times \frac{A_3}{A_{*3}}$$

$$= \frac{2 \times 1{\cdot}633 \times 1{\cdot}0658}{1{\cdot}2932} = 2{\cdot}6917$$

from Table 1, M_4 and p_{04}/p_4 can be obtained by interpolation,

	A/A^*	(diff)	M (diff)
	2·7076		0·22
$A_4/A_{*4} =$	2·6917	(0·0159)	$\therefore M_4 = 0{\cdot}221$
	2·5968	(0·1108)	0·23 (0·01)

$(0{\cdot}221 = 0{\cdot}22 + 0{\cdot}01 \times 159/1108)$

Similarly $p_{04}/p_4 = 1{\cdot}0343 + 0{\cdot}0032 \times 159/1108 = 1{\cdot}0348$

$$p_4 = \frac{p_4}{p_{04}} \times \frac{p_{04}}{p_{01}} \times \frac{p_{01}}{p_1} \times p_1 = \frac{0{\cdot}9607}{1{\cdot}0348} \times 7{\cdot}8244 \times 1 \times 10^5$$

$\therefore$ pressure at exit, $p_4 = 7{\cdot}264 \times 10^5$ N/m² $(M_4 = 0{\cdot}221)$

For thrust calculation, conservation of momentum is considered for the control volume contained by $(A'A'')$ and (BB'); if F_x is the force exerted by the blades on the fluid within the control volume, in the positive X-direction (—+→) then,

momentum out − momentum in = pressure force + F_x

or $0 - (\text{mass flow} \times \text{velocity})_x = (p_1A_1)_x + F_x$

$$\therefore -(\varrho_1 q_1 A_1) q_1 \cos 30° = p_1 A_1 \cos 30° + F_x$$

$(a^2 = \gamma p/\varrho)$, $\therefore \varrho q^2 = \gamma p \dfrac{q^2}{a^2} = \gamma p M^2$

$$\therefore F_x = -p_1 A_1 \cos 30° (\gamma M_1^2 + 1)$$

(per cm depth), $= -1 \times 10^5 \times 0{\cdot}0005 \times (1{\cdot}4 \times 4 + 1)\sqrt{3}/2$

$$\therefore F_x = -285{\cdot}8 \text{ N}$$

$\therefore$ thrust exerted *on* the blades is $+285{\cdot}8$ N per cm depth (——→)

5 FANNO FLOW

5.1 CONCEPT AND ANALYSIS OF ADIABATIC FLOW IN A CONSTANT AREA DUCT WITH FRICTION

The flow of a real compressible fluid in a tube, pipe or duct is always affected by friction between the solid boundaries and the fluid. For steady flow under such conditions a velocity profile describes the variation in flow velocity from zero at the boundary to a maximum at the centre line of a straight symmetrical duct of constant cross-section.

For purposes of analysis mean values of velocity and of the fluid properties are assumed and the relationships developed give good agreement with experimental results. The flow model, as shown in Figure 5.1, also assumes that the fluid flows through a pipe of constant circular cross-section although the equivalent 'hydraulic radius' may be used for ducts of non-circular cross-section. With no transfers of work energy or of heat energy across the system boundaries, the stagnation enthalpy or temperature (assuming a perfect gas) is constant. The flow process is adiabatic but irreversible and therefore non-isentropic: it is referred to as Fanno Flow*.

The skin friction coefficient, C_f, represents the ratio of friction force at the flow boundary to the convective inertia force of the fluid flow and is expressed as,

$$C_f = \frac{\tau_\omega}{\frac{1}{2}\varrho q^2} = f/4 \tag{5.1}$$

where $\tau_\omega = \eta(du/dy)_{y=0}$ is the shear stress at the pipe wall (η is the coefficient of viscosity assumed constant, $(du/dy)_{y=0}$ is the velocity gradient at the pipe wall), f is referred to as the friction factor and ϱ and q represent mean values of density and velocity respectively.

Entry length effects are not taken into account in the following analysis,

continuity, mass flow, $m = \varrho q A = \text{constant}$ (5.2)

$$\therefore \text{ for constant area, } \frac{d\varrho}{\varrho} + \frac{dq}{q} = 0 \tag{5.3}$$

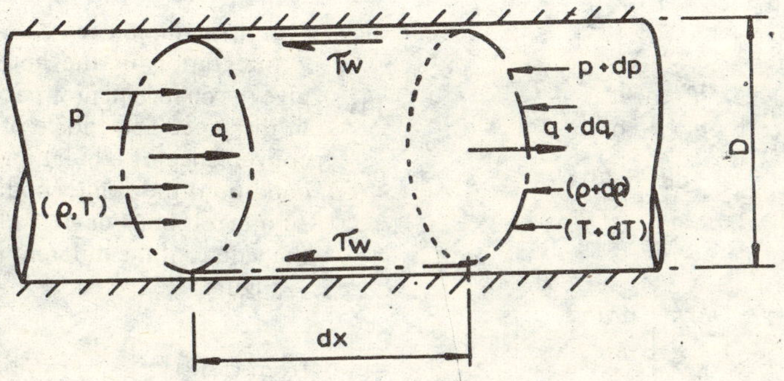

Fig. 5.1 Fanno flow control volume

momentum, $(A\varrho q)\,dq = pA-(p+dp)A-\tau_\omega A_s$ (5.4)

where $A = \dfrac{\pi D^2}{4}$ is the cross-sectional area

* Fanno wrote a thesis in 1904—but there is very little information about the work for which he received recognition (Stodola, A. "Steam and gas turbines" 1945, *1*, 61– Peter Smith).

and $A_s = \pi D\,dx$ is the surface area of the fluid element

from (5.1) and (5.4), $$-\frac{C_f}{2}\left(\frac{4}{D}\right)dx = \frac{dq}{q}+\frac{dp}{\varrho q^2} \quad (5.5a)$$

$$\left(a^2 = \frac{\gamma p}{\varrho}\right), \therefore -\frac{C_f}{2}\left(\frac{4}{D}\right)dx = \frac{dq}{q}+\frac{1}{\gamma M^2}\frac{dp}{p} \quad (5.5b)$$

energy, $$h_0 = h+\frac{q^2}{2} = \text{constant} \quad (5.6)$$

differential, $$c_p\,dT + q\,dq = 0 \quad (5.7)$$

$(a^2 = \gamma RT)$, $$\frac{dT}{T}+(\gamma-1)M^2\frac{dq}{q} = 0 \quad (5.8)$$

Mach number, $$M = \frac{q}{a} = \frac{q}{(\gamma RT)^{0.5}} \quad (5.9)$$

logarithmic differential,

$$\frac{dM}{M} = \frac{dq}{q}-\frac{1}{2}\frac{dT}{T} \quad (5.10)$$

from (5.8) and (5.10),

$$\frac{dq}{q} = \left[\frac{1}{1+\frac{\gamma-1}{2}M^2}\right]\frac{dM}{M}$$

$$\text{or, } \frac{dq}{q} = \frac{1}{2}\left[\frac{1}{M^2}-\frac{\frac{\gamma-1}{2}}{1+\frac{\gamma-1}{2}M^2}\right]dM^2 \quad (5.11)$$

$$= -\frac{d\varrho}{\varrho}$$

$$\text{and, } \frac{dT}{T} = -\left[\frac{(\gamma-1)M^2}{1+\frac{\gamma-1}{2}M^2}\right]\frac{dM}{M}$$

$$= -\left[\frac{\frac{\gamma-1}{2}}{1+\frac{\gamma-1}{2}M^2}\right]dM^2 \quad (5.12)$$

state, $p = \varrho RT$

log arithmic differential, $$\frac{dp}{p} = \frac{d\varrho}{\varrho}+\frac{dT}{T} \quad (5.13)$$

from (5.3), $$\frac{dp}{p} = -\frac{dq}{q}+\frac{dT}{T}$$

from (5.11) and (5.12),

$$\frac{dp}{p} = -\left[\frac{1+(\gamma-1)M^2}{1+\frac{\gamma-1}{2}M^2}\right]\frac{dM}{M}$$

$$\frac{dp}{p} = -\left[\frac{1}{M}+\frac{\frac{\gamma-1}{2}M}{1+\frac{\gamma-1}{2}M^2}\right]dM \quad (5.14)$$

Substitution for dq/q (5.11) and dp/p (5.14) in (5.5b) gives,

$$C_f\left(\frac{4}{D}\right)dx = \frac{2}{\gamma M^3}\left[\frac{1-M^2}{1+\frac{\gamma-1}{2}M^2}\right]dM \quad (5.15)$$

In general, the skin-friction coefficient, C_f, is a function of the diameter Reynolds number, R_D, and of the Mach number, M. For subsonic flow C_f may be assumed to be independent of M; for supersonic flow, $1 < M < 5$, the dependence on M increases only slightly and the effect may be neglected without appreciable error. For constant mass flow R_D is constant and C_f is assumed to be constant. From (5.15),

subsonic flow $M < 1$, dM/dx is positive and M *increases* with length because of friction, (5.15a)

supersonic flow $M > 1$, dM/dx is negative and M *decreases* with length because of friction (5.15b)

In each of the above cases the limiting condition is $M = 1$, the sonic or choking condition.

The change in entropy may be obtained by considering the thermodynamic relation (1.20a)

$$ds = c_p\frac{dT}{T}-R\frac{dp}{p}$$

$$\therefore ds = R\left[\frac{1-M^2}{1+\frac{\gamma-1}{2}M^2}\right]\frac{dM}{M}$$

$$= R\left[\frac{1}{M}-\frac{\frac{\gamma+1}{2}M}{1+\frac{\gamma-1}{2}M^2}\right]dM \quad (5.16)$$

Note: when $M = 1$, $ds = 0$ and the choking condition also represents a maximum entropy condition.

The effect of friction in fluid flow is to create a boundary layer at the pipe wall, which increases in thickness with distance in the flow direction. The undisturbed flow area decreases so that, effectively, dA/dx is negative for flow in a constant area pipe. By comparison with the considerations of isentropic flow through variable area ducts (Chapter 2),

when $M < 1$ the flow speed will *increase* with *decrease* in area—c.f. (5.15a)

when $M > 1$ the flow speed will *decrease* with *increase* in area—c.f. (5.15b)

and, again, in each case the limiting condition is $M = 1$.

Integration of the above differential relationships will give the change in flow parameters between any two stations in the pipe. The choice of limits is quite arbitrary but, for convenience of tabulation and for consistency of reference condition, the sonic condition is used as one of the limits of integration. For $M = 1$, the suffix ($_*$) is used to identify the values of the different parameters at the sonic condition —Figure 5.2.

from (5.12), $$\int_*\frac{dT}{T} = \log_e\frac{T}{T_*}$$

$$= \log_e\left[\frac{1+\frac{\gamma-1}{2}(M_*^2 = 1)}{1+\frac{\gamma-1}{2}M^2}\right] \quad (5.17a)$$

$$\therefore \frac{T}{T_*} = \frac{\gamma+1}{2+(\gamma-1)M^2} \quad (5.17b)$$

from (5.13), $$\int_*\frac{dp}{p} = \log_e\frac{p}{p_*} = \log_e\frac{(M_* = 1)}{M}$$

$$+\tfrac{1}{2}\log_e\left[\frac{1+\frac{\gamma-1}{2}(M_*^2=1)}{1+\frac{\gamma-1}{2}M^2}\right] \quad (5.18a)$$

$$\therefore \frac{p}{p_*} = \frac{1}{M}\left[\frac{\gamma+1}{2+(\gamma-1)M^2}\right]^{0.5} \quad (5.18b)$$

Note: The relationship (5.17b) may be derived in a different way, e.g.

$$T_0/T = 1+(\gamma-1)M^2/2 \quad (2.14)$$

and $$T_0/T_* = (\gamma+1)/2 \quad (2.17a)$$

dividing, $$T/T_* = (\gamma+1)/(2+(\gamma-1)M^2) \quad (5.17b)$$

This approach is used in the following derivation of a relationship between p_0/p_{0*} and M.

Since the flow is not isentropic, the irreversibilities due to friction cause a reduction in total head pressure. At any point in the flow,

$$\therefore \frac{q}{q_*} = M\left[\frac{\gamma+1}{2+(\gamma-1)M^2}\right]^{0.5} = \frac{\varrho_*}{\varrho} \quad (5.20b)$$

from (5.15), $C_f\left(\frac{4}{D}\right)\int_* dx = C_f\left(\frac{4}{D}\right)(x-x_*)$

$$= \frac{1}{\gamma}\left[\frac{1}{(M_*=1)}-\frac{1}{M^2}\right]+\frac{\gamma+1}{2\gamma}\log_e\left[\frac{(M_*^2=1)}{M^2}\right]$$

$$+\frac{\gamma+1}{2\gamma}\log_e\left[\frac{1+\frac{\gamma-1}{2}M^2}{1+\frac{\gamma-1}{2}(M_*^2=1)}\right] \quad (5.21a)$$

$$\therefore 4C_f\left(\frac{L_*}{D}\right) = \frac{1}{\gamma}\left[\frac{1-M^2}{M^2}\right] + \frac{\gamma+1}{2\gamma}\log_e\left[\frac{(\gamma+1)M^2}{2+(\gamma-1)M^2}\right] \quad (5.21b)$$

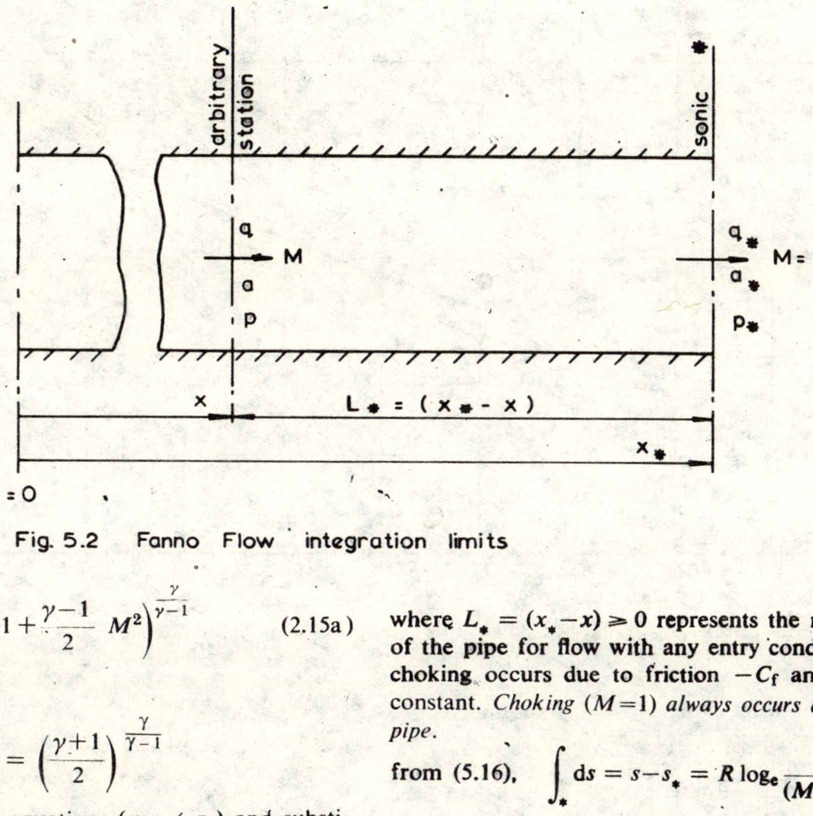

Fig. 5.2 Fanno Flow integration limits

$$\frac{p_0}{p} = \left(1+\frac{\gamma-1}{2}M^2\right)^{\frac{\gamma}{\gamma-1}} \quad (2.15a)$$

at the sonic condition,

$$\frac{p_{0*}}{p_*} = \left(\frac{\gamma+1}{2}\right)^{\frac{\gamma}{\gamma-1}}$$

Combining these two equations ($p_{0*} \neq p_0$) and substituting for $\frac{p}{p^*}$ from (5.18b), the change of total head pressure between any arbitrary station in the flow and the station at which $M = 1$, is given by,

$$\frac{p_0}{p_{0*}} = \frac{1}{M}\left[\frac{2+(\gamma-1)M^2}{\gamma+1}\right]^{\frac{\gamma+1}{2(\gamma-1)}} \quad (5.19)$$

from (5.11), $$\int_* \frac{dq}{q} = \log_e\frac{q}{q_*} = \frac{1}{2}\log_e\left[\frac{M_*^2=1}{M^2}\right]$$

$$+\frac{1}{2}\log_e\left[\frac{1+\frac{\gamma-1}{2}(M_*^2=1)}{1+\frac{\gamma-1}{2}M^2}\right] \quad (5.20a)$$

where $L_* = (x_*-x) \geq 0$ represents the maximum length of the pipe for flow with any entry condition, M, before choking occurs due to friction $-C_f$ and D both being constant. *Choking* ($M=1$) *always occurs at the end of the pipe.*

from (5.16), $$\int_* ds = s-s_* = R\log_e\frac{M}{(M_*=1)}$$

$$+\frac{R}{2}\frac{\gamma+1}{\gamma-1}\log_e\left[\frac{1+\frac{\gamma-1}{2}(M_*^2=1)}{1+\frac{\gamma-1}{2}M^2}\right] \quad (5.22a)$$

$$\therefore \frac{s_*-s}{R} = \log_e\frac{1}{M}\left[\frac{2+(\gamma-1)M^2}{\gamma+1}\right]^{\frac{\gamma+1}{2(\gamma-1)}} \quad (5.22b)$$

from (2.11) or from (5.19),

$$\frac{s_*-s}{R} = \log_e\left(\frac{p_0}{p_{0*}}\right) \quad (5.23)$$

Note: The entropy increases with decrease in total head pressure ($s_* > s$, $p_0 > p_{0*}$) for all adiabatic processes.

5.2 FANNO LINE

If the temperature parameter T/T^* (5.17b) is plotted against the entropy parameter $(s-s_*)/R$ (5.22b), the graph shown in Figure 5.3 is obtained. For a perfect gas, with constant specific heats, the enthalpy parameter h/h_* is equivalent to T/T_* and the intercept on this axis corresponds to the 'no flow' stagnation condition, T_0/T_* or h_0/h_*. The curve plotted in Figure 5.3 is referred to as a 'universal' Fanno

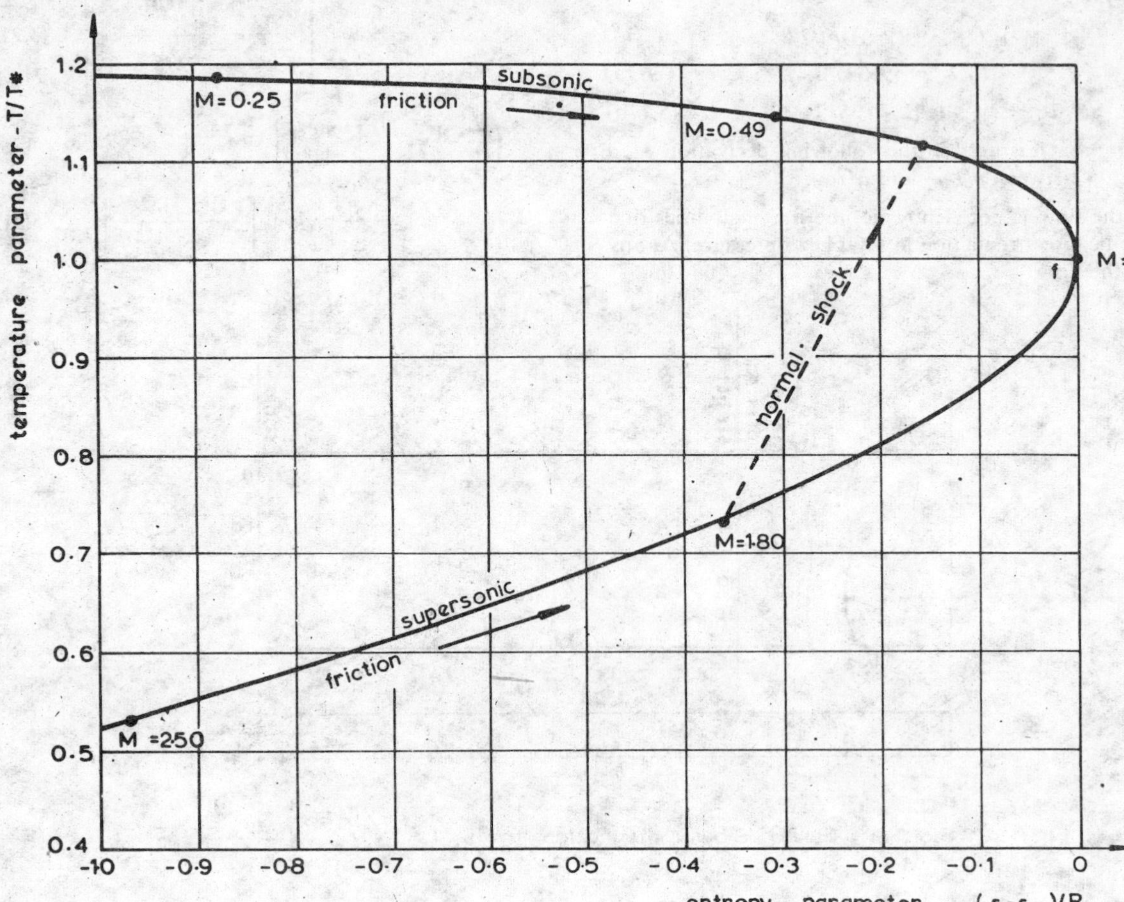

Fig. 5.3 universal Fanno line

line independent of mass flow; a family of lines, for different mass flows, is shown in Figure 5.4.

continuity,
$$\frac{m}{A} = \varrho q = G \tag{5.24}$$

where G is the mass flow per unit area and is constant for a constant area duct.

energy,
$$h_0 = h + \frac{q^2}{2} = \text{constant} \tag{5.6}$$

combining (5.6) and (5.24),
$$h = h_0 - \frac{G^2}{2\varrho^2} \tag{5.25}$$

Since any one property of a perfect gas can be expressed as a function of any two other properties, e.g., $\varrho = \varrho(s, h)$ then, for a constant value of $h_0(T_0)$, different values of G will give different Fanno lines as shown in Figure 5.4.

The maximum entropy condition may also be established by considering the energy equation (5.6) in differential form.

$$dh + q\,dq = 0 \tag{5.7}$$

and *continuity*,
$$\varrho\,dq + q\,d\varrho = 0 \tag{5.3}$$

also, from (1.18),
$$T\,ds = dh - \frac{dp}{\varrho}$$

at f in Figure 5.2, $ds = 0$
$$\therefore dh = \frac{dp}{\varrho} \tag{5.26}$$

combining (5.7) and (5.26),
$$q\,dq = -dp/\varrho \tag{5.27}$$

from (5.3),
$$q^2 = dp/d\varrho = a^2$$

or
$$M^2 = M = 1$$

which confirms the result obtained from (5.16) that the entropy is a maximum when the flow becomes sonic.

from (5.12) and (5.16),

$$\frac{dT}{ds} = -\frac{T}{R}\left[\frac{(\gamma-1)M^2}{1+\frac{\gamma-1}{2}M^2}\right]\left[\frac{1+\frac{\gamma-1}{2}M^2}{1-M^2}\right]$$

$$= \frac{T}{R}\left[\frac{(\gamma-1)M^2}{M^2-1}\right] \tag{5.28}$$

i.e., when $M > 1$, dT/ds is positive (lower half of Fanno line)
when $M < 1$, dT/ds is negative (upper half of Fanno line)

Since the flow with friction is adiabatic the entropy change is always positive (2nd Law) and the direction, in Figure 5.3, of a Fanno flow process from any initial condition (h, s) is always from left to right.

If the duct is long enough the flow will eventually reach a choking condition. If the proposed length of duct is greater than the critical length (for choking) appropriate to specified subsonic entry conditions, the choking condition will be established at the end of the duct with a consequent reduction in flow rate.

For supersonic entry conditions at entry to a duct of length greater than the critical length, a shock wave will occur (Figure 5.3) before the choking condition which will then be reached by a subsonic flow process as described above.

from (5.20b), $$\frac{q}{q_*} = M\left[\frac{6}{5+M^2}\right]^{0\cdot5} = \frac{\varrho_*}{\varrho} \qquad (5.31)$$

from (5.19), $$\frac{p_0}{p_{0*}} = \frac{1}{M}\left[\frac{5+M^2}{6}\right]^3 \qquad (5.32)$$

from (5.21b),
$$4C_f\left(\frac{L_*}{D}\right) = \frac{5}{7}\left[\frac{1}{M^2}-1\right] + \frac{6}{7}\log_e\left[\frac{6M^2}{5+M^2}\right] \qquad (5.33)$$

from (5.22b), $$\frac{s-s_*}{R} = -\log_e\frac{1}{M}\left[\frac{5+M^2}{6}\right]^3 \qquad (5.34)$$

from (5.32), $$= -\log_e(p_0/p_{0*}) \qquad (5.35)$$

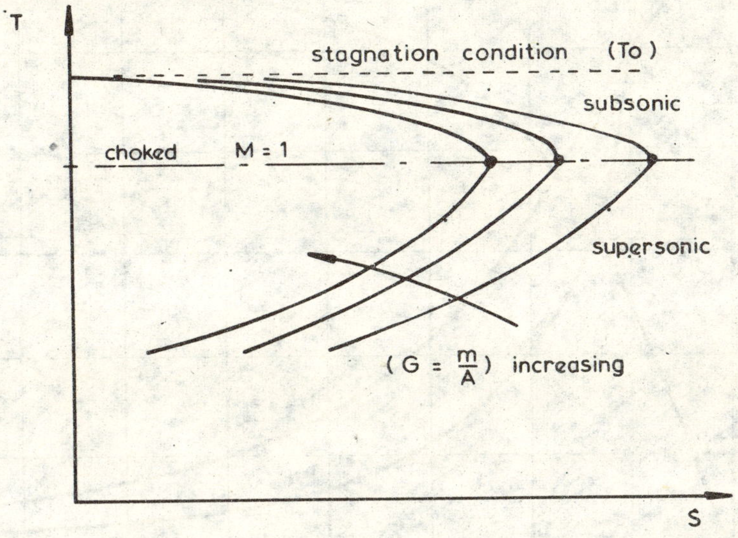

Fig. 5.4 Fanno lines for different mass flows

Summary of Fanno Flow effects:

Flow Property	*Frictional Effects for Initial Conditions* Subsonic ($M < 1$)	Supersonic ($M > 1$)
q m/s	increases	decreases
M	increases	decreases
p N/m^2	decreases	increases
T K	decreases	increases
ϱ kg/m^3	decreases	increases
p_0 N/m^2	decreases	decreases
T_0 K	constant	constant

5.3 SUMMARY OF EQUATIONS

The following are the equations used in Table 3 for steady adiabatic flow of air ($\gamma^a = 7/5$) in a constant area duct with friction and are plotted in Figures 5.5 and 5.6.

from (5.17b), $$\frac{T}{T_*} = \frac{6}{5+M^2} \qquad (5.29)$$

from (5.18b), $$\frac{p}{p_*} = \frac{1}{M}\left[\frac{6}{5+M^2}\right]^{0\cdot5} \qquad (5.30)$$

EXAMPLE 5.1 An insulated pipe, diameter 2·5 cm, has air passing through it. At a station (1) in the pipe, at a distance of 11·452 metres from the open end, the pressure (p_1) and temperature (t_1) are 2×10^5 N/m^2 and 27°C respectively and the Mach number, $M_1 = 0{\cdot}25$.

If the friction coefficient is a constant, $C_f = 0{\cdot}004$, determine the mass flow, the outlet conditions of pressure, temperature and flow velocity and the change in total head pressure over the pipe length.

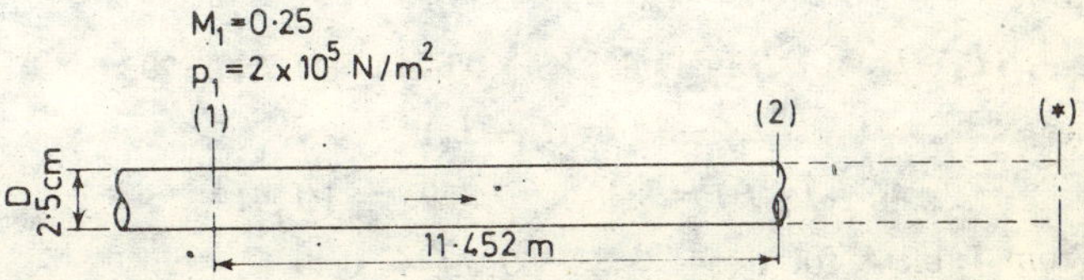

Fig. E5·1

Calculate the mass flow at which choking will occur assuming that ambient conditions and total head temperature are unchanged. For this new condition determine the Mach number at station (1) and the change in total head pressure over the pipe length.

At station (1) Figure E5.1, $(a^2 = \gamma RT)$, $T_1 = 300$ K

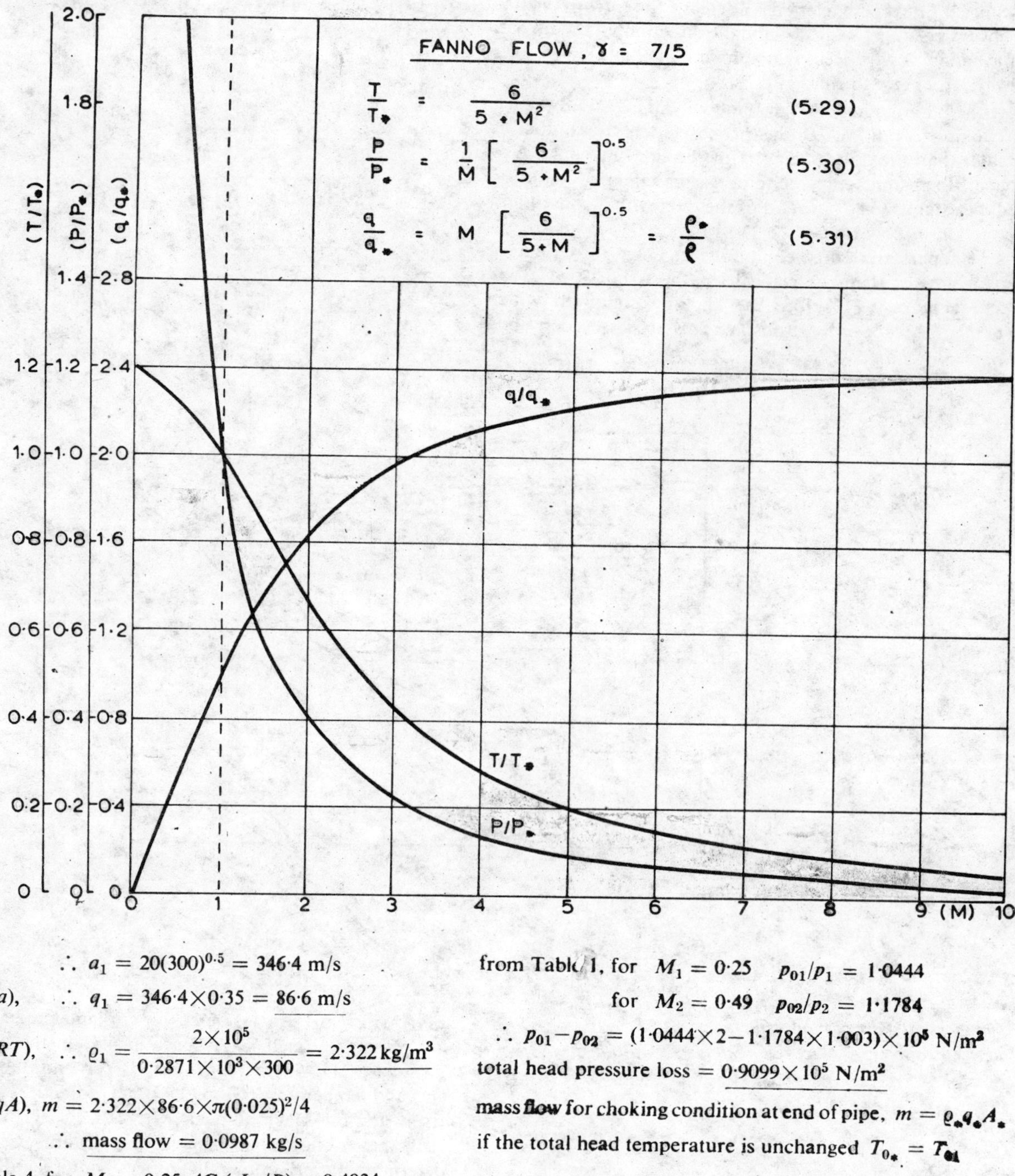

Fig. 5·5

$\therefore\ a_1 = 20(300)^{0.5} = 346{\cdot}4$ m/s

$(M = q/a)$, $\quad \therefore\ q_1 = 346{\cdot}4 \times 0{\cdot}35 = \underline{86{\cdot}6 \text{ m/s}}$

$(p = \varrho RT)$, $\quad \therefore\ \varrho_1 = \dfrac{2\times 10^5}{0{\cdot}2871\times 10^3 \times 300} = \underline{2{\cdot}322 \text{ kg/m}^3}$

$(m = \varrho q A)$, $\quad m = 2{\cdot}322 \times 86{\cdot}6 \times \pi(0{\cdot}025)^2/4$

$\therefore$ mass flow $= \underline{0{\cdot}0987 \text{ kg/s}}$

from Table 4, for $M_1 = 0{\cdot}25$, $4C_f({}_1L_*/D) = 8{\cdot}4834$

$p_1/p_* = 4{\cdot}3546$ and $T_1/T_* = 1{\cdot}1852$ (see Figure 5.3)

now, $\quad L_1 - L_2 = 11{\cdot}452$ m

$\therefore\ (L_1 - L_2)4C_f/D = 11{\cdot}452 \times 4 \times 0{\cdot}004/0{\cdot}025 = 7{\cdot}3293$

${}_1L_* - {}_2L_* = (L_1 - L_2)$

$\therefore\ 4C_f({}_2L_*/D) = 8{\cdot}4834 - 7{\cdot}3293 = 1{\cdot}1541$

From Table 4, for $1{\cdot}1541 \approx 1{\cdot}1539$, $M_2 = 0{\cdot}49$

and, $\quad p_2/p_* = 2{\cdot}1838$, $T_2/T_* = 1{\cdot}1450$ (Figure 5.3)

$$\therefore\ T_2 = \frac{T_2}{T_*} \times \frac{T_*}{T_1} \times T_1 = \frac{1{\cdot}1450}{1{\cdot}1852} \times 300 = \underline{289{\cdot}8 \text{ K}}$$

and, $\quad a_2 = 20(289.8)^{0.5} = 340{\cdot}5$ m/s

$q_2 = 0{\cdot}49 \times 340{\cdot}5 = \underline{166{\cdot}8 \text{ m/s}}$

$$p_2 = \frac{p_2}{p_*} \times \frac{p_*}{p_1} \times p_1 = \frac{2{\cdot}1838}{4{\cdot}3546} \times 2\times 10^5 = \underline{1{\cdot}003 \times 10^5 \text{ N/m}^2}$$

(ambient pressure)

from Table 1, for $M_1 = 0{\cdot}25$ $\quad p_{01}/p_1 = 1{\cdot}0444$

for $M_2 = 0{\cdot}49$ $\quad p_{02}/p_2 = 1{\cdot}1784$

$\therefore\ p_{01} - p_{02} = (1{\cdot}0444 \times 2 - 1{\cdot}1784 \times 1{\cdot}003) \times 10^5$ N/m²

total head pressure loss $= \underline{0{\cdot}9099 \times 10^5 \text{ N/m}^2}$

mass flow for choking condition at end of pipe, $m = \varrho_* q_* A_*$

if the total head temperature is unchanged $T_{0*} = T_{01}$

from Table 1, for $M_1 = 0{\cdot}25$ $T_{01}/T_1 = 1{\cdot}0125$

$\therefore\ T_{01} = 1{\cdot}0125 \times 300 = 303{\cdot}75$ K

for choking ($M = 1$), $\quad T_0/T = 1{\cdot}2$

$\therefore\ T_* = 253{\cdot}1/1{\cdot}2 = 253{\cdot}1$ K

$\therefore\ a_* = 20(253{\cdot}1)^{0.5} = 318{\cdot}2 \text{ m/s} = q_*$

ambient pressure $p_* = 1{\cdot}003 \times 10^5$ N/m² (at outlet)

$(p = \varrho RT)$, $\quad \therefore\ \varrho_* = \dfrac{1{\cdot}003 \times 10^5}{0{\cdot}2871 \times 10^3 \times 253{\cdot}1} = 1{\cdot}3803 \text{ kg/m}^3$

$\therefore$ mass flow $= 1{\cdot}3803 \times 318{\cdot}2 \times 4{\cdot}909 \times 10^{-4} = \underline{0{\cdot}2156 \text{ kg/s}}$

for this choking condition $\quad 4C(L_*/D) = 7{\cdot}3293$

and, interpolating in Table 4, gives

$M_1 = 0{\cdot}265$ and $p_{01}/p_{0*} = 2{\cdot}2786$

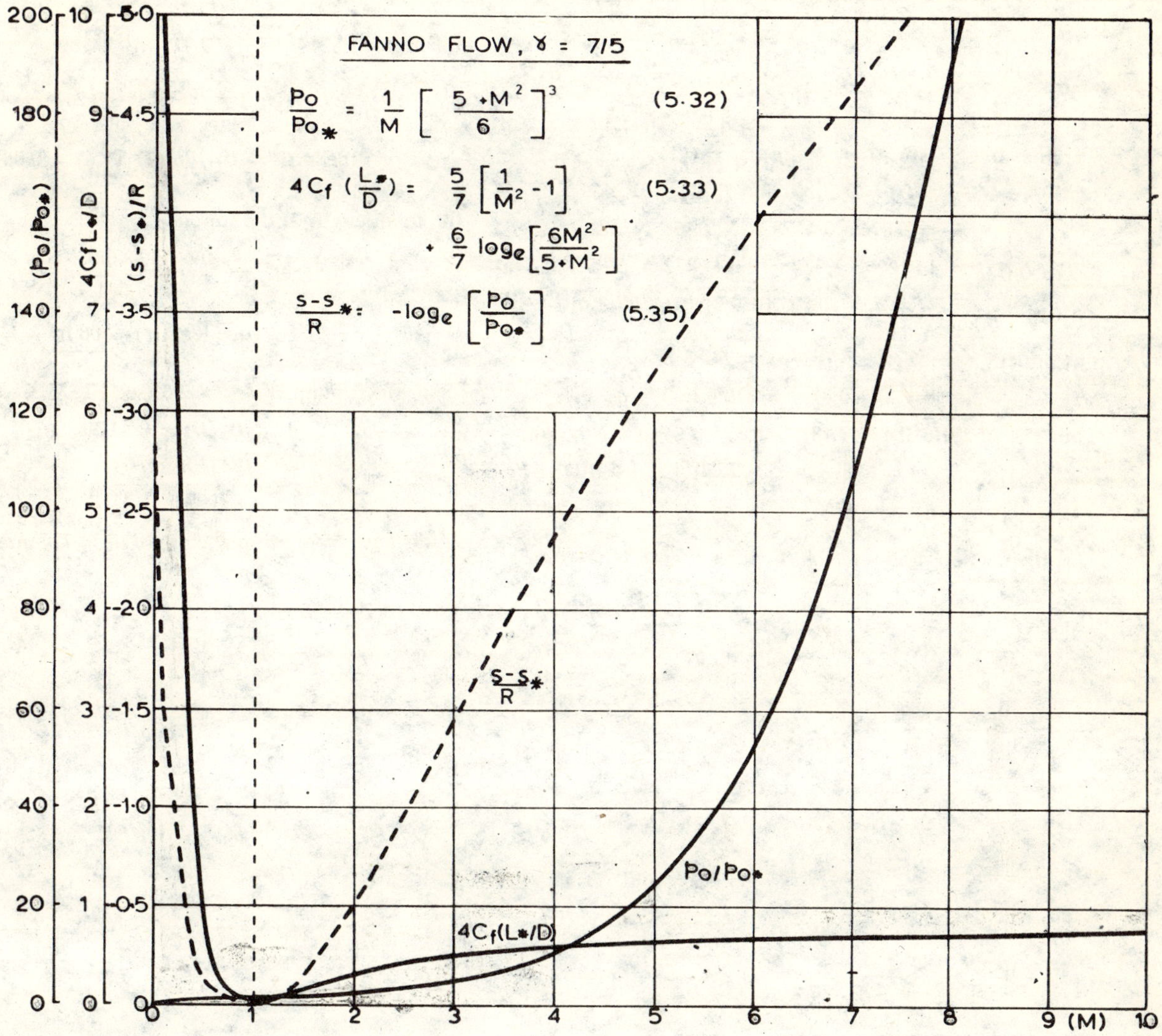

Fig. 5·6

$4C_f(L_*/D)$	(diff.)	M	p_0/p_{0*}
7·6876		0·26	2·3173
$4C_f(_1L_*/D) =$ 7·3293	(0·3583)	0·265	2·2786
6·9832	(0·7044)	0·27	2·2385

$$\therefore p_{01} - p_{0*} = (2{\cdot}2786 - 1) \times p_{0*}$$

$$(p_{0*}/p_* = 1{\cdot}7147), \quad = 1{\cdot}2786 \times 1{\cdot}7147 \times 1{\cdot}003 \times 10^5$$

$\therefore$ total head pressure loss $= \underline{2{\cdot}1989 \times 10^5\ \text{N/m}^2}$

EXAMPLE 5.2 **Figure E5.2 shows a convergent-divergent nozzle with a tube which is attached to the outlet and which** has the same diameter (39 mm) as the nozzle exit diameter.

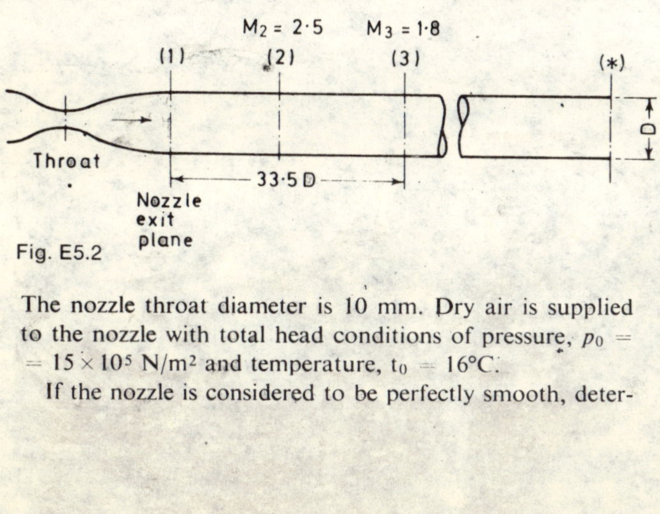

Fig. E5.2

The nozzle throat diameter is 10 mm. Dry air is supplied to the nozzle with total head conditions of pressure, $p_0 = 15 \times 10^5$ N/m² and temperature, $t_0 = 16$°C.

If the nozzle is considered to be perfectly smooth, determine the mass flow of air and the friction drag force on the tube if the flow in the tube is supersonic and choked at the end of the tube.

For this choked condition, at a position of 33.5 diameters from the nozzle exit plane, the Mach number in the tube is 1:8. Determine an average value for the friction coefficient, the total length of the tube and the position in the tube at which the Mach number is 2·5.

For the smooth nozzle it is assumed that the flow of air is isentropic; if the exit plane is referred to as station 1 (Figure E5.2) then,

$$\frac{A_1}{A_*} = \frac{d_1^2}{d_*^2} = (3{\cdot}9)^2 = 15{\cdot}21$$

from Table 1, $M_1 = 4.4$, $p_{01}/p_1 = 255{\cdot}256$, $T_{01}/T_1 = 4{\cdot}872$

and mass flow number $MN_1 = \dfrac{m}{A_1} \cdot \dfrac{a_0}{p_0} = 0{\cdot}0533$

$$A_1 = \pi(0{\cdot}039)^2/4 = 11{\cdot}95 \times 10^{-4}\ \text{m}^2$$

$(a^2 = \gamma RT)$ $\qquad \therefore a_0 = 20(289)^{0.5} = 340$ m/s

$\therefore$ **mass flow,** $m = 0{\cdot}0533 \times (11{\cdot}95 \times 10^{-4}) \times (15 \times 10^5)/340$

$$= \underline{0{\cdot}281\ \text{kg/s}}$$

For friction drag on the tube, conservation of momentum must be considered with F_x the force acting on the *fluid* due to friction ($\pm$).

thus, $$m(q_* - q_1) = (p_1 - p_*)A + F_x \qquad \text{(A)}$$

$$p_{01}/p_1 = 255{\cdot}256$$

$$\therefore p_1 = 15\times10^5/255{\cdot}256 = 5\,876\times5 \text{ N/m}^2$$

from Table 3, for $M_1 = 4{\cdot}4$, $p_1/p_* = 0{\cdot}1128$

also, $4C_f({}_1L_*/D) = 0{\cdot}66149$ and $T_1/T_* = 0{\cdot}2463$

$$\therefore p_* = 5\,876{\cdot}5/0{\cdot}1128 = 52\,096{\cdot}2 \text{ N/m}^2$$

$$T_{01}/T_1 = 4{\cdot}8720 \quad \therefore T_1 = 289/4{\cdot}872 = 59{\cdot}3 \text{ K}$$

$$\therefore q_1 = M_1\times a_1 = 4{\cdot}4\times20(59{\cdot}3)^{0{\cdot}5} = 677{\cdot}8 \text{ m/s}$$

at outlet, $$T_* = T_1/0{\cdot}2463 = 240{\cdot}8 \text{ K}$$

$$\therefore a_* = 20(240{\cdot}8)^{0.5} = 310{\cdot}4 \text{ m/s} = q_*$$

in (*A*), $$0{\cdot}281(310{\cdot}4 - 677{\cdot}8)$$

$$= (5\,876{\cdot}5 - 52\,096{\cdot}2)\times11{\cdot}9\times10^{-4} + F_x$$

$$\therefore F_x = -48N$$

i.e. the friction drag force *on* the tube is +48 N (→)

for $M_3 = 1{\cdot}8$, $4C_f({}_3L_*/D) = 0{\cdot}24189$

$$\therefore \frac{4\bar{C}_f}{D}({}_1L_* - {}_3L_*) = 0{\cdot}66149 - 0{\cdot}24189 = 0{\cdot}4196$$

$$\therefore \text{mean } C_f = 0{\cdot}4196/(33{\cdot}5\times4) = \underline{0{\cdot}00313}$$

the total length of the tube is

$${}_1L_* = \frac{0{\cdot}66149\times0.039}{4\times0{\cdot}00313}$$

∴ tube length is $\underline{2{\cdot}06 \text{ m}}$

for a Mach number $M_2 = 2{\cdot}5$ (see Fig. 5.3)

$$4C_f({}_2L_*/D) = 0{\cdot}43198$$

$$\therefore {}_2L_* = \frac{0{\cdot}43198\times0{\cdot}039}{4\times0{\cdot}00313} = 1{\cdot}346 \text{ m}$$

∴ distance from exit plane of nozzle = <u>714 mm</u> station (2).

6 RAYLEIGH FLOW

6.1 CONCEPT AND ANALYSIS OF FRICTIONLESS FLOW IN A CONSTANT AREA DUCT WITH HEAT ENERGY TRANSFER

The flow processes of a compressible fluid, which have previously been considered, have all been assumed to be adiabatic: no heat energy is transferred across the system boundary and there is no change in the stagnation temperature or stagnation enthalpy.

With flow systems designed for heat exchangers and with flow systems incorporating combustion processes an analysis of thermodynamic and flow characteristics can be made by assuming the flow to be frictionless in a constant area pipe or duct. The mass flow is assumed constant as is the ratio of specific heats but both of these assumptions become suspect for systems involving processes (such as combustion) in which liquid evaporation (or other phase changes) cause variations in mass flow and composition.

Figure 6.1 shows the elemental control volume considered as the flow model for establishing differential relationships

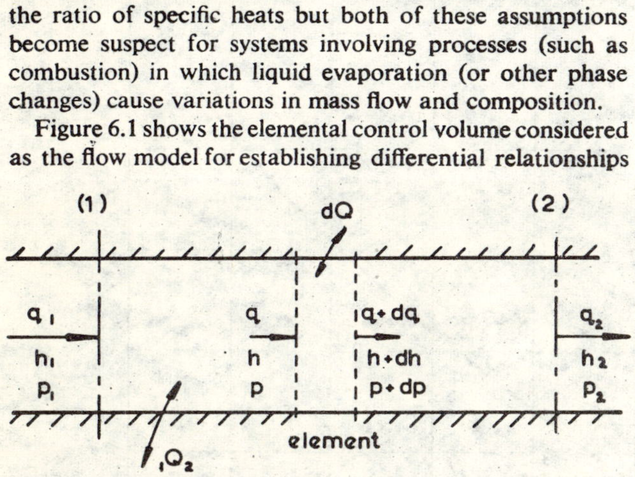

Fig. 6.1 elemental control volume in a constant area duct.

between the flow parameters and the Mach number of the incident flow, for purposes of integration and tabulation.

continuity, $$m = \varrho q A = \text{constant} \qquad (6.1a)$$

$\therefore$ for constant area, $$\frac{d\varrho}{\varrho} + \frac{dq}{q} = 0 \qquad (6.1b)$$

energy, $$h + \frac{q^2}{2} + dQ = (h + dh) + \left(\frac{q + dq}{2}\right)^2 \qquad (6.2a)$$

or, $$dQ = dh + (dq)^2 + q\,dq$$

neglecting second order small quantities,

$$dQ = c_p dT + q\,dq \qquad (6.2b)$$

alternatively, $$h_0 = h + \frac{q^2}{2} \neq \text{constant}$$

and, $$dh_0 = dQ = dh + q\,dq \qquad (6.2b)$$

dividing (6.2b) by $h = c_p T$ and using $a^2 = \gamma RT$,

$$\frac{dQ}{c_p T} = \frac{dQ}{h} = \frac{dT}{T} + (\gamma - 1)M^2 \frac{dq}{q} = \frac{dT_0}{T} \qquad (6.3)$$

momentum, $$(\varrho q A)dq = pA - (p + dp)A$$

or, $$\varrho q\,dq = -dp \qquad (6.4)$$

$a^2 = \dfrac{\gamma p}{\varrho}$ $$\therefore \frac{dp}{p} = -\gamma M^2 \frac{dq}{q} \qquad (6.5)$$

state, $$p = \varrho RT \qquad (6.6)$$

log differential, $$\frac{dp}{p} = \frac{d\varrho}{\varrho} + \frac{dT}{T} \qquad (6.7a)$$

from (6.1b), $$\frac{dp}{p} = -\frac{dq}{q} + \frac{dT}{T} \qquad (6.8)$$

Mach number, $$M^2 = \frac{q^2}{a^2} = \frac{q^2}{\gamma RT} \qquad (6.9)$$

log differential, $$\frac{dM^2}{M^2} = 2\frac{dq}{q} - \frac{dT}{T} \qquad (6.10)$$

with (6.5) and (6.8), $$\frac{dM^2}{M^2} = (1 + \gamma M^2)\frac{dq}{q} \qquad (6.11)$$

and, $$\frac{dM^2}{M^2} = \left(\frac{1 + \gamma M^2}{1 - \gamma M^2}\right)\frac{dT}{T} \qquad (6.12)$$

in (6.3), $$\frac{dQ}{h} = \frac{dM^2}{M^2}\left[\frac{1 - \gamma M^2}{1 + \gamma M^2} + \frac{(\gamma - 1)M^2}{1 + \gamma M^2}\right]$$

or,

$$\frac{dQ}{h} = \left(\frac{1-M^2}{1+\gamma M^2}\right)\frac{dM^2}{M^2} \tag{6.13}$$

$$\left(ds = \frac{dQ}{T}\right), \quad \therefore \frac{ds}{R} = \frac{\gamma}{\gamma-1}\left(\frac{1-M^2}{1+\gamma M^2}\right)\frac{dM^2}{M^2} \tag{6.14}$$

from (6.12) and (6.13),

$$\frac{dT}{T} = \left(\frac{1-\gamma M^2}{1-M^2}\right)\frac{dQ}{h} = \frac{1-\gamma M^2}{1+\gamma M^2}\frac{dM^2}{M^2} \tag{6.15}$$

from (6.11) and (6.13),

$$\frac{dq}{q} = \left(\frac{1}{1-M^2}\right)\frac{dQ}{h} = \left(\frac{1}{1+\gamma M^2}\right)\frac{dM^2}{M^2} = -\frac{d\rho}{\rho} \tag{6.16}$$

from (6.7b) and (6.15),

$$\frac{dp}{p} = \left(\frac{-\gamma M^2}{1-M^2}\right)\frac{dQ}{h} = \left(\frac{-\gamma M^2}{1+\gamma M^2}\right)\frac{dM^2}{M^2} \tag{6.17}$$

Differential equations are useful for identifying the sense of the changes of the flow parameters for subsonic or supersonic flow and for transfer of heat energy to the fluid (dQ positive) and from the fluid (dQ negative).

Figure 6.2 anticipates a discussion of the Rayleigh line and shows the variation of a temperature parameter with an entropy parameter. As with Fanno flow, when $M = 1$, $ds = 0$ and this choking condition (point r) represents a maximum entropy (or isentropic) condition. The effect of adding heat energy to the subsonic flow of a compressible fluid is to accelerate the fluid towards the sonic condition while, conversely, extraction of heat energy cause the fluid to decelerate.

When heat energy is added to the supersonic flow of a compressible fluid the Mach number reduces, theoretically, to $M = 1$ and cooling causes the Mach number to increase.

In contrast to Fanno flow, there is another maximum condition (indicated by (6.15)) when $M = 1/\gamma^{0.5}$, $dh = 0 = dT$ and this corresponds to a maximum enthalpy (or isothermal) condition, point (t) in Figure 6.2.

In addition to the maxima conditions it is useful to consider the changes in gradient of the Rayleigh line:

from (6.14) and (6.15),

$$\frac{dT}{ds} = \frac{\gamma-1}{\gamma}\frac{T}{R}\left(\frac{1-\gamma M^2}{1-M^2}\right) \tag{6.18}$$

i.e., when $M > 1$, $\frac{dT}{ds}$ is positive (lower half ($z \rightarrow r$) of Rayleigh line)

when $\frac{1}{\gamma^{0.5}} < M < 1$, $\frac{dT}{ds}$ is negative (upper half ($t \rightarrow r$) of Rayleigh line)

when $M < \frac{1}{\gamma^{0.5}}$, $\frac{dT}{ds}$ is positive (upper half ($a \rightarrow t$) of Rayleigh line)

The above effects of heat energy transfer to the flow of a compressible fluid, on initial flow conditions as specified by Mach number, are summarised overleaf:

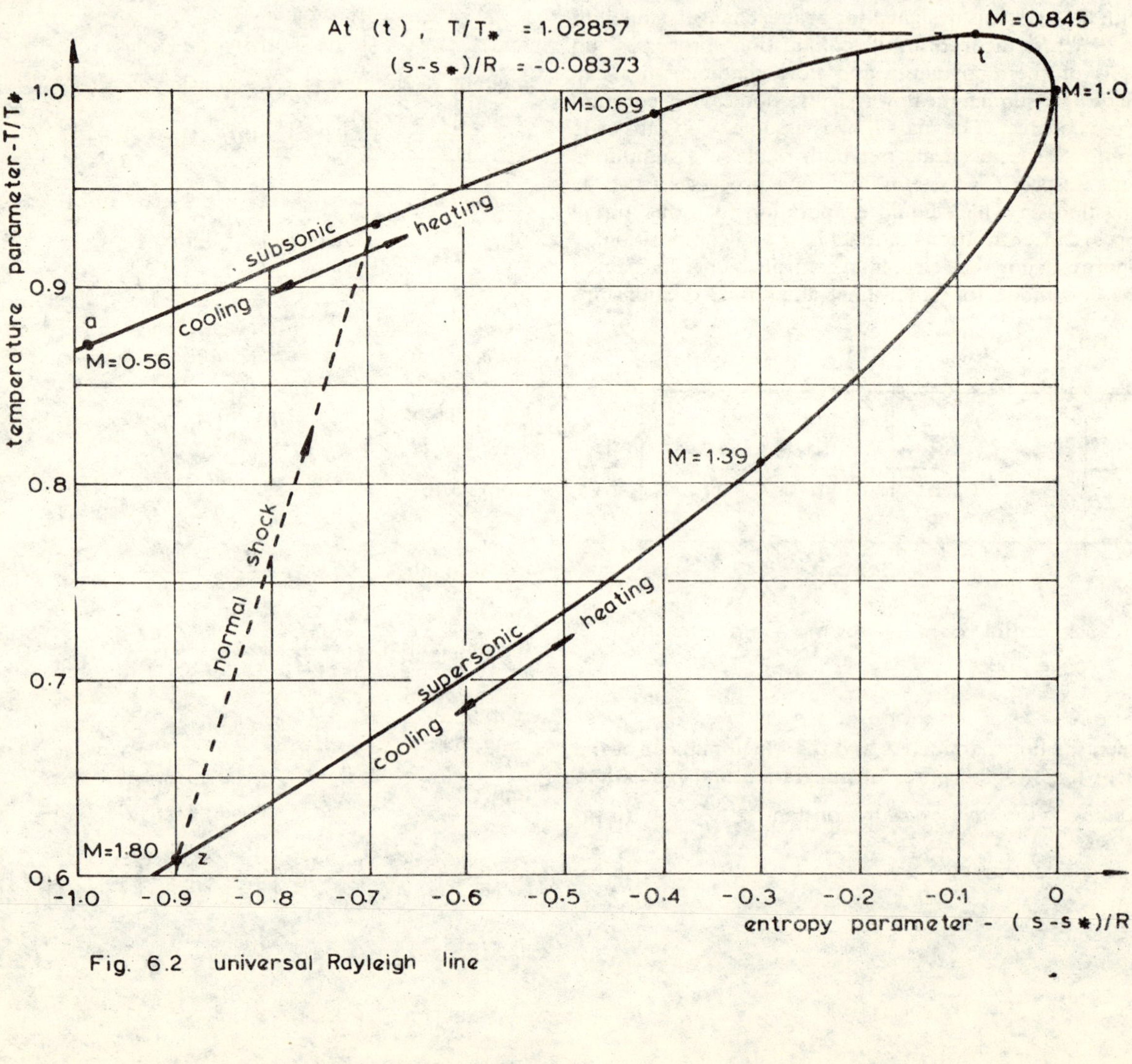

Fig. 6.2 universal Rayleigh line

	Heat Energy Added Heating, $dQ > 0$ Subsonic ($M < 1$)	Supersonic ($M > 1$)	*Heat Energy Removed* Cooling, $dQ < 0$ Subsonic ($M < 1$)	Supersonic ($M > 1$)
$\left(\frac{dq}{q}\right)$ q m/s	increases	decreases	decreases	increases
$\left(\frac{dM^2}{M^2}\right)$ M	increases	decreases	decreases	increases
$\left(\frac{dp}{p}\right)$ p N/m²	decreases	increases	increases	decreases
$\left(\frac{d\varrho}{\varrho}\right)$ ϱ kg/m³	decreases	increases	increases	decreases
$\left(\frac{ds}{R}\right)$ s kJ/kg K	increases	increases	decreases	decreases
(dh_0) T_0 K	increases	increases	decreases	decreases
(see 6.22b) p_0 N/m²	decreases	decreases	increases	increases
$\left(\frac{dT}{T}\right)$ (h) T K	increases $M < \frac{1}{\gamma^{0.5}}$	increases	decreases $M < \frac{1}{\gamma^{0.5}}$	decreases
(equation (6.15)	decreases, $1 > M > \frac{1}{\gamma^{0.5}}$		increases, $1 > M > \frac{1}{\gamma^{0.5}}$	

Integration of the above differential relationships gives the change in characteristics of the fluid between any two stations in the flow. The choice of limits is quite arbitrary but, for convenience of tabulation, the sonic condition is chosen as one of the limits of the integration and, for $M = 1$, the suffix ($_*$) is used to denote the values of the different parameters at sonic condition.

from (6.15), $$\int_* \frac{dT}{T} = \log_e \frac{T}{T_*} = \log_e \left[\frac{M^2}{(M_*^2 = 1)}\right] + 2\log_e \left[\frac{1+\gamma(M_*^2 = 1)}{1+\gamma M^2}\right] \qquad (6.19a)$$

$$\therefore \frac{T}{T_*} = \left[\frac{(1+\gamma)M}{1+\gamma M^2}\right]^2 \qquad (6.19b)$$

from (6.17), $$\int_* \frac{dp}{p} = \log_e \frac{p}{p^*} = \log_e \left[\frac{1+\gamma(M_*^2 = 1)}{1+\gamma M^2}\right] \qquad (6.20a)$$

$$\therefore \frac{p}{p_*} = \frac{1+\gamma}{1+\gamma M^2} \qquad (6.20b)$$

from (6.16), $$\int_* \frac{dq}{q} = \log_e \frac{q}{q_*} = \log_e \left[\frac{M^2}{(M_*^2 = 1)}\right] + \log_e \left[\frac{1+\gamma(M_*^2 = 1)}{1+\gamma M^2}\right] \qquad (6.21b)$$

$$\therefore \frac{q}{q_*} = \frac{(1+\gamma)M^2}{1+\gamma M^2} = \frac{\varrho_*}{\varrho} \qquad (6.21b)$$

With Rayleigh flow the transfers of heat energy cause changes of both the total head pressure and total head temperature conditions. At any point in the flow,

$$\frac{p_0}{p} = \left(1+\frac{\gamma-1}{2}M^2\right)^{\frac{\gamma}{\gamma-1}} \qquad (2.15a)$$

at the sonic condition, $$\frac{p_{0*}}{p_*} = \left(\frac{\gamma+1}{2}\right)^{\frac{\gamma}{\gamma-1}}$$

and, $$\frac{p_0}{p_{0*}} = \frac{p_0}{p} \times \frac{p_*}{p_{0*}} \times \frac{p}{p_*}$$

In this form, by combining the above equations with (6.20b) the change of total head pressure between any arbitrary station in the flow and the station at which $M = 1$, is given by:

$$\frac{p_0}{p_{0*}} = \left[\frac{1+\gamma}{1+\gamma M^2}\right]\left[\frac{2+(\gamma-1)M^2}{\gamma+1}\right]^{\frac{\gamma}{\gamma-1}} \qquad (6.22a)$$

Note: By taking logs and differentiating (2.15a) above and by using (6.17) the differential form of (6.22) can be obtained as:

$$\frac{dp_0}{p_0} = \frac{\gamma M^2(M^2-1)}{(1+\gamma M^2)(2+(\gamma-1)M^2)} \frac{dM^2}{M^2} \qquad (6.22b)$$

The corresponding change in total head temperature as a function of the flow Mach number at any station in the flow, is obtained in a similar manner,

$$\frac{T_0}{T} = 1+\frac{\gamma-1}{2}M^2 \qquad (2.14)$$

and, $$\frac{T_0}{T_{0*}} = \frac{T_0}{T} \times \frac{T_*}{T_{0*}} \times \frac{T}{T_*}$$

using (6.19b), $$\frac{T_0}{T_{0*}} = \frac{M^2(1+\gamma)(2+(\gamma-1)M^2)}{(1+\gamma M^2)^2} \qquad (6.23)$$

The change in entropy for the irreversible processes of heating or cooling is obtained from (6.14),

$$\frac{1}{R}\int_{*} ds = \frac{s-s_*}{R} = \frac{\gamma}{\gamma-1}\log_e\left[\frac{M^2}{(M_* = 1)^2}\right]$$

$$+\frac{\gamma}{\gamma-1}\log_e\left[\frac{1+\gamma(M_*^2 = 1)}{1+\gamma M^2}\right]^{\frac{\gamma+1}{\gamma}} \quad (6.24a)$$

or,

$$\frac{s-s_*}{R} = \frac{\gamma}{\gamma-1}\log_e M^2\left[\frac{\gamma+1}{1+\gamma M^2}\right]^{\frac{\gamma+1}{\gamma}} \quad (6.24b)$$

6.2 RAYLEIGH LINE

Figure 6.2 shows the graph obtained by plotting the temperature parameter, T/T_* (6.19b), against the entropy parameter, $(s-s_*)/R$, (6.24b). For a perfect gas with constant specific heats the enthalpy parameter, h/h_*, corresponds to T/T_*. The stagnation condition, T_0/T_* or h_0/h_* is not constant. The curve shown in Figure 6.2 is referred to as the 'universal' Rayleigh Line independent of mass flow; for different mass flow conditions, there is the following analysis:

continuity $$\frac{m}{A} = \varrho q = G \quad (5.23)$$

where G is the mass flow per unit area and is constant for a constant area duct.

momentum, $$pA+(\varrho qA)q = F$$

or, $$p+\varrho q^2 = \frac{F}{A} \quad (6.25)$$

where F/A is a constant in the absence of friction and area change; the momentum equation is used instead of the energy equation since $h_0 \neq$ constant.

combining (5.23) and (6.25), $$p+\frac{G^2}{\varrho} = \frac{F}{A} \quad (6.26)$$

Since any one property of a perfect gas can be expressed as a function of any two other properties, i.e., $\varrho = \varrho(s, h)$ and $p = p(s, h)$ then, for a constant value of F/A, different values of G will give different Rayleigh lines.

The maxima conditions of a Rayleigh line may be established as follows:

6.2.1 Maximum entropy or isentropic condition, ds=0

energy, $$dQ = dh_0 = d\left(h+\frac{q^2}{2}\right) \quad (6.2)$$

$$\therefore\ Tds = dQ = dh+qdq \quad (6.27)$$

also, $$Tds = dh-\frac{dp}{\varrho} \quad (1.18)$$

at r in Figure 6.2, $ds = 0$ $$\therefore\ dh = \frac{dp}{\varrho} = -qdq \quad (6.28a)$$

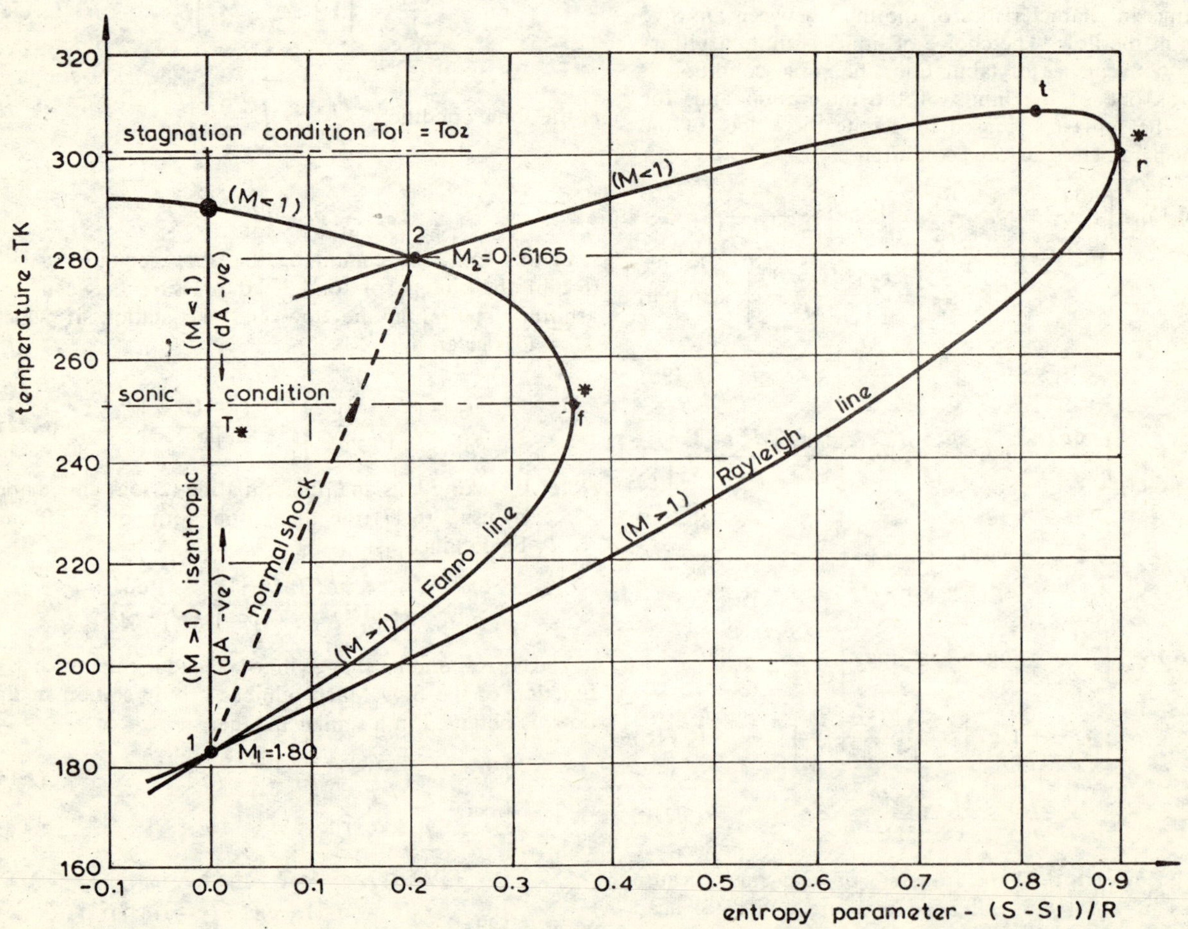

Fig.6.3 one-dimensional compressible flow processes

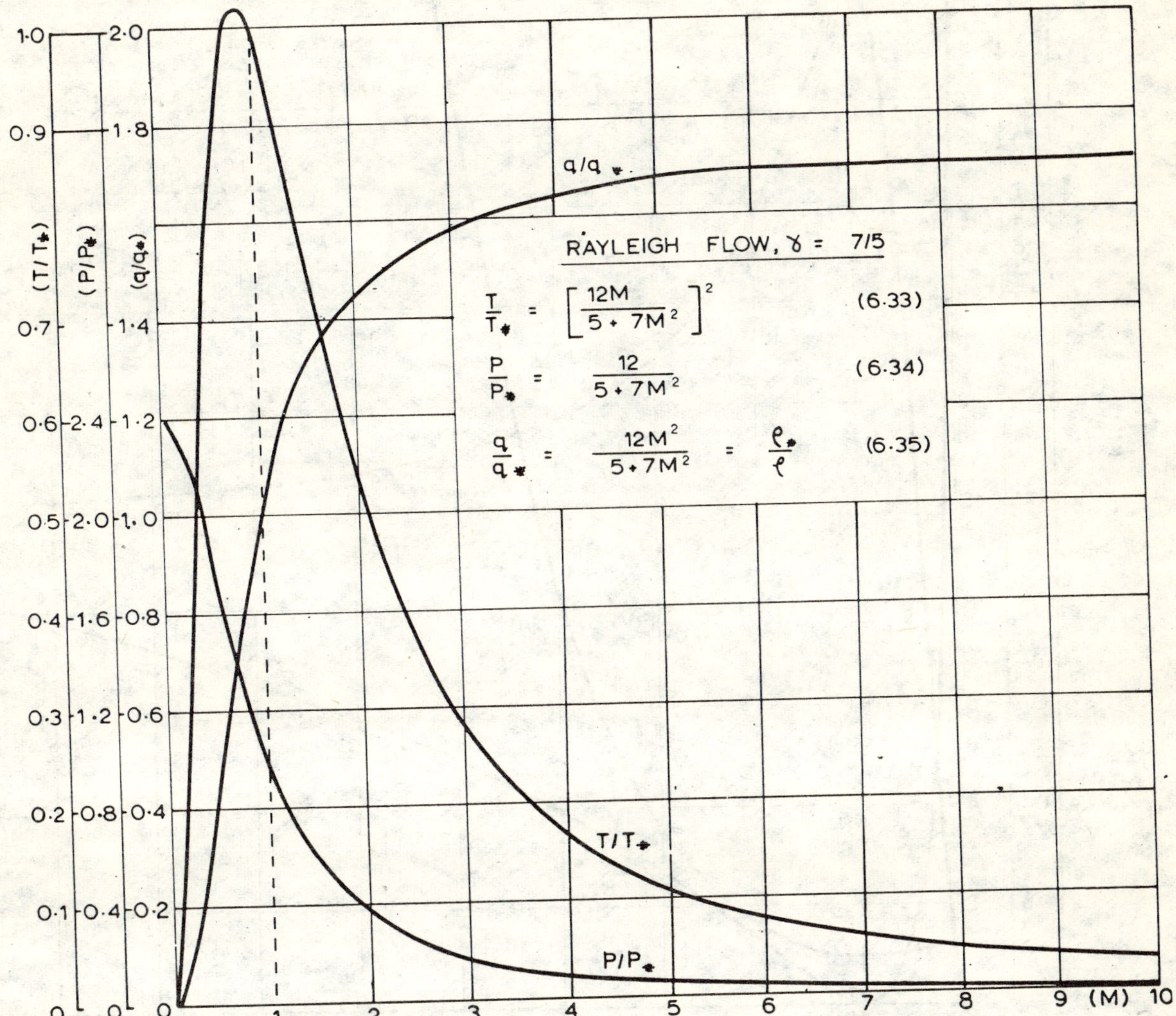

Fig. 6·4

$$\therefore\ \mathrm{d}p = -\varrho q \mathrm{d}q \qquad (6.28b)$$

continuity, $\varrho \mathrm{d}q + q\mathrm{d}\varrho = 0$ (6.1b)

in (6.27b), $q^2 = \left(\frac{\mathrm{d}p}{\mathrm{d}\varrho}\right)_{s=\mathrm{const.}} = a^2$

or, $M^2 = 1 = M$

which confirms the result obtained in (6.14) that the entropy change is a maximum when the flow becomes sonic.

6.2.2 Maximum enthalpy (temperature) or isothermal condition, dh=0=dT

from (6.27) and (1.18), for $\mathrm{d}h = 0 = \mathrm{d}T$ at t in Figure 6.2

$$q\mathrm{d}q = -\frac{\mathrm{d}p}{\varrho} \qquad (6.29)$$

continuity, $\varrho \mathrm{d}q + q\mathrm{d}\varrho = 0$ (6.1b)

and, in (6.29), $q^2 = \left(\frac{\mathrm{d}p}{\mathrm{d}\varrho}\right)_{T=\mathrm{const.}} \neq a^2$ (6.30)

state, $p = \varrho RT$

differential form, $\mathrm{d}T = 0$, $\left(\frac{\mathrm{d}p}{\mathrm{d}\varrho}\right)_{T=\mathrm{const.}} = \frac{p}{\varrho} = \frac{a^2}{\gamma}$ (6.31)

in (6.30), $q^2 = \frac{a^2}{\gamma}$

or $M = \frac{1}{\gamma^{0.5}}$ (6.32)

which confirms the result obtained in (6.15) that the temperature is a maximum when $M = 1/\gamma^{0.5}$. These maxima conditions are also shown in Figure 6.3 which is a representation of a specific Rayleigh condition (see Section 6.4).

6.3 SUMMARY OF EQUATIONS

The following are the equations used in Table 4 for steady frictionless flow of air ($\gamma = 7/5$) in a constant area duct with heat energy transfer as plotted in Figures 6.4 and 6.5.

from (6.19b), $\frac{T}{T_*} = \left[\frac{12M}{5+7M^2}\right]^2$ (6.33)

from (6.20b), $\frac{p}{p_*} = \frac{12}{5+7M^2}$ (6.34)

from (6.21b), $\frac{q}{q^*} = \left[\frac{12M^2}{5+7M^2}\right] = \frac{\varrho_*}{\varrho}$ (6.35)

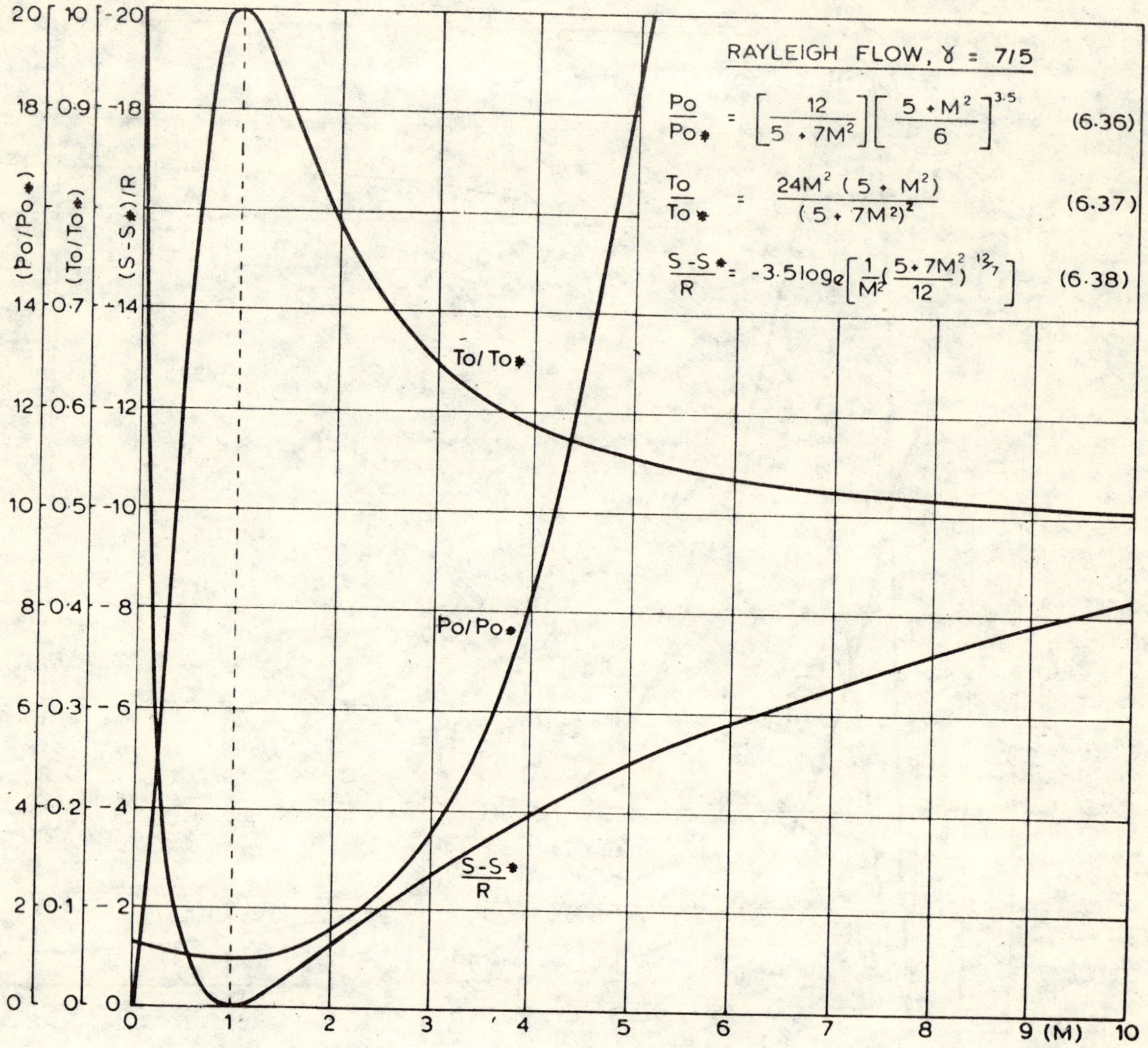

Fig. 6·5

from (6.22), $$\frac{p_0}{p_{0*}} = \left[\frac{12}{5+7M^2}\right]\left[\frac{5+M^2}{6}\right]^{3.5} \qquad (6.36)$$

from (6.23), $$\frac{T_0}{T_{0*}} = \frac{24M^2(5+M^2)}{(5+7M^2)^2} \qquad (6.37)$$

from (6.24b), $$\frac{s-s_*}{R} = -3{\cdot}5\ \log_e\left[\frac{1}{M^2}\left(\frac{5+7M^2}{12}\right)^{12/7}\right] \qquad (6.38)$$

a plane normal shock wave (Chapter 4), Fanno flow (Chapter 5) and Rayleigh flow. The flow equations are those of conservation of mass, of momentum, of energy and the isentropic relation $p/\varrho^{\gamma} = \text{const}$. The equation of state $p = \varrho RT$ has also been used for a perfect gas.

Table 6.1 shows which equations have been used for the appropriate analysis of each of the above flow processes.

The universal Fanno and Rayleigh lines which are shown in Figures 5.3 and 6.2 respectively cannot be compared directly except when the mass flow (per unit area) and initial total head conditions are the same in each case. The $T-s$ relationships which can readily be obtained from Tables 3 and 4 (see EXAMPLE 6.3) are shown plotted in Figure 6.3 for initial conditions of total head temperature $T_{01} = 301{\cdot}1$ K and $M_1 = 1.8$.

Table 6.1

Process	p/ϱ^{γ} = *const.*	*continuity*	*momentum*		*energy*
Isentropic nozzle flow	✓	✓	✓	or	✓
Plane normal shock wave		✓	✓		✓
Fanno flow		✓	✓		
Rayleigh flow		✓			✓

6.4 GRAPHICAL PRESENTATION OF ONE-DIMENSIONAL STEADY FLOW PROCESSES FOR AIR (γ=7/5)

Four fundamental equations have been used in previous sections to examine and develop the analysis of the different processes of isentropic nozzle flow (Chapter 2), flow through

For Fanno flow, the continuity and energy equations were used to develop property relationships such as the $T-s$ relationship which is presented graphically with the absolute static temperature TK as a function of a non-dimensional entropy parameter $(s-s_1)/R$ referred to the initial condition (1) specified above. No friction term is involved in this relation.

For Rayleigh flow the continuity and momentum equations were used to develop a $T-s$ relationship which is presented in a similar manner. No heat energy transfer term is involved in this relation.

Thus the intersections of the two curves at (1) and (2) in Figure 6.3 represent conditions which satisfy the plane normal shock wave relation between (1) and (2) developed from considerations of continuity, momentum and energy.

For completeness the isentropic process for a convergent-divergent nozzle, is also shown in Figure 6.3.

EXAMPLE 6.1 Show that the local Mach number equals $1/\sqrt{\gamma}$ at the point of maximum enthalpy on the Rayleigh line. Identify this point on a Rayleigh line and state the value of M at the point of maximum entropy.

Fuel is injected and burned in air moving with a speed of 87·4 m/s along a straight pipe of uniform diameter. The constant pressure heat of reaction of the fuel is 12 000 kJ/kg, the stagnation temperature of the air at inlet is 202°C and the burned gas stream discharges from the duct into a region where the pressure is 1×10^5 N/m². Duct cross-sectional area is 75 cm².

For a fuel-air ratio of 0·2, calculate the stagnation pressure at the inlet section, the flow speed at outlet and the mass flow.

Determine the change in entropy between the position in the pipe at which $M = 0{\cdot}56$ and the end of the pipe. Calculate the total heat energy transfer rate between these two positions.

(U. Leeds 1970)

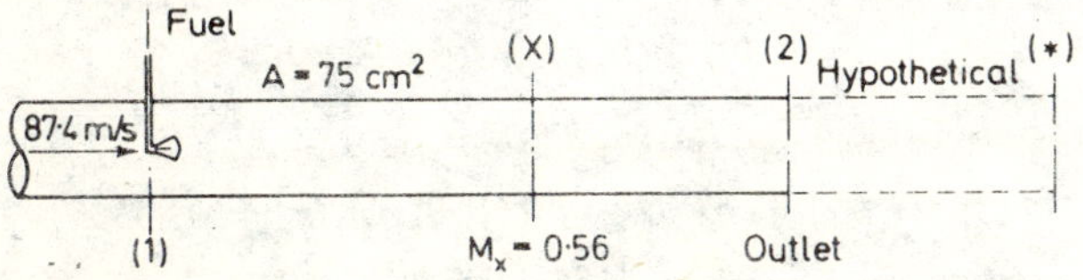

Fig. E6.1

The first part of this problem is dealt with in Sections 6.2.1 and 6.2.2.

Figure E6.1, shows a representation of the second part of the problem with a fuel/air ratio of 0·2. The subsequent analysis will assume that the products of combustion have the properties of dry air as a perfect gas so that the tables may be used.

0·2 kg of fuel produces $(0{\cdot}2\times12\,000)/1{\cdot}2$ kJ/kg of mixture ($\gamma = 1{\cdot}4$)

$$\therefore\ {}_1Q_2 = 2000 \text{ kJ/kg}$$

energy, $\quad h_{02} = h_{01} + {}_1Q_2$

or $\quad T_{02} = T_{01} + 2000/1{\cdot}00485$

where $c_p = 1{\cdot}00485$ kJ/kg K for tabulated data although c_p is a function of TK

$$\therefore\ \underline{T_{02}/T_{01} = 5{\cdot}19}$$

energy, $\quad h_{01} = h_1 + q_1^2/2$

$$\therefore\ T_{01} = 475 = T_1 + 87{\cdot}4^2/(2\times1{\cdot}00485\times10^3)$$

whence, $\quad T_1 = 471{\cdot}2$ K

$$\therefore\ T_{01}/T_1 = 475/471{\cdot}2 = \underline{1{\cdot}008}$$

from Table 1, $M_1 = 0{\cdot}2$

from Table 4, $q_1/q_* = 0{\cdot}0909$, $p_{01}/p_{0*} = 1{\cdot}2346$, $T_{01}/T_{0*} = 0{\cdot}1736$

$$\therefore\ \frac{T_{02}}{T_{0*}} = \frac{T_{02}}{T_{01}}\times\frac{T_{01}}{T_{0*}} = 5{\cdot}19\times0{\cdot}1736 = 0{\cdot}901$$

from Table 4 ($T_{02}/T_{0*} \simeq 0{\cdot}9012$), $M_2 = 0{\cdot}69$

$$q_2/q_* = 0{\cdot}6856,\quad p_{02}/p_0^* = 1{\cdot}046,\quad \frac{s_2-s_*}{R} = -0{\cdot}40913$$

from Table 1, $M_2 = 0{\cdot}69$, $p_{02}/p_2 = 1{\cdot}3748$

$$\therefore\ p_{01} = \frac{p_{01}}{p_{0*}}\times\frac{p_{0*}}{p_{02}}\times\frac{p_{02}}{p_2}\times p_2 = \frac{1{\cdot}2346}{1{\cdot}046}\times1{\cdot}3748\times10^5 \text{ N/m}^2$$

$\therefore$ stagnation pressure at inlet $= \underline{1{\cdot}6227\times10^5 \text{ N/m}^2}$

also, $$q_2 = \frac{q_2}{q_*}\times\frac{q_*}{q_1}\times q_1 = \frac{0{\cdot}6856}{0{\cdot}0909}\times87{\cdot}4 = \underline{659{\cdot}2 \text{ m/s}}$$

from Table 4, $T_1/T_* = 0{\cdot}2066$, $T_2/T_* = 0{\cdot}9874$

$$\therefore\ T_2 = \frac{0{\cdot}9874}{0{\cdot}2066}\times471{\cdot}2 = 2252 \text{ K}$$

and $\quad T_* = 2252/0{\cdot}9874 = 2280{\cdot}7$ K

$(p = \varrho RT)_2 \quad \therefore\ \varrho_2 = (1\times10^5)/(0{\cdot}2871\times10^3\times2280{\cdot}7)$

$$= 0{\cdot}1527 \text{ kg/m}^3$$

$(m = \varrho qA)_2 \quad \therefore$ mass flow $= 0{\cdot}1527\times659{\cdot}2\times0{\cdot}0075$

$$= 0{\cdot}755 \text{ kg/s}$$

from Table 4, for $M_x = 0{\cdot}56$, $T_{0x}/T_{0*} = 0{\cdot}7725$, (see Figure 6.2)

$$\frac{s_x - s_*}{R} = -0{\cdot}98977$$

$$T_x/T_* = 0{\cdot}8723$$

$\therefore$ change in entropy $s_2 - s_x = (-0{\cdot}40913 + 0{\cdot}98977)0{\cdot}2871$

$$= \underline{+0{\cdot}1667 \text{ kJ/kg K}}$$

heat energy transfer $\quad {}_xQ_2 = h_{02} - h_{0x}$ kJ/kg

$$\therefore\ {}_xQ_2 = mc_p(T_{02} - T_{0x}) \text{ kJ/s}$$

where $$T_{02} = \frac{T_{02}}{T_{0*}}\times\frac{T_{0*}}{T_{01}}\times T_{01} = \frac{0{\cdot}901}{0{\cdot}1736}\times475 = 2465{\cdot}3 \text{ K}$$

and $$T_{0x} = \frac{0{\cdot}7725}{0{\cdot}1736}\times475 = 2113{\cdot}7 \text{ K}$$

$$\therefore\ {}_xQ_2 = 0{\cdot}755\times1{\cdot}00485(2465{\cdot}3 - 2113{\cdot}7)$$

$\therefore$ heat energy transfer rate $= \underline{266{\cdot}7 \text{ kJ/s}}$

Note: from Figure 6.2, between $M = 0{\cdot}56$ and $M = 0{\cdot}69$ the graph is almost linear

$$\text{area under graph} = \left(\frac{s_2 - s_x}{R}\right)\left(\frac{T_x/T_* + T_2/T_*}{2}\right)$$

$$= 0{\cdot}53991$$

heat energy transfer $= 0{\cdot}53991\times(T_*\times R)$

$= 353{\cdot}5$ kJ/kg

(rate) $= 266{\cdot}9$ kJ/s

(The difference between the two results may be attributed to the assumption of linearity.)

EXAMPLE 6.2 Rayleigh flow is the term used to describe the effect of heat energy transfer on the one-dimensional flow of a perfect gas in a pipe of constant cross-section.

For the choking condition ($M = 1$, $q_* = a_*$) obtain an expression for $(s-s_*)/R$ in terms of T/T_* and show that the values of those parameters when T/T_* is a maximum, are:

$$T/T_* = (\gamma+1)^2/4\gamma$$

$$\frac{(s-s_*)}{R} = \frac{\gamma}{\gamma-1}\log_e\frac{(\gamma^2+1)^2}{4\gamma} - \log_e\left(\frac{\gamma+1}{2}\right)$$

An appropriate form of entropy relation with which to start the analysis, is that of equation (1.20a)

$$ds = c_p\, dT/T - R\, dp/p \qquad \text{(A)}$$

As in Chapter 6, integration of the above equation is between any arbitrary point in the flow and the sonic or choking condition, $M = 1$, $q_* = a_*$

thus, $$s-s_* = c_p \log_e (T/T_e) - R\log_e (p/p_*) \qquad \text{(B)}$$

from (6.19b) and (6.20b), $$\frac{1}{M^2}\frac{T}{T_*} = \left(\frac{p}{p_*}\right)^2 \qquad \text{(C)}$$

and from (6.20b), $$M^2 = \left(\frac{\gamma+1}{\gamma}\right)\frac{p_*}{p} - \frac{1}{\gamma}$$

substituting in (C) $$\frac{T}{T_*} = \left(\frac{p}{p^*}\right)^2\left[\left(\frac{\gamma+1}{\gamma}\right)\frac{p_*}{p} - \frac{1}{\gamma}\right]$$

or, $$\left(\frac{p}{p_*}\right)^2 - (\gamma+1)\frac{p}{p_*} + \frac{\gamma T}{T_*} = 0 \qquad \text{(D)}$$

solution of this quadratic equation gives

$$\frac{p}{p_*} = \frac{(\gamma+1)\pm[(\gamma+1)^2-4\gamma(T/T_*)]^{0\cdot5}}{2}$$

substituting in (B),

$$s-s_* = \frac{\gamma}{\gamma-1}R\log_e\frac{T}{T_*} - R\log_e\frac{p}{p_*}$$

$$\therefore \frac{s-s_*}{R} = \frac{\gamma}{\gamma-1}\log_e\frac{T}{T_*}$$

$$-\log_e\left[\frac{(\gamma+1)\pm[(\gamma+1)^2-4\gamma(T/T_*)]^{0\cdot5}}{2}\right] \qquad \text{(E)}$$

Figure 6.2 is the graphical description of this relation for the universal Rayleigh line and points (t) and (r) represent maximum values of (T/T_*) and $(s-s_*)/R$ respectively. Clearly, $M = 1$ for maximum entropy and, in the previous example (or from (6.18)), $M = 1/\gamma^{0\cdot5}$ for maximum temperature.

Equation (6.19b) gives, $$\frac{T}{T_*} = \left[\frac{(1+\gamma)M}{1+\gamma M^2}\right]^2$$

$$M = 1/\gamma^{0\cdot5} \qquad \therefore \left(\frac{T}{T_*}\right)_{max} = \frac{(\gamma+1)^2}{4\gamma}$$

substituting in (E) gives

$$\frac{(s-s_*)_{max}}{R} = \frac{\gamma}{\gamma-1}\log_e\frac{(\gamma+1)^2}{4\gamma} - \log_e\left(\frac{\gamma+1}{2}\right)$$

for $\gamma = 7/5$ (Figure 6.2), $M_t = 0{\cdot}845$

$$\left(\frac{T}{T_*}\right)_{max} = 1{\cdot}02857$$

$$\frac{(s-s_*)_{max}}{R} = -0{\cdot}08373$$

EXAMPLE 6.3 Air flows steadily through a pipe of constant cross-section and, at a certain section (1), has temperature $T_1 = 182{\cdot}7$ K, pressure $p_1 = 0{\cdot}4\times10^5$ N/m^2 and Mach number $M_1 = 1{\cdot}8$.

Plot Fanno and Rayleigh lines corresponding to condition (1) on a $(T-s)$ diagram and find from the diagram the temperature, pressure and Mach number downstream of a normal shock occurring at (1).

From Table 1 for $M_1 = 1{\cdot}8$, $T_0/T_1 = 1{\cdot}648$

$\therefore T_0 = 1{\cdot}648\times182{\cdot}7 = 301{\cdot}1$ K

($a^2 = \gamma RT$), $a_1 = 20(182{\cdot}7)^{0\cdot5} = 270{\cdot}3$ m/s

$u_1 = 1{\cdot}8\times270{\cdot}3 = 486{\cdot}5$ m/s

($p = \varrho RT$), $\varrho_1 = (0{\cdot}4\times10^5)/(0{\cdot}2871\times10^3\times182{\cdot}7)$

$= 0{\cdot}7626$ kg/m^3

from Table 3 for *Fanno* flow,

$M_1 = 1{\cdot}8$, $T_1/T_{*f} = 0{\cdot}7282$, $p_1/p_{*f} = 0{\cdot}4741$

$\therefore T_{*f} = 182{\cdot}7/0{\cdot}7282 = 250{\cdot}9$ K

from Table 4 for *Rayleigh* flow,

$M_1 = 1{\cdot}8$, $T_1/T_* = 0{\cdot}6089$, $p_1/p_* = 0{\cdot}4335$

$\therefore T_{*r} = 182{\cdot}7/0{\cdot}6089 = 300$ K

$\therefore T_{*r} \neq T_{*f}$ and a plot of T/T_* against $(s-s_*)/R$ (as in Figures 5.3 and 6.2) would not give a common point for $M_1 = 1{\cdot}8$.

In general, from (1.22a),

$$s-s_1 = c_p\log_e(T/T_1) - R\log_e(p/p_1) \qquad \text{(A)}$$

Fanno Flow

from (5.17b), $$\frac{T}{T_*} = \frac{\gamma+1}{2+(\gamma-1)M^2}$$

and from (5.18b), $$\frac{p}{p_*} = \frac{1}{M}\left[\frac{\gamma+1}{2+(\gamma-1)M^2}\right]^{0\cdot5}$$

arranging, $$\frac{p}{p_1} = \frac{M_1}{M}\left(\frac{T}{T_1}\right)^{0\cdot5} \qquad \text{(B)}$$

now, $$T_0/T = 1+\frac{\gamma-1}{2}M^2$$

$$\therefore \frac{M_1}{M} = \left(\frac{T_0-T_1}{T_0-T}\cdot\frac{T}{T_1}\right)^{0\cdot5}$$

substituting in (B), $\dfrac{p}{p_1} = \dfrac{T}{T_1}\left(\dfrac{T_0-T_1}{T_0-T}\right)^{0.5}$

and in (A),

$$s-s_1 = c_p \log_e (T/T_1)+\frac{R}{2} \log_e \left(\frac{T_0-T}{T_0-T_1}\right) - R \log \frac{T}{T_1}$$

or, $$\frac{s-s_1}{R} = \frac{1}{\gamma-1} \log_e\left(\frac{T}{T_1}\right)+\frac{R}{2}\log_e\left(\frac{T_0-T}{T_0-T_1}\right) \quad \text{(C)}$$

Rayleigh Flow,

from (6.19b), $$\frac{T}{T_*} = \left[\frac{(1+\gamma)M}{1+\gamma M^2}\right]^{0.5}$$

and from (6.20b), $$\frac{p}{p_*} = \frac{1+\gamma}{1+\gamma M^2}$$

arranging, as above, $$\frac{p}{p_1} = \frac{M_1}{M}\left(\frac{T}{T_1}\right)^{0.5} \quad \text{(B)}$$

but, $$\frac{M_1}{M} = \left(\frac{T_{01}-T_1}{T_0-T}\cdot\frac{T}{T_1}\right)^{0.5}$$

(T_0 is only constant for adiabatic processes) giving,

$$\frac{s-s_1}{R} = \frac{1}{\gamma-1}\log_e\left(\frac{T}{T_1}\right)+\frac{R}{2}\log_e\left(\frac{T_0-T}{T_{01}-T_1}\right) \quad \text{(D)}$$

Comparing (C) and (D) it is evident that $T_0 = T_{01} = T_{02}$ will represent two common points or solutions of these equations and the corresponding intersections of Fanno and Rayleigh lines are shown in Figure 6.3.

In (A), $s_2-s_1 = c_p \log_e (T_2/T_1) - R \log_e (p_2/p_1)$ which also corresponds to a plane normal shock wave process, shown as a dashed line (having no intermediate equilibrium condition) in Figure 6.3.

The Fanno and Rayleigh lines are plotted by assuming point (1) to have $M_1 = 1.8$ and $\dfrac{(s-s_1)}{R} = 0$

The following tabulations of data taken from Tables 3 and 4 illustrate the point.

Fanno Flow–Table 3

$$\left(T_{*f} = 250.9\ K,\ \frac{(s_1-s_*)}{R} = -0.3639\right)$$

M	$(s-s_*)/R$	T/T_{*f}	$(s-s_1)/R$	TK
2·0	−0·5233	0·6667	−0·1594	167·3
1·8	−0·3639	0·7282	0 (1)	182·7 (1)
1·6	−0·2233	0·7937	+0·1406	199·1
1·4	−0·1088	0·8621	+0·2551	216·3
1·2	−0·0300	0·9317	+0·3339	233·8
1·0	0	1·0000	+0·3639	250·9
0·8	−0·0375	1·0638	+0·3264	266·9
			+0·21 (2)	280 (2) solution
0·6	−0·1724	1·1194	+0·1915	280·9
0·4	−0·4638	1·1628	−0·0999	291·7
0·2	−1·0864	1·1905	−0·7225	298·7

Rayleigh Flow–Table 4

$$\left(T_{*r} = 300\ \text{K}_2 = \frac{(s_1-s_*)}{R} = -0.9003\right)$$

M	$(s-s_*)/R$	T/T_{*r}	$(s-s_1)/R$	TK
2·0	−1·2176	0·5289	−0·3173	158·7
1·8	−0·9003	0·6089	0 (1)	182·7 (1)
1·6	−0·5926	0·7017	+0·3077	210·5
1·4	−0·3128	0·8054	+0·5875	241·6
1·2	−0·0945	0·9118	+0·8058	273·5
1·0	0	1·0000	+0·9003	300·0
0·9	−0·0327	1·0245	+0·8676	307·5
0·8	−0·1477	1·0255	+0·7526	307·4
			+0·21 (2)	280 (2) solution
0·6	−0·7717	0·9167	+0·1286	275·0

From Figure 6.3 the temperature downstream of a normal shock occurring at (1) is $T_2 = 280$ K and the change in entropy across the shock is $(s_2-s_1)/R = 0.21$

from (5.17b), $\gamma = 7/5$, $\dfrac{T_2}{T_{*f}} = \dfrac{6}{5+M_2^2}$

$$\therefore M_2^2 = (6\times 250.9)/280-5 = 0.3764$$

$$\therefore M_2 = 0.6135$$

$(a^2 = \gamma RT)$, $\therefore a_2 = 20(280)^{0.5} = 334.7$ m/s

$$\therefore u_2 = 0.6135\times 334.7 = 205.3 \text{ m/s}$$

continuity, mass flow $= \varrho_1 u_1 A = \varrho_2 u_2 A$

$$\therefore \varrho_2 = 0.7626\times 486.5/205.3 = 1.8071 \text{ kg/m}^3$$

$(p = \varrho RT)$, $\therefore p_2 = 1.8071\times 0.2871\times 10^3\times 280$

$$= 1.453\times 10^5 \text{ N/m}^2$$

Normal Shock—Table 2 (figures in brackets refer to graphical solution)

for $M_1 = 1.8$, $\quad M_2 = 0.6165(0.6135)$

$p_2/p_1 = 3.6133$

$\therefore p_2 = 1.445\times 10^5$ N/m² $(1.453\times 10^5$ N/m²)

$T_2/T_1 = 1.5316$ $\therefore T_2 = 279.8$ K (280 K)

also, $\quad (s_2-s_1)/R = 0.2074$ (0·21)

EXAMPLE 6.4 A perfect gas flows along a non-adiabatic, straight duct of constant cross-sectional area A and perimeter λ. Show that the condition that the Mach number M of the flow remains constant along the duct is

$$\frac{dT_0}{dx} = -\frac{\gamma M^2 T_0}{(1+\gamma M^2)}\frac{\lambda}{A}C_f$$

where T_0 is the local stagnation temperature of the gas, γ is the ratio of the principal specific heat capacities of the gas, C_f is the friction coefficient for the pipe and x is measured along the pipe in the direction of the flow.

Air flows at a constant Mach number of 0·5 along a pipe of diameter 100 mm. At a particular station (1) of the pipe, $T_0 = 550$ K. Assuming that air is a perfect gas and $C_f = 0.01$, estimate the static temperature T at a point 5 m downstream of (1).

(U. London 1969)

The required analysis examines the effects of friction and of heat energy transfer and an appropriate control volume (similar to Figure 6.1) is shown in Figure E 6.4

continuity, $m = \varrho q A = \text{constant}$

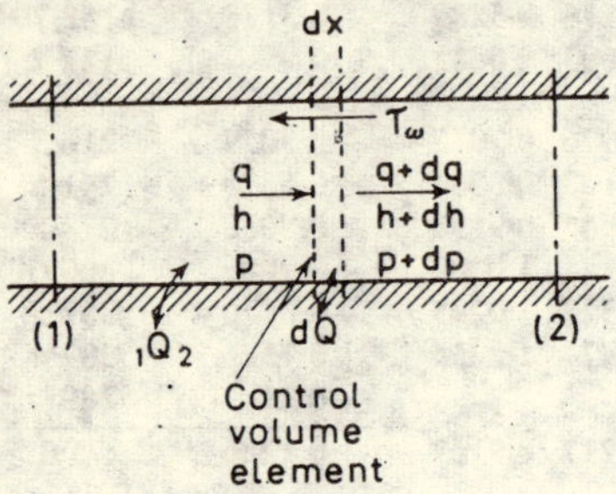

Fig. E6.4

$A = \text{constant}, \quad \therefore \mathrm{d}(\varrho q A) = 0 = \dfrac{\mathrm{d}\varrho}{\varrho} + \dfrac{\mathrm{d}q}{q} + \left(\dfrac{\mathrm{d}A}{A} = 0\right)$ (A)

momentum, $(\varrho q A)\,\mathrm{d}q = [p-(p+\mathrm{d}p)]A - \tau_\omega \lambda\,\mathrm{d}x$

$$\therefore \mathrm{d}p + \varrho q\,\mathrm{d}q + \frac{1}{2}\varrho q^2\left(\frac{\lambda}{A}C_\mathrm{f}\right)\mathrm{d}x = 0$$

$(a^2 = \gamma p/\varrho)$, $\therefore \dfrac{\mathrm{d}p}{p} + \gamma M^2\dfrac{\mathrm{d}q}{q} + \dfrac{\gamma M^2}{2}\left(\dfrac{\lambda C_\mathrm{f}}{A}\right)\mathrm{d}x = 0$ (B)

now $q = M(\gamma R T)^{0\cdot5}$ $\therefore \dfrac{\mathrm{d}q}{q} = \dfrac{\mathrm{d}M}{M} + \dfrac{1}{2}\dfrac{\mathrm{d}T}{T}$ (C)

substituting (C) in (B),

$$\frac{\mathrm{d}p}{p} + \frac{\gamma M^2}{2}\left(\frac{\lambda C_\mathrm{f}}{A}\right)\mathrm{d}x + \gamma M^2\frac{\mathrm{d}M}{M} + \frac{\gamma M^2}{2}\frac{\mathrm{d}T}{T} = 0 \quad \text{(D)}$$

energy, ${}_1Q_2 = h_{02} - h_{01}$

or $\mathrm{d}Q = \mathrm{d}h_0 = c_p\,\mathrm{d}T_0$

where, $T_0 = T\left(1 + \dfrac{\gamma-1}{2}M^2\right)$ (2.14)

$$\therefore \mathrm{d}Q/c_p = \mathrm{d}T_0 = \mathrm{d}T\left(1+\frac{\gamma-1}{2}M^2\right) + T(\gamma-1)M\,\mathrm{d}M \quad \text{(E)}$$

for a perfect gas $p = \varrho RT$, $\therefore \dfrac{\mathrm{d}p}{p} = \dfrac{\mathrm{d}\varrho}{\varrho} + \dfrac{\mathrm{d}T}{T}$

from (A), $\dfrac{\mathrm{d}p}{p} = -\dfrac{\mathrm{d}q}{q} + \dfrac{\mathrm{d}T}{T}$

from (C), $\dfrac{\mathrm{d}p}{p} = -\dfrac{\mathrm{d}M}{M} + \dfrac{1}{2}\dfrac{\mathrm{d}T}{T}$

substituting in (D) $(\gamma M^2 - 1)\dfrac{\mathrm{d}M}{M} + \dfrac{(\gamma M^2+1)}{2}\dfrac{\mathrm{d}T}{T} +$

$$\frac{\gamma M^2}{2}\left(\frac{\lambda C_\mathrm{f}}{A}\right)\mathrm{d}x = 0 \quad \text{(F)}$$

from (2.14) and (E), $\dfrac{\mathrm{d}T_0}{T_0} = \dfrac{\mathrm{d}T}{T} + \dfrac{(\gamma-1)M}{1+\dfrac{\gamma-1}{2}M^2}\mathrm{d}M$ (G)

substituting in (F), $\dfrac{\mathrm{d}T_0}{T_0}\dfrac{(\gamma M^2+1)}{2} + \dfrac{\gamma M^2}{2}\left(\dfrac{\lambda C_\mathrm{f}}{A}\right)\mathrm{d}x$

$$= \left[\frac{(\gamma M^2+1)(\gamma-1)M^2}{2+(\gamma-1)M^2} - (\gamma M^2-1)\right]\frac{\mathrm{d}M}{M} \quad \text{(H)}$$

for constant Mach number $\mathrm{d}M = 0$,

$$\therefore \frac{\mathrm{d}T_0}{\mathrm{d}x} = -\frac{\gamma M^2 T_0}{(1+\gamma M^2)}\frac{\lambda}{A}C_\mathrm{f}$$

In the above result (H), $\lambda = \pi D$, $A = \pi D^2/4$

$\therefore \lambda/A = 4/D$

$$\therefore \frac{\mathrm{d}T_0}{T_0} = -\left(\frac{1\cdot4\times0\cdot25}{1+1\cdot4\times2\cdot5}\right)\left(\frac{4\times0\cdot01}{0\cdot1}\right)\int_1^2 \mathrm{d}x$$

$$\therefore \log_e\frac{T_{02}}{T_{01}} = -\frac{0\cdot14}{1\cdot35}\times5 = -0\cdot51852$$

or, $T_{01}/T_{02} = 1\cdot6845$

from (G) or (2.14), for $\mathrm{d}M = 0$, $\dfrac{T_{02}}{T_{01}} = \dfrac{T_2}{T_1}$

from (2.14) $T_{01} = T_1(1+0\cdot2\times0\cdot25)$

$\therefore T_1 = 550/1\cdot05 = 523\cdot8\ \mathrm{K}$

$\therefore T_2 = T_1/1\cdot6845 = 311\ \mathrm{K}$

Note: for a general solution of equation (H) when the Mach number is variable under the effects of friction and heat energy transfer, numerical procedures are required. A good approximation however, may be obtained by considering the effects of friction and of heat energy transfer separately and then adding the changes (e.g. of pressure).

7 THE PLANE OBLIQUE SHOCK WAVE

In Section 3.2, the concept of a Prandtl–Meyer (isentropic) compression process was discussed and is illustrated in Figure 7.1. The incident flow (M_1) gradually changes direction though a convex bend before continuing with uniform flow at a final condition ($M_2 < M_1$). Intermediate Mach lines of varying gradient indicate the gradual change in flow characteristics, which may be considered to be isentropic. The convergence of such Mach lines, however, creates a region in which there is a significant finite change in the fluid flow properties and the process is thus not reversible but is still adiabatic.

bend. For a sharp corner, the plane oblique shock wave is assumed to extend from the corner in a straight line with the flow properties uniform on either side. Other boundaries are assumed to have no influence on the flow model.

Since the oblique shock wave is associated with a change of flow direction, plane oblique shock flow is two-dimensional and suitable components of velocity are selected for analysis. As will be shown later, the two limiting conditions of plane oblique shock flow are one-dimensional isentropic flow and plane normal shock flow, both of which involve no change of flow direction.

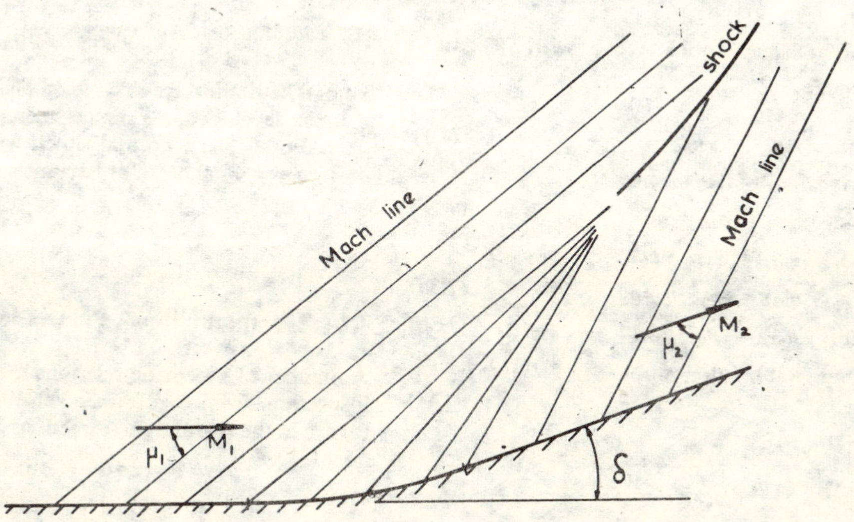

Fig. 7.1 oblique shock wave development

This discontinuity in pressure, temperature, etc., is treated as a shock and the curvature of the shock line in Figure 7.1 is an indication that the shock strength varies between the inlet and outlet conditions. To simplify the analysis the flow model shown in Figure 7.2, with a corner replacing the bend, is used to represent the deflection of an incident supersonic flow through the same angle as the

7.1 ANALYSIS – BUSEMANN SHOCK POLAR DIAGRAM

Figure 7.3 represents the steady uniform adiabatic flow of a perfect gas deflected through an angle (δ) by a plane oblique shock wave inclined at angle (θ) to the incident flow direction. Velocity components (q_n) and (q_t) are respectively normal and parallel to the oblique shock wave.

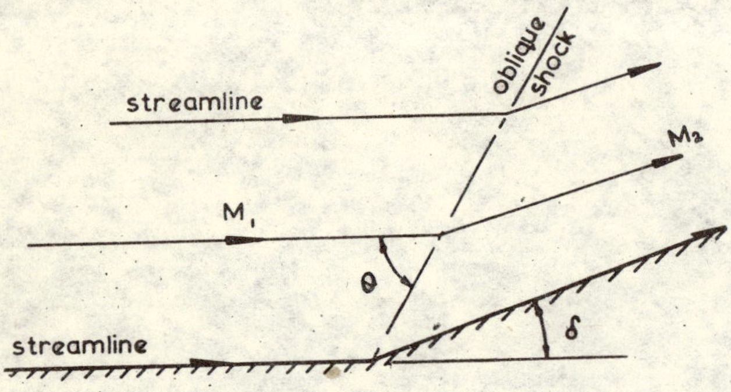

Fig. 7.2 plane oblique shock wave at sharp corner.

Since the velocity component (q_t) is aligned with the shock it is not involved with mass or momentum transfer *across* the shock from condition (1) to condition (2). (q_t) is therefore not associated with the changes in flow characteristics.

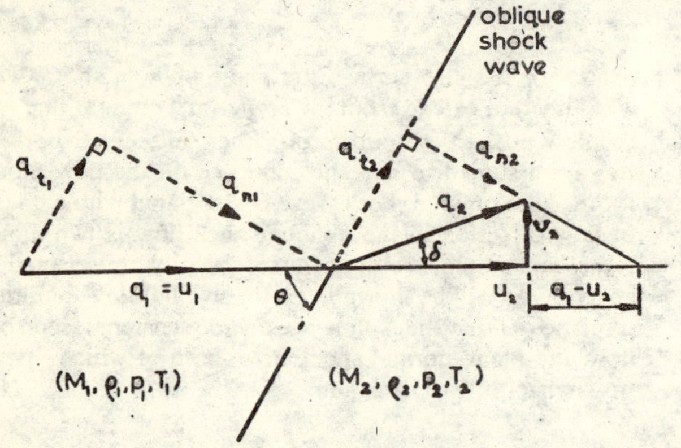

Fig. 7.3 plane oblique shock wave with kinematic changes.

continuity, $A\rho_1 q_{n1} = \rho_2 q_{n2} = \rho q_n$ (7.1)

momentum, $(A\varrho q_n)_1 q_{t1} = (A\varrho q_n)_2 q_{t2}$ (7.2)

from (7.1), $q_{t1} = q_{t2} = q_t$ (7.2a)

also, $A\varrho q_n(q_{n2}-q_{n1}) = A(p_1-p_2)$ (7.3)

or, $q_{n1}(q_{n1}-q_{n2}) = (p_2-p_1)/\varrho_1$ (7.3a)

or, $q_{n2}(q_{n1}-q_{n2}) = (p_2-p_1)/\varrho_2$ (7.3b)

adding (7.3a) and (7.3b), $(q_{n1}^2-q_{n2}^2) = (p_2-p_1)(1/\varrho_1+1/\varrho_2)$ (7.4)

energy, $$\frac{q_1^2}{2}+\frac{\gamma}{\gamma-1}\frac{p_1}{\varrho_1} = \frac{q_2^2}{2}+\frac{\gamma}{\gamma-1}\frac{p_2}{\varrho_2} = \frac{\gamma}{\gamma-1}\frac{p_0}{\varrho_0} = \frac{\gamma+1}{\gamma-1}\frac{q_*^2}{2} = \frac{c^2}{2} \quad (7.5)$$

or, $(a^2 = \gamma p/\varrho)$, $$\frac{q_{n1}^2+q_{t1}^2}{2}+\frac{\gamma}{\gamma-1}\frac{p_1}{\varrho_1} = \frac{q_{n2}^2+q_{t2}^2}{2}+\frac{\gamma}{\gamma-1}\frac{p_2}{\varrho_2} = \frac{a_0^2}{\gamma-1} \quad (7.5a)$$

from (7.2a), $$\frac{q_{n1}^2}{2}+\frac{\gamma}{\gamma-1}\frac{p_1}{\varrho_1} = \frac{q_{n2}^2}{2}+\frac{\gamma}{\gamma-1}\frac{p_2}{\varrho_2} \quad (7.5b)$$

or, $$(q_{n1}^2-q_{n2}^2) = \frac{2\gamma}{\gamma-1}\left(\frac{p_2}{\varrho_2}-\frac{p_1}{\varrho_1}\right) \quad (7.5c)$$

Equations (7.3), (7.4) and (7.5c) are exactly similar (with $q_n \equiv q$) to equations (3.1), (3.2) and (3.4) which were used to develop the Rankine-Hugoniot (property) relations and the changes in flow characteristics across a plane normal shock wave—for a normal shock, $q_t = 0$.

Since the property changes are not affected by the velocity component variables the Rankine-Hugoniot relations are the same for any plane (oblique) shock wave, thus,

$$\frac{T_2}{T_1} = \frac{p_2}{p_1}\left[\frac{\gamma+1}{\gamma-1}+\frac{p_2}{p_1}\right]\Big/\left[\frac{\gamma+1}{\gamma-1}\frac{p_2}{p_1}+1\right] \quad \text{c.f. (4.5)}$$

7.1.1 Prandtl's Relation

from (7.5a), $$q_{n1}+\frac{2\gamma}{\gamma-1}\frac{p_1}{\varrho_1 q_{n1}} = \frac{2}{\gamma-1}\frac{a_0^2}{q_{n1}}-\frac{q_t^2}{q_{n1}} \quad (7.6a)$$

and, $$q_{n2}+\frac{2\gamma}{\gamma-1}\frac{p_2}{\varrho_2 q_{n2}} = \frac{2}{\gamma-1}\frac{a_0^2}{q_{n2}}-\frac{q_t^2}{q_{n2}} \quad (7.6b)$$

(7.6a) minus (7.6b), $$q_{n1}-q_{n2}+\frac{2\gamma}{\gamma-1}\left(\frac{p_1}{\varrho_1 q_{n1}}-\frac{p_2}{\varrho_2 q_{n2}}\right) = \left(\frac{2}{\gamma-1}a_0^2-q_t^2\right)\left(\frac{1}{q_{n1}}-\frac{1}{q_{n2}}\right) \quad (7.7)$$

from (7.1) and (7.3), $$q_{n1}-q_{n2} = \frac{p_2}{\varrho_2 q_{n2}}-\frac{p_1}{\varrho_1 q_{n1}} \quad (7.8)$$

substituting in (7.7), $$1-\frac{2\gamma}{\gamma-1} = -\left(\frac{2a_0^2}{\gamma-1}+q_t^2\right)\frac{1}{q_{n1}q_{n2}}$$

or, $$q_{n1}q_{n2} = \frac{2a_0^2}{\gamma+1}-\frac{\gamma-1}{\gamma+1}q_t^2 \quad (7.9a)$$

From (7.5), $$q_{n1}q_{n2} = q_*^2-\frac{\gamma-1}{\gamma+1}q_t^2 \quad (7.9b)$$

Note: The Prandtl Relation for a plane normal shock wave is developed in a similar manner in EXAMPLE 4.5. For a plane normal shock wave, $q_t = 0$, $q_{n1} = q_1$, $q_{n2} = q_2$ and (7.9b) becomes,

$$q_1 q_2 = q_*^2 \quad (7.9c)$$

7.1.2 Busemann Shock Polar Diagram

A number of kinematic relations may be written down, by inspection, from Figure 7.3. Velocity components (u) and (v) are in the (X) and (Y) directions respectively.

$$u_2 = q_{t2}\cos\theta+q_{n2}\sin\theta \quad (7.10)$$

$$v_2 = q_{t2}\sin\theta-q_{n2}\cos\theta \quad (7.11)$$

$$q_t = q_{t1} = q_{t2} = q_1\cos\theta \quad (7.12)$$

$$(v_1 = 0),\ q_{n1} = q_1\sin\theta = u_1\sin\theta \quad (7.13)$$

$$\tan\theta = (q_1-u_2)/v_2 \quad (7.14)$$

$$\sin(\theta-\delta) = q_{n2}/q_2 \quad (7.15a)$$

from (7.10), $$q_{n2} = (u_2-q_{t2}\cos\theta)/\sin\theta \quad (7.15b)$$

substituting (7.12), (7.13) and (7.14) in (7.9b),

$$q_1(u_2-q_1\cos^2\theta) = q_*^2-q_1^2\cos^2\theta\ (\gamma-1)/(\gamma+1) \quad (7.16a)$$

or, re-arranging, $$(u_2 q_1-q_*^2)\sec^2\theta = \frac{2}{\gamma+1}q_1^2 \quad (7.16b)$$

from (7.14) and using $\sec^2\theta = 1+\tan^2\theta$

$$v_2^2 = \frac{(u_2 q_1 - q_*^2)(q_1-u_2)^2}{\frac{2}{\gamma+1} q_1^2 - u_2 q_1 + q_*^2} \qquad (7.17a)$$

or, $$\left(\frac{v_2}{q_*}\right)^2 = \left[\left(\frac{u_2}{q_*}-\frac{q_*}{q_1}\right)\left(\frac{q_1}{q_*}-\frac{u_2}{q_*}\right)^2\right] \Big/ \left[\frac{2}{\gamma+1}\frac{q_1}{q_*} - \frac{u_2}{q_*} + \frac{q_*}{q_1}\right] \qquad (7.17b)$$

For a given or assumed value of the non-dimensional incident velocity parameter (q_1/q_*) a series of arbitrary values of (u_2/q_*) may be substituted to give corresponding $(\pm)$ values of (v_2/q_*). A typical graph, described as a Cartesian Oval , is shown in Figure 7.4. which represents the Busemann Shock Polar Diagram.

The Cartesian Oval is a strophoid which has the particular characteristics of a folium of Descartes (1596–2650), the Latin version of whose name is identified with Cartesian coordinates.

7.1.3 Interpretation of the Shock Polar Diagram

Before examining the detailed use of the shock polar it is useful to consider the range from which arbitrary values of (u_2/q_*) are selected.

The Cartesian Oval makes two intercepts with the horizontal axis (Figure 7.4),

$$v_2/q_* = 0, \quad u_2 = q_2.$$

from (7.17b), $$\left(\frac{u_2}{q_*}-\frac{q_*}{q_1}\right)\left(\frac{q_1}{q_*}-\frac{u_2}{q_*}\right)^2 = 0 \qquad (7.20)$$

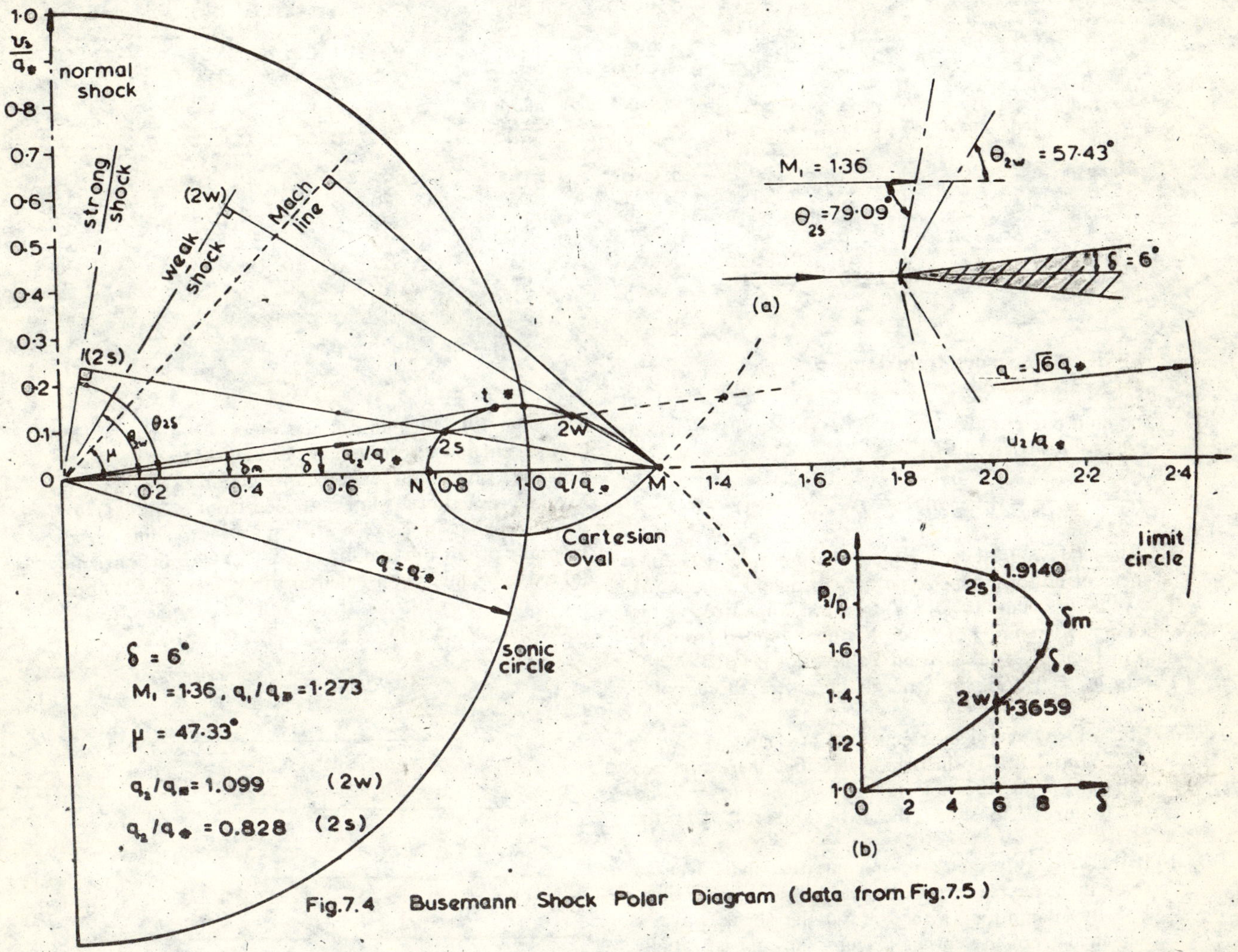

Fig.7.4 Busemann Shock Polar Diagram (data from Fig.7.5)

The Busemann Shock Polar contains a great deal of information and is conveniently used as a family of curves for a range of values of M_1, or $(q_1/q_*) = f(M_1)$ thus,

from (7.5), $$\frac{q_1^2}{2} + \frac{\gamma}{\gamma-1}\frac{p_1}{\varrho_1} = \frac{\gamma+1}{\gamma-1}\frac{q_*^2}{2}$$

or, $$\left(a^2 = \frac{\gamma p}{\varrho}\right) \quad \frac{q_1^2}{2} + \frac{q_1^2}{(\gamma-1)M_1^2} = \frac{\gamma+1}{\gamma-1}\frac{q_*^2}{2} \qquad (7.18)$$

whence, $$\left(\frac{q_1}{q_*}\right)^2 = \frac{(\gamma+1)M_1^2}{2+(\gamma-1)M_1^2} \qquad (7.19)$$

and $$\frac{u_2}{q_*}\left(=\frac{q_2}{q_*}\right) = \frac{q_*}{q_1} \qquad (7.20a)$$

or, $$\frac{u_2}{q_*}\left(=\frac{q_2}{q_*}\right) = \frac{q_1}{q_*} \qquad (7.20b)$$

(7.20a) represents point N in Figure 7.4 and corresponds to the normal shock wave condition as represented by (7.9c), $q_1 q_2 = q_*^2$.

(7.20b) represents point M in Figure 7.4 and corresponds to the incident isentropic flow Mach wave or infinitely weak shock condition, $q_2 = q_1$.

thus, the range is. $\dfrac{q_*}{q_1} \leqslant \dfrac{u_2}{q_*} \leqslant \dfrac{q_1}{q_*}$ (7.20c)

(Mathematically, the shock polar may be plotted for values of $\dfrac{u_2}{u_*} > \dfrac{q_1}{q_*}$ (dashed line in Figure 7.4) but has no physical significance).

In Figure 7.4, the sonic circle $\left(M = 1, \dfrac{q}{q_*} = 1\right)$ is plotted. The sonic circle has a radius of unity and within this circle ($q_1 < q_*$) and no shock exists with an incident subsonic flow. Point N for the normal shock wave exit condition lies well within the subsonic region and represents a limit condition of oblique shock flow.

The tangency point always lies inside the sonic circle and the *strong* shock solution (2S) always produces subsonic flow downstream; it therefore corresponds to the effect of a plane normal shock wave (point N).

The proximity of the tangency point to the sonic point indicates that the flow downstream of a *weak* shock (2W) is usually supersonic but can theoretically be subsonic.

The extract (Figure 7.5) from table 5 for $M_1 = 1{\cdot}36$ ($q_1/q_* = 1{\cdot}273$) illustrates the above points.

Note: (a) The top and bottom lines of data refer to the isentropic (point M) and normal shock (point N) conditions respectively.
(b) For $\delta = 8°$ both the strong and weak shock solutions are subsonic ($M_2 < 1$).

M_1 (q_1/q_*)	δ	θ	T_2/T_1	p_2/p_1	p_{02}/p_{01}	M_2	q_2/q_*	u_2/u_1	V_2/q_*	$(s_2-s_1)/R$
1.36	0.0	47.332	1.0000	1.0000	1.0000	1.3600	1.273	1.273	0.000	0.0000
(1.273)	1.0	48.678	1.0142	1.0504	1.0000	1.3244	1.248	1.229	0.021	+1.2136$_{10}$ -5
	2.0	50.114	1.0287	1.1039	0.9999	1.2879	1.223	1.186	0.041	+9.6435$_{10}$ -5
	4.0	53.355	1.0593	1.2225	0.9992	1.2108	1.166	1.103	0.077	+8.2658$_{10}$ -4
	6.0	57.430	1.0941	1.3659	0.9969	1.1226	1.099	1.020	0.107	+3.0857$_{10}$ -3
	8.0	64.286	1.1439	1.5830	0.9901	0.9940	0.995	0.919	0.129	+9.9194$_{10}$ -3
$\delta_* =$ →	7.9411	63.935	1.1416	1.5746	0.9905	1.0000	1.000	0.923	0.129	+9.5023$_{10}$ -3
$\delta_m =$ →	8.3253	68.307	1.1680	1.6964	0.9851	0.9298	0.940	0.876	0.128	+1.4962$_{10}$ -2
	8.0	72.305	1.1881	1.7918	0.9801	0.8747	0.892	0.844	0.119	+2.0031$_{10}$ -2
	6.0	79.094	1.2133	1.9140	0.9727	0.8033	0.828	0.807	0.085	+2.7635$_{10}$ -2
	4.0	83.246	1.2230	1.9614	0.9696	0.7751	0.802	0.794	0.055	+3.0853$_{10}$ -2
	2.0	86.740	1.2276	1.9842	0.9681	0.7614	0.790	0.787	0.027	+3.2461$_{10}$ -2
	1.0	88.383	1.2287	1.9895	0.9677	0.7582	0.787	0.786	0.014	+3.2836$_{10}$ -2
	0.0	90.000	1.2290	1.9912	0.9676	0.7572	0.786	0.786	0.000	+3.2959$_{10}$ -2

Fig 7·5 Extract from table 5

The limit circle is also shown in the shock polar diagram and the radius is given by,

$$\frac{q}{q_*} = \frac{c}{q_*} = \left(\frac{\gamma+1}{\gamma-1}\right)^{0.5} = \sqrt{6} \quad \text{for air } (\gamma = 7/5)$$

Figure 7.4a represents a typical sharp corner deflecting an incident uniform supersonic flow through an angle (δ). The velocity vector (q_2/q_*) is drawn at angle (δ) on the shock polar (Figure 7.4) which has been plotted from the data in Figure 7.5 taken from Table 5 for $\delta = 6°$, $M_1 = 1{\cdot}36$ ($q_1/q_* = 1{\cdot}273$). In general, there are three possibilities:

(a) The velocity vector intersects the Cartesian Oval twice, at (2S) and (2W) which indicates that there are two values of (q_2/q_*) possible for each value of (q_1/q_*) and (δ). Figure 7.4b shows the two values of p_2/p_1 for $\delta = 6°$.

(b) If the velocity vector does not intersect the Cartesian Oval then no plane oblique shock wave is possible. For this condition (with $\delta > \delta_m$) the shock wave is detached from the corner (Figure 7.6) and is initially normal to the upstream boundary before curving away in a downstream direction with diminishing strength.

(c) The velocity vector is tangent to the curve (t) and, for a given (q_1/q_*), represents the maximum deflection (δ_m) of the incident flow through a plane oblique shock wave.

The tangency point (t) is very close to the (sonic) point (*) of intersection of the Cartesian Oval with the sonic circle and effectively divides the two possible solutions into *strong* (point 2S) and *weak* (point 2W) shocks.

The angle (θ) which an oblique shock wave makes with the incident flow direction may be obtained from the shock polar by a simple and general construction as follows:
For the *weak* shock, a line extended from point (M)(q_1/q_*), through point (2W) is at right angles to a line drawn from the origin and making angle (θ_w) with the horizontal axis. A similar construction through point (2S) produces (θ_s) for the *strong* shock and these angles are also shown in Figure 7.4a.

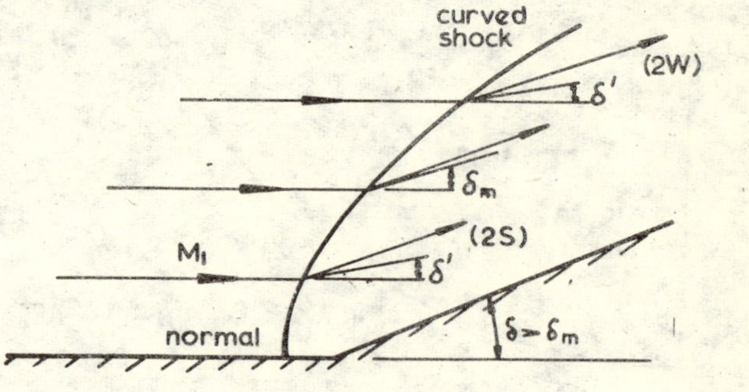

Fig.7.6 curved detached shock.

A tangency condition at point (M) produces, by the same construction, a line passing through the origin at Mach angle (μ). For point (N) the appropriate line passing through the origin is the vertical axis at right angles to the incident flow direction for a plane normal shock wave.

7.2 COMPUTATION OF PLANE OBLIQUE SHOCK WAVE DATA

If velocity (or Mach number) is involved in analysis and development of equations, together with property changes, the plane oblique shock wave may be considered to act as a plane normal shock wave for the flow component at right angles to it. The previous relations (Chapter 4) for a normal shock will thus apply to an oblique shock if M_1 and M_2 are replaced by $M_1 \sin\theta$ and $M_2 \sin(\theta-\delta)$.

from (4.9), $$\frac{T_2}{T_1} = \frac{[2\gamma M_1^2 \sin^2\theta-(\gamma-1)][(\gamma-1)M_1^2\sin^2\theta+2]}{(\gamma+1)^2M_1^2\sin^2\theta} \tag{7.21}$$

from (4.8a), $$\frac{p_2}{p_1} = 1+\frac{2\gamma}{\gamma+1}(M_1^2\sin^2\theta-1) \tag{7.22}$$

$$= \frac{2\gamma M_1^2\sin^2\theta-(\gamma-1)}{\gamma+1} \tag{7.22a}$$

from (4.13),

$$\frac{p_{02}}{p_{01}} = \left[\frac{(\gamma+1)M_1^2\sin^2\theta}{(\gamma-1)M_1^2\sin^2\theta+2}\right]^{\frac{\gamma}{\gamma-1}}\left[\frac{\gamma+1}{2\gamma M_1^2\sin^2\theta-(\gamma-1)}\right]^{\frac{1}{\gamma-1}} \tag{7.23}$$

from (4.10), $$\frac{\varrho_1}{\varrho_2} = \frac{(\gamma-1)M_1^2\sin^2\theta+2}{(\gamma+1)M_1^2\sin^2\theta} = \frac{q_{n2}}{q_{n1}} \tag{7.24}$$

from (4.11), $$M_2\sin(\theta-\delta) = \left[\frac{(\gamma-1)M_1^2\sin^2\theta+2}{2\gamma M_1^2\sin^2\theta-(\gamma-1)}\right]^{0\cdot5} \tag{7.25}$$

As previously noted, the change of entropy for *any* adiabatic process is given by (4.14),

$$s_2-s_1 = R\log_e(p_{01}/p_{02}) \tag{7.26}$$

Velocity ratios for plotting the shock polar diagram may also be obtained as functions of (M_1) and (θ) thus:

from (7.12), $$q_{t2}/q_1 = \cos\theta \tag{7.27}$$

from (7.13) and (7.24), $$\frac{q_{n2}}{q_1} = \frac{(\gamma-1)M_1^2\sin^2\theta+2}{(\gamma+1)M_1^2\sin\theta} \tag{7.28}$$

$$\therefore \frac{q_2}{q_1} = \left[\frac{q_{t2}^2+q_{n2}^2}{q_1^2}\right]^{0\cdot5}$$

$$= \left[1-\frac{4(M_1^2\sin^2\theta-1)(\gamma M_1^2\sin^2\theta+1)}{(\gamma+1)^2M_1^4\sin^2\theta}\right]^{0\cdot5} \tag{7.29}$$

and, from (7.19), $$\frac{q_2}{q_*} = \frac{q_2}{q_1}\times\frac{q_1}{q_*}\ \text{ or }\ \frac{q_2}{q_*} = \frac{q_1}{q_*}\cos\delta \tag{7.30}$$

also, from (7.15a), $$\sin(\theta-\delta) = \frac{q_{n2}}{q_2} = \frac{q_{n2}}{q_1}\times\frac{q_1}{q_2} \tag{7.31}$$

using the results of (7.28), (7.29) with (7.31) in (7.15) it may be shown that,

$$M_2 = \left[\frac{(\gamma+1)^2M_1^4\sin^2\theta-4(M_1^2\sin^2\theta-1)(\gamma M_1^2\sin^2\theta+1)}{[(2\gamma M_1^2\sin^2\theta-(\gamma-1)][(\gamma-1)M_1^2\sin^2\theta+2)}\right]^{0\cdot5} \tag{7.32}$$

A relationship between (θ), (δ) and M_1 is more convenient if the variables (θ) and (δ) are separated:

from Figure 7.3, $$\tan\delta = \frac{v_2}{u_2} = \frac{v_2}{q_*}\times\frac{q_*}{u_2} \tag{7.33}$$

from (7.14), $$\tan\theta = (q_1-u_2)/v_2$$

combining, $$\frac{u_2}{q_1} = \frac{1}{1+\tan\delta\tan\theta} \tag{7.34}$$

from (7.10), (7.27) and (7.28),

$$\frac{u_2}{q_1} = 1-\frac{2(M_1^2\sin^2\theta-1)}{(\gamma+1)M_1^2} \tag{7.35}$$

from (7.11), (7.27) and (7.28),

$$\frac{v_2}{q_1} = \frac{2\cot\theta(M_1^2\sin^2\theta-1)}{(\gamma+1)M_1^2} \tag{7.36}$$

from (7.34) and (7.35),

$$\tan\delta = \frac{2\cot\theta(M_1^2\sin^2\theta-1)}{2+M_1^2(\gamma+1-2\sin^2\theta)} \tag{7.37}$$

thus, knowing (δ), (q_2/q_*), (u_2/q_*) and (v_2/q_*) may be determined.

All of the above parameters can be expressed in terms of (M_1) and (θ) but the most convenient way of presenting plane oblique shock wave data is in terms of the incident Mach number (M_1) and the flow deflection angle (δ). The following extract is from NACA Report No. 1135 (1):

Parameters M_1 and δ

"No convenient explicit relations exist. However, the value of $\sin^2\theta$ can be found by solving the following cubic equation (2):

$$\sin^6\theta+b\sin^4\theta+c\sin^2\theta+d = 0 \tag{7.38}$$

where, $$b = -\left[\frac{M_1^2+2}{M_1^2}\right]-\gamma\sin^2\delta$$

$$c = \left[\frac{2M_1^2+1}{M_1^4}\right]+\left[\frac{(\gamma+1)^2}{4}+\frac{\gamma-1}{M_1^2}\right]\sin^2\delta$$

$$d = -\frac{\cos^2\delta}{M_1^4}$$

The smallest of the three roots corresponds to a decrease in entropy and should be disregarded."

Since the structure of the shock polar and the nature of the graphical solution were known, the computation technique was based on the point of tangency. The tangency condition for maximum deflection angle (δ_m) is found directly by differentiating (7.37),

$\frac{d(\tan\delta)}{d\theta} = 0$ produces the following quadratic equation in $\sin^2\theta$,

$$2\gamma M_1^4\sin^4\theta+M_1^2(4-(\gamma+1)M_1^2)\sin^2\theta-(\gamma+1)M_1^2-2 = 0 \tag{7.39}$$

(1) NACA Report N°. 1135, 'Equations, Tables and Charts for Compressible Flow'

(2) THOMPSON, M. J., 'A Note on the Calculation of Oblique Shock-Wave Characteristics', Journal of Aeronautical Science **17**, 11, 744 Nov. 1950

whence, $\sin^2 \theta_m = \dfrac{1}{4\gamma M_1^2} \times$

$$\left[(\gamma+1)M_1^2 - 4 + [(\gamma+1)[(\gamma+1)M_1^4 + 8(\gamma-1)M_1^2 + 16]]^{0\cdot5} \right] \quad (7.40)$$

The required starting value of (δ_{max}) is then determined from (7.37) and appropriate values of $\delta < \delta_{max}$ used in (7.38) for subsequent computation, by the Newton-Raphson method, of the required parameters for the *weak* shock and *strong* shock solutions. The calculated data is then presented in terms of (M_1) and (δ).

The sonic point (δ_*, θ_*) is readily obtained by putting $M_2 = 1$ in (7.25) to produce another quadratic equation in $\sin^2 \theta$,

$$2\gamma M_1^4 \sin^4 \theta + M_1^2[(3-\gamma)-(\gamma+1)M_1^2] \sin^2 \theta - 2 = 0 \quad (7.41)$$

whence, $\sin^2 \theta_* = \dfrac{1}{4\gamma M_1^2} \{(\gamma+1)M_1^2 - (3-\gamma)$

$$+ [(\gamma+1)[(\gamma+1)M_1^4 - 2(3-\gamma)M_1^2 + (\gamma-9)]]^{0\cdot5}\} \quad (7.42)$$

7.3 SUMMARY OF EQUATIONS

The following are the equations used in Table 5 for steady adiabatic flow of air ($\gamma = 7/5$) through a plane oblique shock wave and are plotted in Figures 7.7, 7.8 and 7.9

from (7.19),

$$\frac{q_1}{q_*} = \left[\frac{6M_1^2}{5+M_1^2} \right]^{0\cdot5} \quad (7.43)$$

Note: (7.43) corresponds exactly to (2.36) which is plotted in Figure 2.11

from (7.40),

$$\sin^2 \theta_m = [3M_1^2 - 5 + (9M_1^4 + 12M_1^2 + 60)^{0\cdot5}]/7M_1^2 \quad (7.44)$$

from (7.42),

$$\sin^2 \theta_* = [3M_1^2 - 2 + (9M_1^4 - 12M_1^2 + 39)^{0\cdot5}]/7M_1^2 \quad (7.45)$$

from (7.37),

$$\tan \delta = \frac{\cot \theta (M_1^2 \sin^2 \theta - 1)}{1 + M_1^2(1\cdot2 - \sin^2 \theta)} \quad (7.46)$$

from (7.21),

$$\frac{T_2}{T_1} = \frac{(7M_1^2 \sin^2 \theta - 1)(M_1^2 \sin^2 \theta + 5)}{36 M_1^2 \sin^2 \theta} \quad (7.47)$$

from (7.22),

$$\frac{p_2}{p_1} = \frac{7M_1^2 \sin^2 \theta - 1}{6} \quad (7.48)$$

from (7.23),

$$\frac{p_{02}}{p_{01}} = \left(\frac{6M_1^2 \sin^2 \theta}{M_1^2 \sin^2 \theta + 5} \right)^{3\cdot5} \left(\frac{6}{7M_1^2 \sin^2 \theta - 1} \right)^{2\cdot5} \quad (7.49)$$

from (7.32),

$$M_2 = \left[\frac{36M_1^4 \sin^2 \theta - 5(M_1^2 \sin^2 \theta - 1)(7M_1^2 \sin^2 \theta + 1)}{(7M_1^2 \sin^2 \theta - 1)(M_1^2 \sin^2 \theta + 5)} \right]^{0\cdot5} \quad (7.50)$$

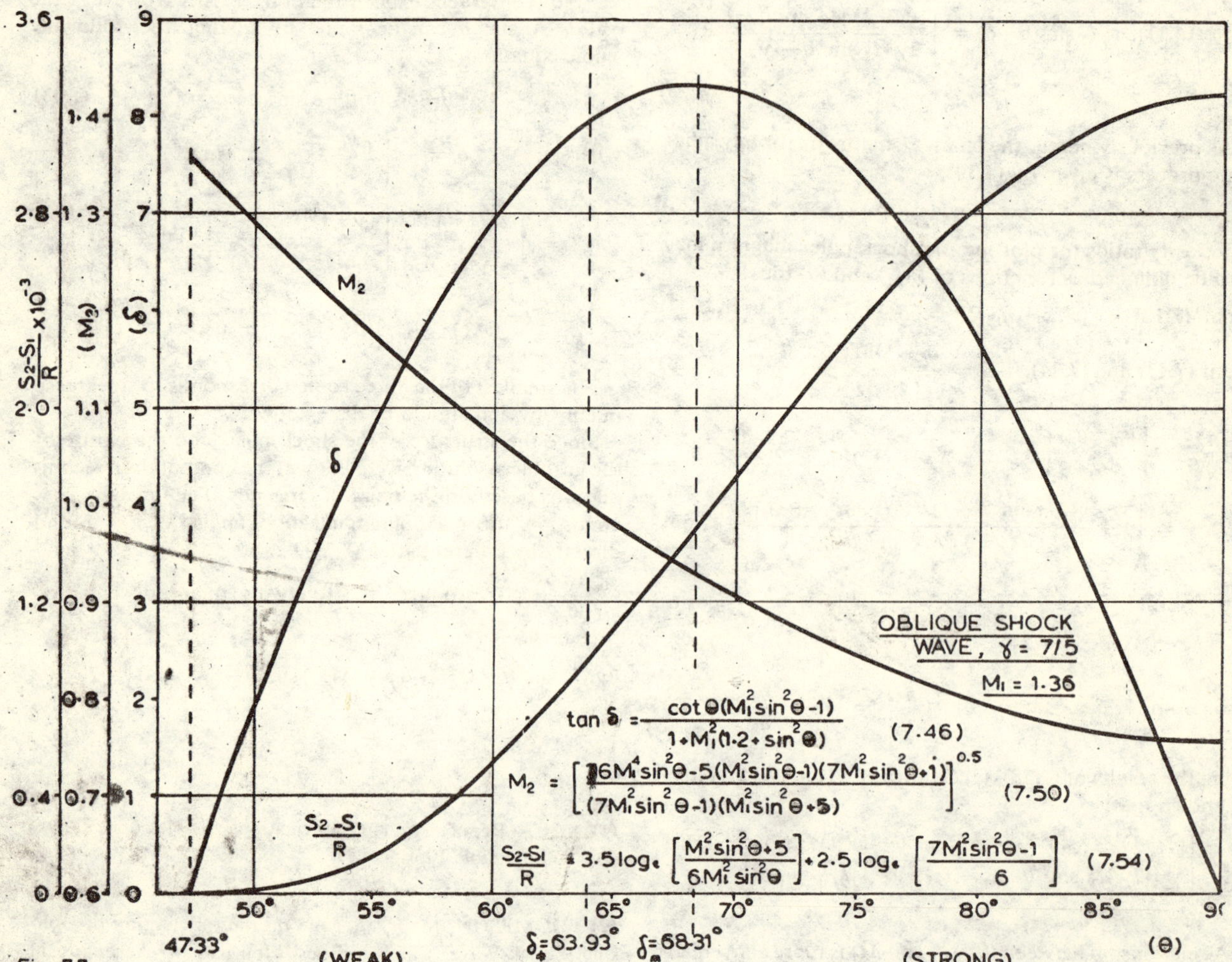

Fig. 7·7

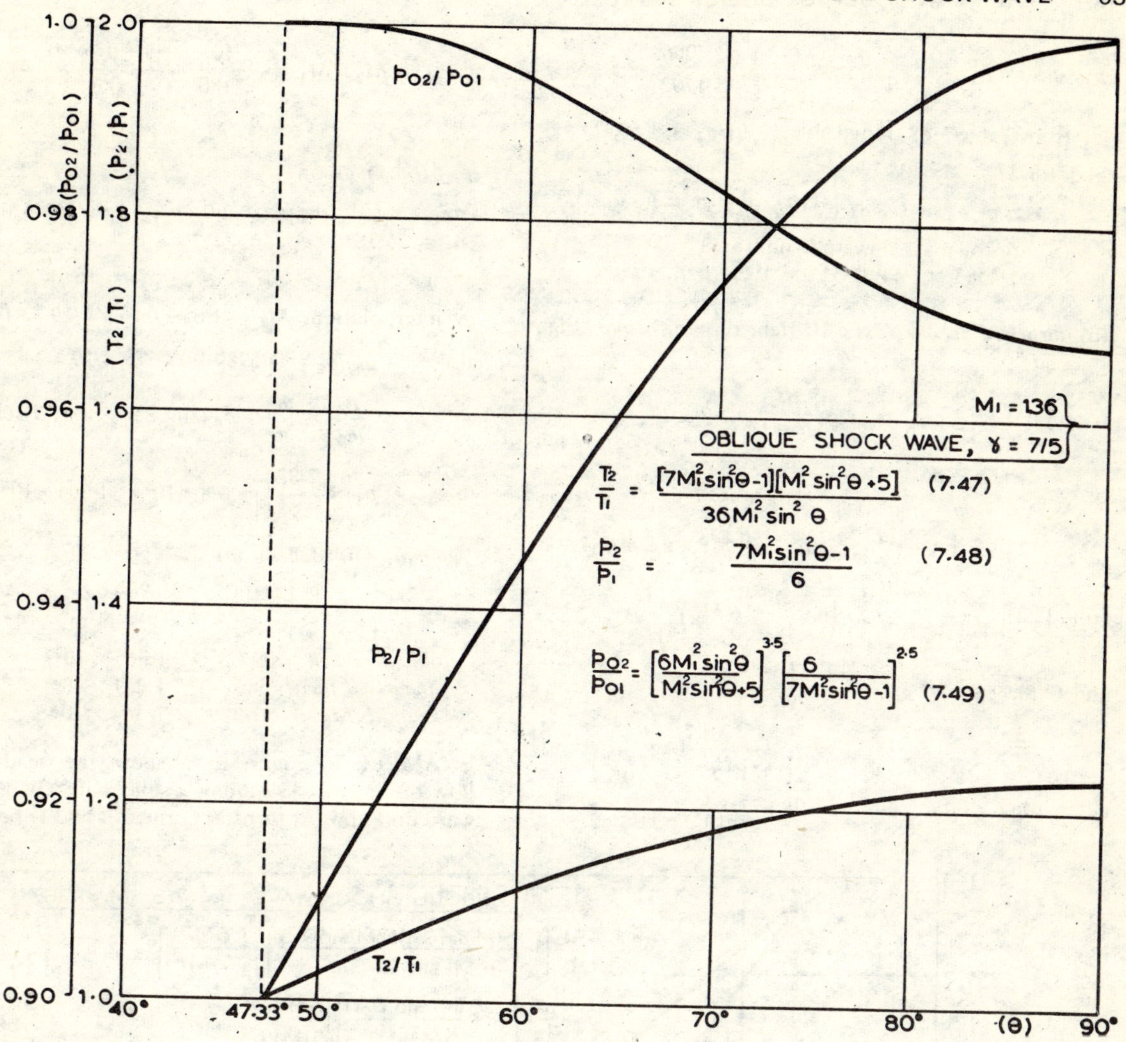

Fig. 7·8

from (7.29) and (7.40),

$$\frac{q_2}{q_*} = \left[1 - \frac{5(M_1^2\sin^2\theta-1)(7M_1^2\sin^2\theta+5)}{36M_1^4\sin^2\theta}\right]^{0.5}\left[\frac{6M_1^2}{5+M_1^2}\right]^{0.5} \tag{7.51}$$

from (7.35) and (7.43)

$$\frac{u_2}{q_*} = \left[1 - \frac{5(M_1^2\sin^2\theta-1)}{6M_1^2}\right]\left[\frac{6M_1^2}{5+M_1^2}\right]^{0.5} \tag{7.52}$$

from (7.36) and (7.43),

$$\frac{v_2}{q_*} = \left[\frac{5\cot\theta(M_1^2\sin^2\theta-1)}{6M_1^2}\right]\left[\frac{6M_1^2}{5+M_1^2}\right]^{0.5} \tag{7.53}$$

from (7.26) and (7.49),

$$\frac{s_2-s_1}{R} = 3{\cdot}5\log_e\left[\frac{M_1^2\sin^2\theta+5}{6M_1^2\sin^2\theta}\right] + 2{\cdot}5\log_e\left[\frac{7M_1\sin^2\theta-1}{6}\right] \tag{7.54}$$

Figures 7.11 and 7.12 show the variations of some of the above parameters for the condition of maximum deflection (δ_m). The particular relationships with M_1 are obtained by substituting the result of (7.44) in (7.47), (7.48), (7.49) and (7.54) for Figure 7.10 and in (7.46) and (7.50) for Figure 7.11. Corresponding values are tabulated in Table 5.

EXAMPLE 7.1 A uniform supersonic stream of air with a pressure (p) of $0{\cdot}8\times10^5$ N/m² and a Mach number (M) of 2·0 approaches a sharp 10° corner at which a weak oblique shock wave occurs.

(a) Determine the final Mach number, the final pressure and the ratio of the final mass flow per unit area to the initial mass flow per unit area;

(b) Compare the results of (a), for final Mach number and final pressure with the corresponding results for a 10° turn based on linearised theory and for a 10° turn based on Prandtl–Meyer theory.

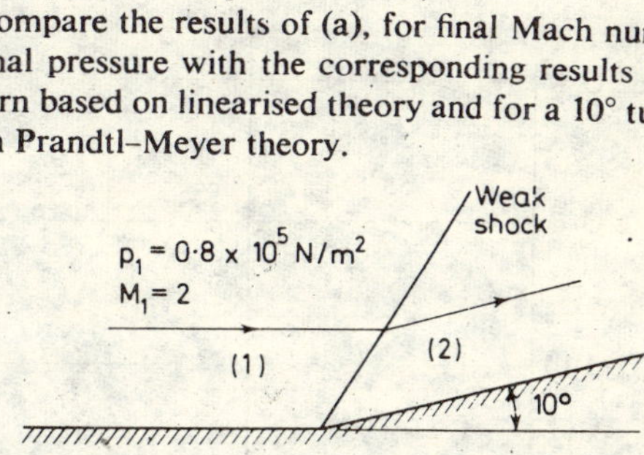

Fig. E7.1

Figure E7.1 shows the deflection through 10° of an incident uniform flow with $M_1 = 2{\cdot}0$

(a) from Table 5 for $\delta_2 = 10°$, $M_2 = 1{\cdot}6405$

also, $p_2/p_1 = 1{\cdot}7066$

$\therefore\ p_2 = 1\ 7066\times0{\cdot}8\times10^5 = 1{\cdot}365\times10^5$ N/m²

and, $T_2/T_1 = 1{\cdot}1702$

mass flow per unit area $= \varrho q$ $\therefore$ ratio $= \dfrac{\varrho_2 q_2}{\varrho_1 q_1}$

$p = \varrho RT \quad \therefore \frac{\varrho_2}{\varrho_1} = \frac{p_2}{p_1} \times \frac{T_1}{T_2} = \frac{1{\cdot}7066}{1{\cdot}1702} = 1{\cdot}4583$

from Table 5 or from Table 1 $q_1/a_* = 1{\cdot}6330$

and for $M_2 = 1{\cdot}6405$,

$$q_2/a_* = 1{\cdot}4487 + 0{\cdot}05 \times 0{\cdot}0057 = 1{\cdot}449$$

$\therefore$ ratio of mass flows per unit area =

$$1{\cdot}4583 \times 1{\cdot}449/1{\cdot}633 = 1{\cdot}294$$

(b) ***linearised** theory* (Type II. right-running wave, Figure 3·3*d*)

from (3.5), $\frac{dp}{p} = \frac{\gamma M^2}{+(M^2-1)^{0{\cdot}5}} d\delta$

or $\frac{p_2 - p_1}{p_1} = \frac{10\pi}{180} \times \frac{1{\cdot}4 \times 4}{\sqrt{3}} = 0{\cdot}5643$

$\therefore p_2 = 1{\cdot}5643 \times 0{\cdot}8 \times 10^5 = 1{\cdot}251 \times 10^5 \text{ N/m}^2$

from (3.7), $\frac{dM}{M} = \frac{dq}{q}\left(1 + \frac{\gamma - 1}{2} M^2\right)$

and (3.5), $-\gamma M^2 \frac{dq}{q} = \frac{\gamma M^2}{+(M^2-1)^{0{\cdot}5}} d\delta$

eliminating dq/q, $dM = -d\delta \frac{M\left(1 + \frac{\gamma-1}{2} M^2\right)}{+(M^2-1)^{0{\cdot}5}}$

or $M_2 - M_1 = -\frac{10\pi}{180} \cdot \frac{2(1 + 0{\cdot}2 \times 4)}{\sqrt{3}} = -0{\cdot}363$

$\therefore M_2 = 1{\cdot}637$

Prandtl–Meyer theory

from Table 1, for $M_1 = 2{\cdot}0$, $p_{01}/p_1 = 7{\cdot}8244$ and ${}_*\delta_1 = 26{\cdot}38°$

$\therefore {}_*\delta_2 = {}_*\delta_1 - {}_1\delta_2 = 16{\cdot}38°$

by interpolation, $M_2 = 1{\cdot}65 + 0{\cdot}01 \times 4/29 = 1{\cdot}651$

also, $p_{02}/p_2 = 4{\cdot}5789 + 0{\cdot}069 \times 4/29 = 4{\cdot}5884$

$p_2 = \frac{p_2}{p_{02}} \times \frac{p_{01}}{p_1} \times p_1$ (since $p_{02} = p_{01}$, isentropic)

$\therefore p_2 = \frac{7{\cdot}8244}{4{\cdot}5844} \times 0{\cdot}8 \times 10^5 = 1{\cdot}364 \times 10^5 \text{ N/m}^2$

Summary (10° deflection)

	Pressure (P_2)–N/m²	*Mach number* M_2
oblique shock wave	$1{\cdot}365 \times 10^5$	1·6405
linearised theory	1.251×10^5	1·637
Prandtl–Meyer theory	$1{\cdot}364 \times 10^5$	1·651

EXAMPLE 7.2 Figure E 7.2 shows the steady supersonic flow of air ($\gamma = 1{\cdot}4$) along a solid wall ABCD, the downward and upward turns at corners B and C being 7·1° and

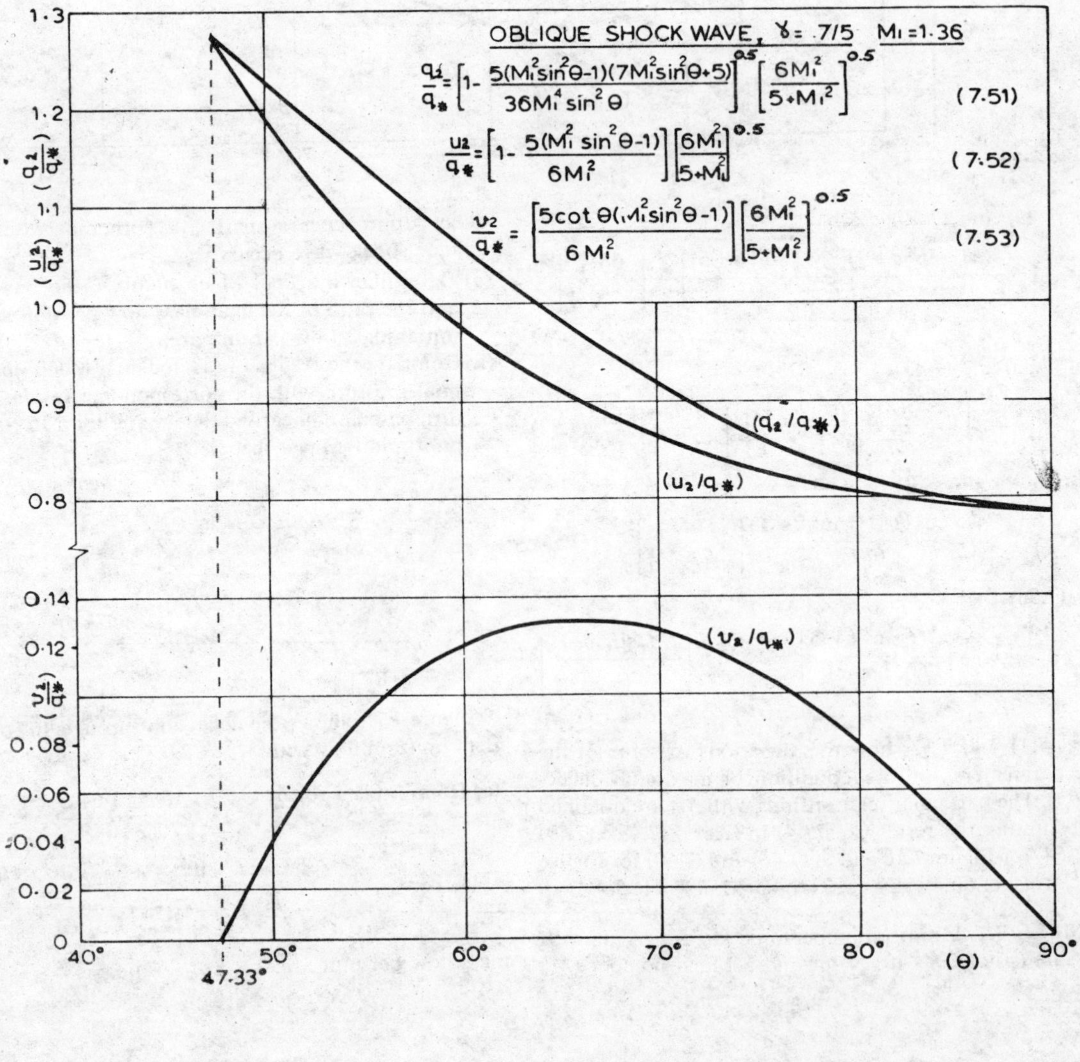

Fig. 7·9

12° respectively. If the Mach number (M_3) and the pressure (p_3), in region (3), are respectively 1·71 and 1·42×10^5 N/m^2, find the Mach number (M_1) and pressure (p_1) in region (1) and also the angular extent (α) of the expansion fan from corner B.

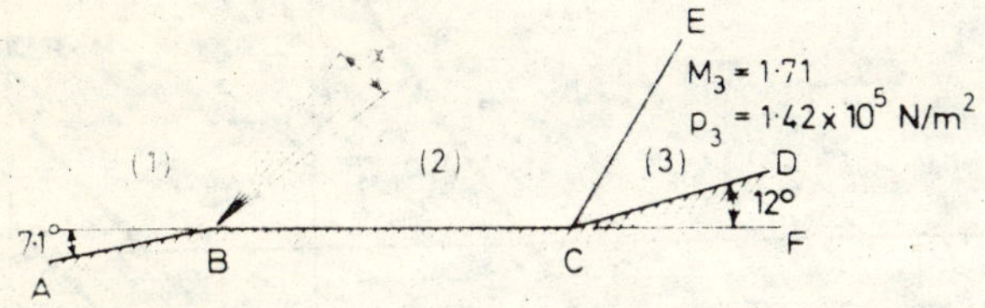

Fig. E7.2

If, with all upstream conditions unchanged, the upward turn at C were progressively increased, at what magnitude of the angle DCF would shock detachment occur? What would be the percentage change in recovery factory $R_{31} = p_{03}/p_{01}$ from the above condition for DCF = 12°?

(U. Birmingham 1964).

Table 5 has to be searched until a value of M_2 = 1·71 corresponding to 12° has been found. This provides a value of M_1 = 2·16 which then is used as M_2 according to the specification of the problem with regions (2) and (3) being separated by a weak oblique shock wave.

$\therefore$ for ${}_2\delta_3 = 12°$ then $M_2 = 2{\cdot}16$ and $M_3 = 1.71$

(Table 5) from Table 1, ${}_*\delta_2 = 30{\cdot}69°$

$\therefore {}_*\delta_1 = {}_*\delta_2 - {}_1\delta_2 = 30{\cdot}69 - 7{\cdot}1 = 23{\cdot}59$

whence, M_1 = 1·9 for a 7·1° turn in the flow direction,

also, $\dfrac{p_{01}}{p_1} = 6{\cdot}7006$ and $\dfrac{p_{03}}{p_3} = 5{\cdot}0111$ (for $M_3 = 1{\cdot}71$)

from Table 5, recovery factor $R_{13} = \dfrac{p_{03}}{p_{01}} = 0{\cdot}9707$

$$p_1 = \frac{p_1}{p_{01}} \times \frac{p_{01}}{p_{03}} \times \frac{p_{03}}{p_3} \times p_3$$

$$\therefore \quad p_1 = \frac{5{\cdot}0111}{6{\cdot}7006 \times 0{\cdot}9707} \times 1{\cdot}42 \times 10^5 = 1{\cdot}094 \times 10^5 \text{ N/m}^2$$

the expansion fan angle, $\alpha = \mu_1 + {}_1\delta_2 - \mu_2$

from Table 1, $\mu_1 = 31{\cdot}76°$ and $\mu_2 = 27{\cdot}58°$

$$\therefore \quad \alpha = 31{\cdot}76 + 7{\cdot}1 - 27{\cdot}58 = 11{\cdot}28°$$

when shock detachment occurs the limiting angle for shock wave deflection of flow is $\delta_m = 25{\cdot}5245°$ (Table 5)

at this condition $\left(\dfrac{p_{03}}{p_{01}}\right)_m = 0{\cdot}7435 = R_{31_m}$

$$\therefore \quad \% \text{ change} = \frac{0{\cdot}9707 - 0{\cdot}7435}{0{\cdot}9707} = 30{\cdot}6\%$$

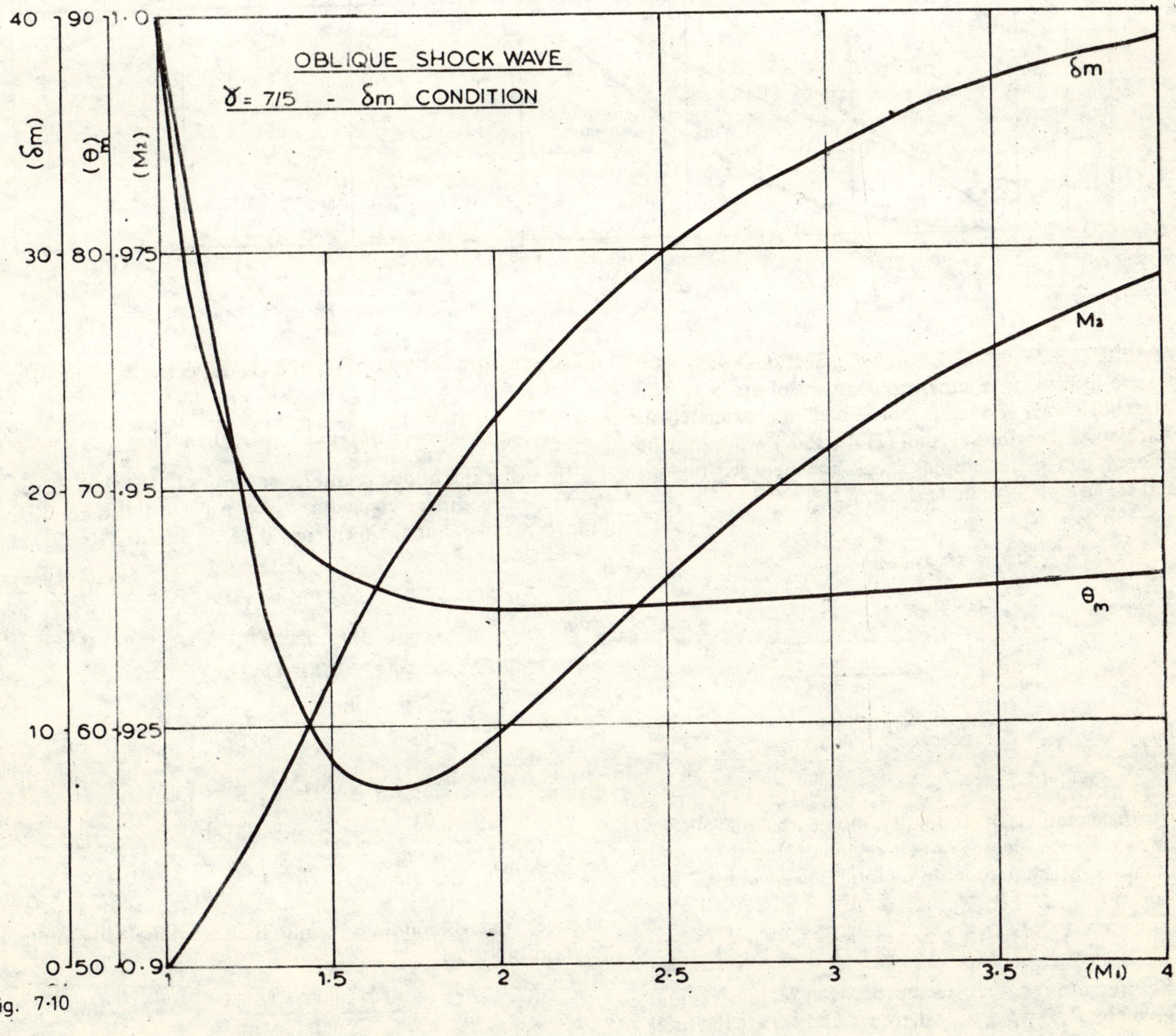

Fig. 7·10

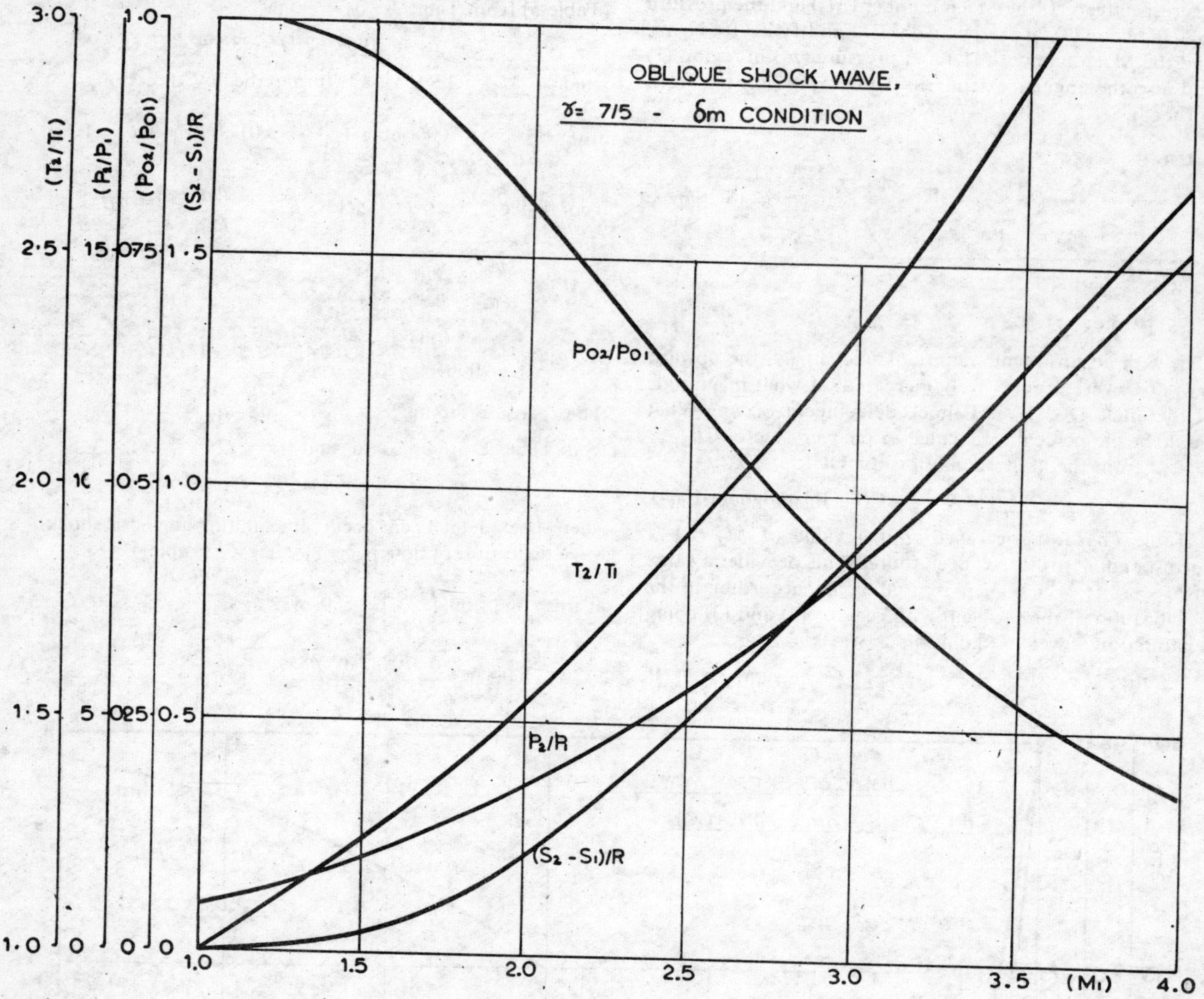

Fig. 7·11

•EXAMPLE 7.3 Figure E 7.3 shows a sectional view of a Pitot tube inserted in a supersonic stream of air ($\gamma = 7/5$) and the shock wave formed is represented approximately as shown. Between regions (1) and (2) the shock wave may be considered as a plane normal shock and between regions (1) and (3) as a plane oblique shock wave.

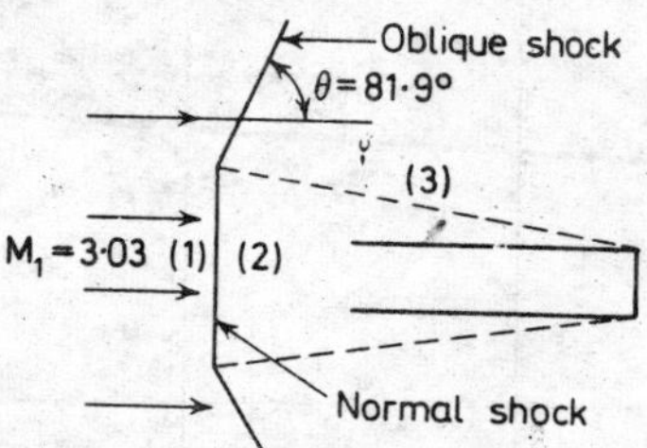

Fig. E7.3

Write the equations of continuity, momentum and energy appropriate to the flow between regions (1) and (2) and show that the Mach number in region (2) is given by,

$$M_2^2 = \frac{(\gamma-1)M_1^2+2}{2\gamma M_1^2-(\gamma-1)}$$

If the pressure recovery factor between regions (1) and (2) is given by $R_{21} = p_{02}/p_{01}$, where p_{02} and p_{01} are the total head pressures in regions (2) and (1) respectively, show that,

$$R_{21} = \left[\frac{6}{7M_1^2-1}\right]^{2\cdot5}\left[\frac{1\cdot2M_1^2}{1+0\cdot2M_1^2}\right]^{3\cdot5}$$

By using the above relations or by use of the flow tables provided, compare R_{21} with $R_{31} = p_{03}/p_{01}$ and determine the respective changes in entropy if $M_1 = 3{\cdot}03$ and $\theta = 81{\cdot}9°$

(U. Leeds 1966)

continuity, $\varrho_1 q_1 = \varrho_2 q_2$

momentum, $\varrho q(q_2 - q_1) = (p_1 - p_2)$

energy, $\frac{q_1^2}{2} + \frac{\gamma}{\gamma-1}\frac{p_1}{\varrho_1} = \frac{q_2^2}{2} + \frac{\gamma}{\gamma-1}\frac{p_2}{\varrho_2}$

giving, $\frac{p_2}{p_1} - 1 = \frac{\varrho_1 q_1^2}{p_1}\left(1 - \frac{q_2}{q_1}\right)$

$(a^2 = \gamma p/\varrho)$. $\therefore \frac{p_2}{p_1} - 1 = \gamma M_1^2\left(1 - \frac{q_2}{q_1}\right)$

whence, using continuity and the Rankine-Hugoniot relationship (4.6)

$$\frac{p_2}{p_1} = \frac{2\gamma M_1^2 - (\gamma-1)}{\gamma+1}$$

and, $\dfrac{p_1}{p_2} = \dfrac{2\gamma M_2^2-(\gamma-1)}{\gamma+1}$

whence, $M_2^2 = \dfrac{(\gamma-1)M_1^2+2}{2\gamma M_1^2-(\gamma-1)}$

now, $\dfrac{p_{02}}{p_2} = \left(1+\dfrac{\gamma-1}{2}M_2^2\right)^{\frac{\gamma}{\gamma-1}}$

eliminating p_2 and M_2 gives,

$$\frac{p_{02}}{p_1} = \left[\frac{\gamma+1}{2}M_1^2\right]^{\frac{\gamma}{\gamma-1}}\left[\frac{\gamma+1}{2\gamma M_1^2-(\gamma-1)}\right]^{\frac{\gamma}{\gamma-1}}$$

and, since $\dfrac{p_{01}}{p_1} = \left[1+\dfrac{\gamma-1}{2}M_1^2\right]^{\frac{\gamma}{\gamma-1}}$

then, $\dfrac{p_{02}}{p_{01}} = \left[\dfrac{(\gamma+1)M_1^2}{2+(\gamma-1)M_1^2}\right]^{\frac{\gamma}{\gamma-1}}\left[\dfrac{\gamma+1}{2\gamma M_1^2-(\gamma-1)}\right]^{\frac{1}{\gamma-1}}$

$(\gamma = 7/5)$, $R_{21} = \dfrac{p_{02}}{p_{01}} = \left[\dfrac{1{\cdot}2M_1^2}{1+0{\cdot}2M_1^2}\right]^{3{\cdot}5}\left[\dfrac{6}{7M_1^2-1}\right]^{2{\cdot}5}$

$M_1 = 3{\cdot}03$, $\theta = 81{\cdot}9°$, $\sin\theta = 0{\cdot}99002$ $\therefore M_1 \sin\theta = 3$

from Table 2, $R_{21} = p_{02}/p_{01} = 0{\cdot}32$

and, $\dfrac{s_2-s_1}{R} = +1{\cdot}13952$

Plane normal shock wave relations may be applied directly to plane oblique shock waves by substituting $M_1 \sin\theta$ for M_1 where $M_1 \sin\theta$ is effectively the flow component normal to the plane oblique shock wave.

thus,

$$R_{31} = \frac{p_{03}}{p_{01}} = \left[\frac{1{\cdot}2M_1^2\sin^2\theta}{1+0{\cdot}2M_1^2\sin^2\theta}\right]^{3{\cdot}5}\left[\frac{6}{7M_1^2\sin^2\theta-1}\right]^{2{\cdot}5}$$

$$= (27/7)^{3.5}(6/62)^{2.5}$$

$$R_{31} = 0{\cdot}3283$$

$\therefore$ change in recovery factor $= \dfrac{R_{31}-R_{21}}{R_{21}} = +2{\cdot}59\%$

for all adiabatic processes with ideal and compressible fluids

$$\frac{s_3-s_1}{R} = +\log_e\left(\frac{p_{01}}{p_{03}}\right) = \log_e(1/0{\cdot}3282) = 1{\cdot}11573$$

change in entropy change $= \dfrac{1{\cdot}11573-1{\cdot}13952}{1{\cdot}13952} = -2{\cdot}09\%$

EXAMPLE 7.4 Figure E7.4 shows two different types of intake ducts, for supersonic air, for which the performances are to be compared for inlet conditions at (1) of Mach number, $M_1 = 1{\cdot}9$, static pressure, $p_1 = 0{\cdot}7\times10^5$ N/m² and static temperature, $t_1 = 7°$C.

(a) represents an intake design which has one normal shock (1)–(4);

(b) represents an intake design which has two oblique shocks (1)–(2), (2)–(3) and a normal shock (3)–(4).

Compare the intake performances by comparing the pressure recovery factor, $R_{14} = p_{04}/p_{01}$ and the total change in entropy (s_4-s_1) for (a) and (b). If the minimum area in each case is 1·5 m², calculate the mass flow for each intake.

For (b) determine the angles which the oblique shock waves (1)–(2) and (2)–(3) make with the free stream. At what flow speed (m/s) does the initial shock become detached?

(U. Leeds 1968)

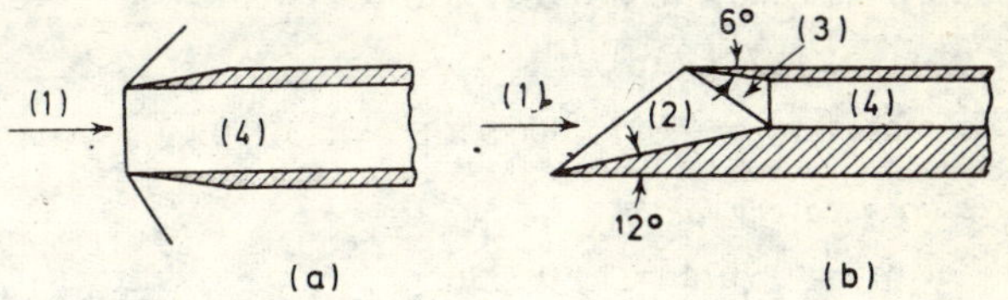

Fig. E7.4

(a) plane normal shock wave intake.

from Table 2, for $M_1 = 1{\cdot}9$,

$$p_{04}/p_{01} = 0{\cdot}7674 \text{ i.e., } (1-0{\cdot}2326)$$

$$M_2 = 0{\cdot}5956 \text{ and } (s_4-s_1)/R = +0{\cdot}264\,804$$

also, $p_4/p_1 = 4{\cdot}0450$

$\therefore$ $p_4 = 4{\cdot}045\times0{\cdot}7\times10^5 = 2{\cdot}8315\times10^5$ N/m²

and, $T_4/T_1 = 1{\cdot}6079$ $\therefore$ $T_4 = 1{\cdot}6079\times280 = 450{\cdot}2$ K

$(a^2 = \gamma RT)_4$, $\therefore a_4 = 20(T_4)^{0{\cdot}5} = 20(450{\cdot}2)^{0{\cdot}5} = 424{\cdot}4$ m/s

$\therefore$ $q_4 = 0{\cdot}5956\times424{\cdot}4 = 252{\cdot}8$ m/s

$(p = \varrho RT)_4$,

$\therefore$ $\varrho_4 = (2{\cdot}8315\times10^5)/(0{\cdot}2871\times10^3\times450{\cdot}2) = 2{\cdot}191$ kg/m³

$(m = \varrho qA)_4$,

$\therefore$ mass flow $= 2{\cdot}191\times252{\cdot}8\times1{\cdot}5 = 830{\cdot}7$ kg/s

(b) multi-shock wave intake.

(1)–(2) oblique shock wave,

from Table 5, for $M_1 = 1{\cdot}9$, ${}_1\delta_2 = 4°$

then ${}_1\theta_2 = 35{\cdot}279°$ and $M_2 = 1{\cdot}76$

also, $p_{02}/p_{01} = 0{\cdot}9990$, $(s_2-s_1)/R = +0{\cdot}00028552$

and, $p_2/p_1 = 1{\cdot}2382$, $T_2/T_1 = 1{\cdot}0633$

(2)–(3) oblique shock wave

from Table 5, for $M_2 = 1{\cdot}76$ (incident Mach number),

${}_2\delta_3 = 4°+6° = 10°$ then ${}_2\delta_3 = 45{\cdot}224°$ and $M_3 = 1{\cdot}4096$

also, $p_{03}/p_{02} = 0{\cdot}9871$, $(s_3-s_2)/R = +0{\cdot}003\,714\,3$

$p_3/p_2 = 1{\cdot}6544$, $T_3/T_2 = 1{\cdot}1590$

${}_1\theta_3 = {}_2\theta_3 - {}_1\delta_2 = 54{\cdot}224°-4° = 50{\cdot}224°$

(3)–(4) normal shock wave.

from Table 3, $M_3 = 1{\cdot}4096$ (incident Mach number) by interpolation, $M_4 = 0{\cdot}7355+0{\cdot}04\times0{\cdot}0042 = 0{\cdot}7357$

$$p_{04}/p_{03} = 0{\cdot}9557+0{\cdot}04\times0{\cdot}0025 = 0{\cdot}9558$$

$$(s_4-s_3)/R = (4{\cdot}5383-0{\cdot}04\times0{\cdot}26437)\times10^{-2}$$

$$= +0{\cdot}0452426$$

$$p_4/p_3 = 2{\cdot}1528-0{\cdot}04\times0{\cdot}0328 = 2{\cdot}1515$$

$$T_4/T_3 = 1{\cdot}2612-0{\cdot}04\times0{\cdot}0065 = 1{\cdot}2609$$

total effect (1)–(4).

$$(s_4-s_1)/R = \Sigma\Delta s/R = [(s_4-s_3)+(s_3-s_2)+(s_2-s_1)]/R$$

$$\therefore (s_4-s_1)/R = 0{\cdot}049242$$

$$p_{04}/p_{01} = p_{04}/p_{03}\times p_{03}/p_{02}\times p_{02}/p_{01} = 0{\cdot}9425$$

similarly, $p_4/p_1 = 5{\cdot}1519 \quad \therefore\ p_4 = 0{\cdot}7\times10^5\times5{\cdot}1519$

$= 3{\cdot}606\times10^5\ \text{N/m}^2$

and, $T_4/T_1 = 1{\cdot}5539 \quad \therefore\ T_4 = 1{\cdot}5539\times280 = 435{\cdot}1\ \text{K}$

$\therefore\ a_4 = 20(435{\cdot}1)^{0{\cdot}5} = 417{\cdot}2\ \text{m/s}$

$\therefore\ q_4 = (3{\cdot}606\times10^5)/(0{\cdot}2871\times10^3\times435{\cdot}1) =$

$= 2{\cdot}887\ \ \text{kg/m}^3$

$\therefore$ mass flow $= 2{\cdot}887\times306{\cdot}9\times1{\cdot}5 = 1328{\cdot}9$ kg/s

Pressure recovery factor.

$R_{14(a)} = 0{\cdot}7674 \leftarrow 1-0{\cdot}2326$

$R_{14(b)} = 0{\cdot}9425 = 1-0{\cdot}0575$ (improved design)

$\therefore$ reduction of total head pressure loss is 75·3% by improving the intake design

$(s_4-s_1)/R_{(a)} = +0{\cdot}264804$

$(s_4-s_1)/R_{(b)} = +0{\cdot}049242$ (improved design)

$\therefore$ reduction of increase in entropy is 81·4%

Shock wave detachment. for ${}_1\delta_2 = 4°$ (i.e. 3.9442)

$M_1 = 1{\cdot}20,\ a_1 = 20(280)^{0{\cdot}5} = 334{\cdot}7$ m/s

$\therefore\ q_1 = 1{\cdot}20\times334{\cdot}7 = 401{\cdot}6$ m/s

EXAMPLE 7.5 Figure E7.5a shows an air stream with a Mach number of 2·4 and a stagnation pressure of 12×10^5 N/m², flowing parallel to a plane wall and past a symmetrically placed 20° wedge.

Determine the angle of the shock wave *AB* attached to the point of the wedge and the angle of the reflected wave *BC*. What is the static pressure and Mach number immediately downstream of the reflected wave?

Determine the wedge angle at which the shock wave *AB* will cease to be attached and the wedge angle at which Mach reflection will take place at the wall.

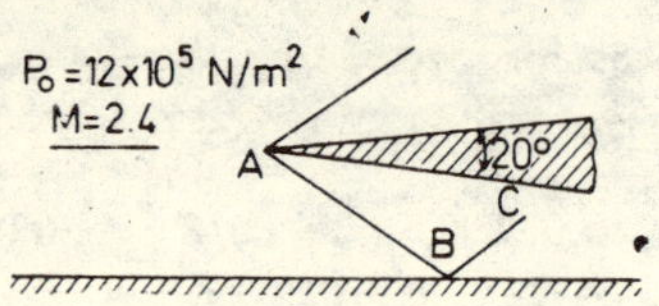

Fig. E7.5a

Reflection of Oblique Shock Waves

In Figure E7.5b the primary shock wave *AB* (1)–(2), deflects the flow downwards towards the plane boundary and parallel to the wedge boundary. If $M_2 > 1$ another oblique shock wave, *BC*, occurs to bring the flow parallel to the plane boundary. This reflected shock must therefore provide an equal and opposite deflection (δ).

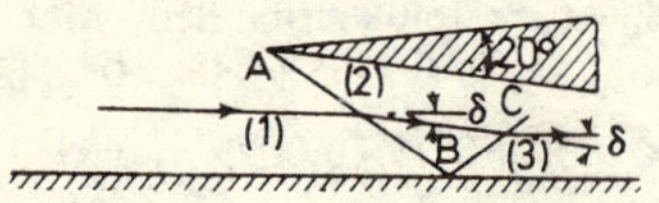

Fig. E7.5b

For (2)–(3) calculations the flow direction of the incident flow (M_2) is taken as datum.

Mach Reflection

If the deflection caused by the primary wave is too large a more complex system of shock waves is required to satisfy boundary conditions downstream (Figure E7.5c).

From Table 5, for $M_1 = 2{\cdot}4$, $\theta_{AB} = 33{\cdot}023$, $p_2/p_1 = 1{\cdot}8292$

from Table 1, $p_{01}/p_1 = 14{\cdot}62$

again, from Table 5, $M_2 \approx 2{\cdot}0(1{\cdot}9994)$ giving $\theta_{BC} = 39{\cdot}314°$

$\therefore$ shock wave *BC* makes an angle of $(39{\cdot}314-10) = 29{\cdot}3°$ with the incident flow (horizontal) direction.

for $M_2 = 2{\cdot}0$, $M_3 = 1{\cdot}64\ (1{\cdot}6405)$, $p_3/p_2 = 1{\cdot}7066$

$$\therefore \text{ static pressure } p_3 = \frac{p_3}{p_2}\times\frac{p_2}{p_1}\times\frac{p_1}{p_{01}}\times p_{01}$$

$$\therefore\ p_3 = \frac{1{\cdot}7066\times1{\cdot}8292}{14{\cdot}62}\times12\times10^5 = 2{\cdot}562\times10^5\ \text{N/m}^2$$

wedge angle for detachment $= 2\times\delta_m$ for $M_1 = 2{\cdot}4$

$\therefore$ wedge angle $= 2\times28{\cdot}68(14) = 57{\cdot}36$

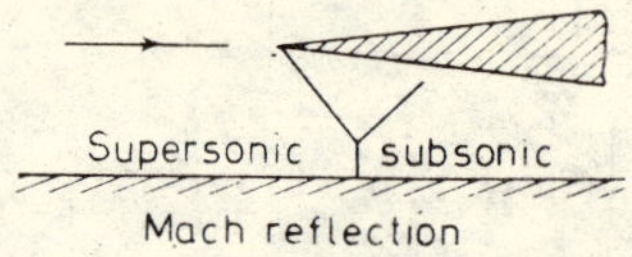

Fig. E7.5c

The exact wedge half-angle for Mach reflection will be that angle which will produce a value of M_2 for which it will be equal to the maximum deflection δ_m for M_2.

The solution is readily obtained graphically by plotting values of δ (from Table 5 for $M_1 = 2{\cdot}4$) and δ_m for an appropriate range of values of M_2, against M_2. These values are tabulated as follows and plotted in Figure E7.5d.

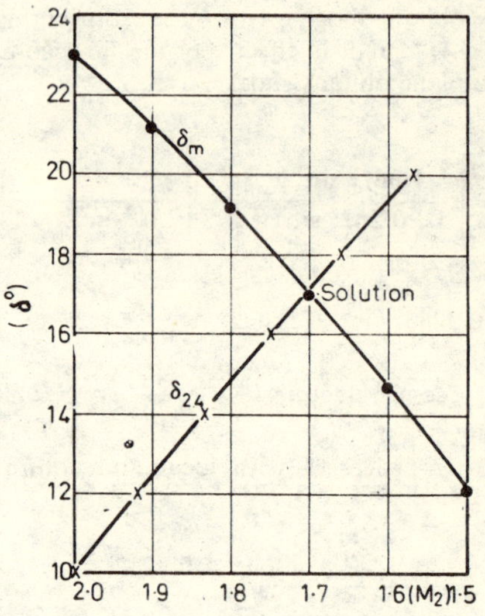

Fig. E7.5d

The intersection of the two graphs $\delta_{2{\cdot}4}$ and δ_m give $M_2 \approx 1{\cdot}7$ for which $\delta_m = 17{\cdot}0119$ giving the wedge angle for Mach reflection as $(2\times17) = 34°$. Data for graphical solution of Mach reflection:

Table 5 ($M_1 = 2{\cdot}4$)		($M_2 \equiv$ incident flow)	
δ	M_2	(M_2)	$\delta_m(M_2)$
10	1·9994	2·0	22·9735
12	1·9181	1·9	21·1675
14	1·8350	1·8	19·1833
16	1·7497	1·7	17·0119
18	1·6613	1·6	14·6515
20	1·5689	1·5	15·1127

EXAMPLE 7.6 A two-dimensional wedge having a total wedge angle of 16° is used to measure the Mach number of the flow in a supersonic wind tunnel. Determine the value of the Mach number when the shock waves attached to the wedge make an angle of 38° to the direction of the flow upstream of the wedge.

If the wind tunnel walls are parallel, find the angle the reflected shock wave makes with the wall.

What is the smallest Mach number the wedge could be used to measure, and at what Mach number will simple reflection cease at the wall?

When the Mach number in the tunnel is $M_1 = 1{\cdot}60$ Mach reflection takes place at the wall. Calculate the Mach numbers in the subsonic flow on both sides of the vortex sheet arising from the triple point?

(U. Leeds 1969)

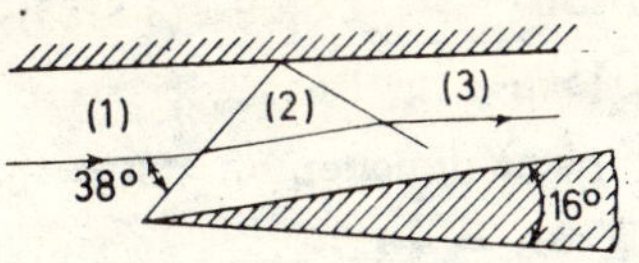

Fig. E7.6a

Figure E7.6a, shows the oblique shock wave attached to the wedge at an angle of 38° to the incident flow (for the upper half only).

From Table 5, for $\delta = 8°$ and $\theta = 38°$ $M_1 = 1{\cdot}96$ and $M_2 = 1{\cdot}6761$,

for the oblique shock wave reflected between regions (2) and (3), interpolation for $M_2 = 1{\cdot}6761$ in Table 5 gives ${}_2\theta_3 = 45{\cdot}298°$

$\therefore$ angle the reflected wave makes with wall (i.e. horizontal direction) is

$$ {}_2\theta_3 - {}_1\theta_2 (= \delta) = 45{\cdot}3 - 8° = 37{\cdot}3° $$

The smallest Mach number which may be measured by a wedge causing a flow deflection of 8° is found in Table 5 by inspection of δ_m the maximum deflection angle. From Figure 7.5, $\delta_m = 8{\cdot}3253°$ and the corresponding Mach number of the incident flow is 1·36.

For simple reflection to cease at the wall, $M_2 < 1{\cdot}36$ which corresponds to an incident flow Mach number of approximately 1·64. For lower Mach numbers, the simple wave reflection is replaced by the Mach reflection referred to in the previous example and illustrated in Figure E7.6b.

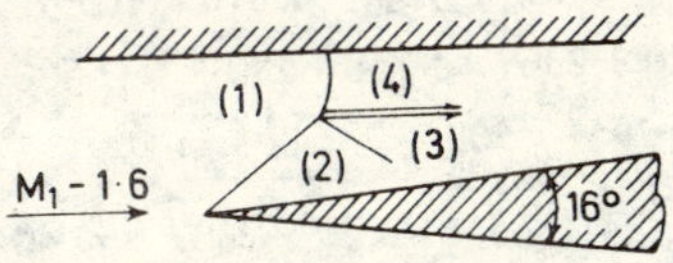

Fig. E7.6b

The strong shock wave between regions (1) and (4) is normal to the wall but the area for analysis is local to the point at which there is a slip plane downstream of the Mach reflection. This slip plane is caused by the velocities q_4 and q_3 having different speeds but the same direction; also $p_3 = p_4$.

When $M_1 = 1{\cdot}6$, $p_2/p_1 = 1{\cdot}4843$, $M_2 = 1{\cdot}3195 \approx 1{\cdot}32$ which becomes the incident flow Mach number for the oblique shock wave process (2) → (3) and a graphical solution is used:

for flow directions in (3) and (4) to be the same,

$$ \delta_{14} = \delta_{12} - \delta_{23} \quad \text{(positive upwards)} $$
$$ = 8 - \delta_{23} $$

also,
$$ \frac{p_4}{p_1} = \frac{p_3}{p_1} = \frac{p_3}{p_2} \times \frac{p_2}{p_1} = \frac{p_3}{p_2} \times 1{\cdot}4843 $$

$\therefore$ from Table 5 and using the strong shock wave solutions for $M_1 = 1{\cdot}60$ (1 → 4) and $M_2 = 1{\cdot}32$ (2 → 3) the following table is established,

($M_1 = 1{\cdot}60$)		($M_2 = 1{\cdot}32$)		
δ_{14}	p_4/p_1	δ_{23}	p_3/p_2	p_3/p_1
2	2·8166	6°	1·7634	2·6174
4°	2·8059	4°	1·8294	2·7154
6°	2·7870	2°	1·8577	2·7574
8°	2·7576	0°	1·8661	2·7699

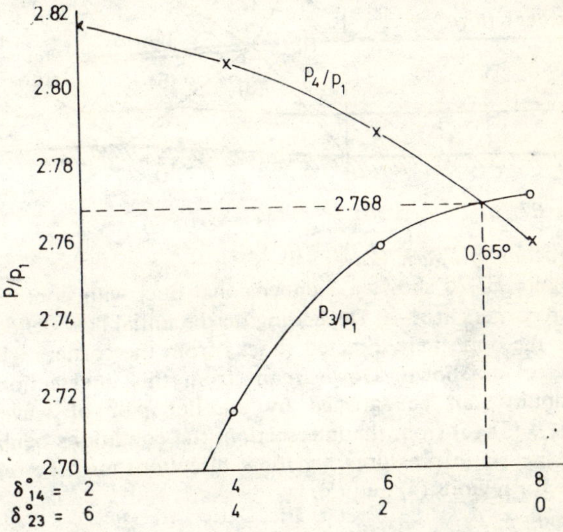

Fig. E7.6c

from the graph, Figure E7.6c, the intersection of the two curves satisfies the condition that $p_4 = p_3$ for a flow direction of 0·65°. From Table 5, by interpolation, $M_3 = 0{\cdot}777$ and $M_4 = 0{\cdot}697$.

EXAMPLE 7.7 Figure E7.7a shows a uniform two-dimensional supersonic airstream (1) ($t_1 = 7°C$) emerging to atmosphere from a duct, the lower edge of the jet being turned upwards through 14 degrees at exit by a deflector

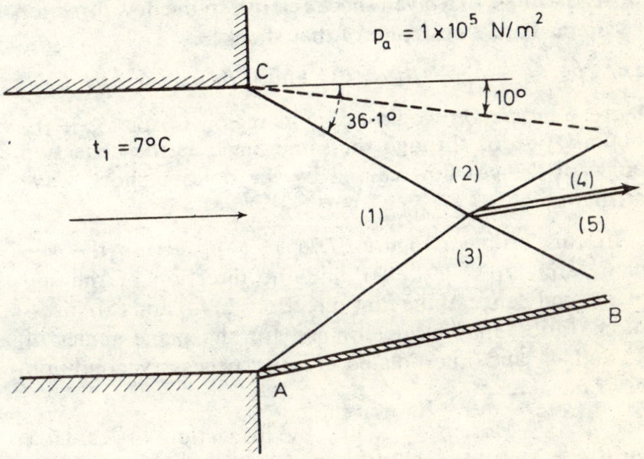

Fig. E7.7a

plate AB. The pressure at (1) being below atmospheric, an oblique plane shock develops at the upper lip (C) and, making an angle of 36·1 degrees with the initial flow direction, turns the upper edge of the jet downwards through 10 degrees. If the atmospheric pressure is 1×10^5 N/m^2, use the flow tables to determine the Mach numbers and static pressures in regions (1), (2) and (3).

Determine the angle at which the slip plane, separating regions (4) and (5), slopes upwards relative to the initial flow direction given that the value of this angle is between 2 and 4 degrees. What are the magnitudes of flow velocities (q_4) and (q_5).

(U. Birmingham 1957)

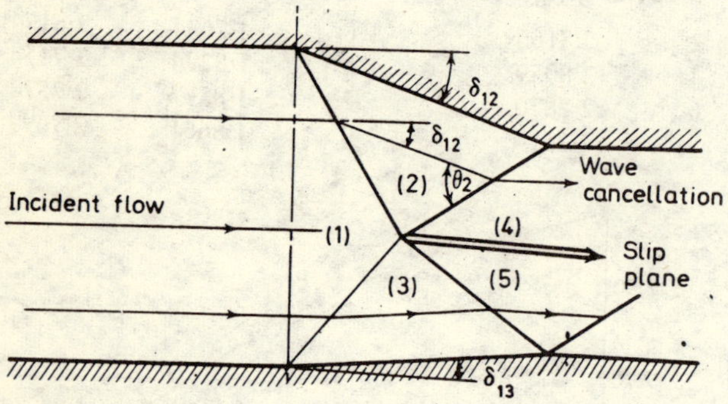

Fig. E7.7b

Interaction of Shock Waves

Figure E7.7b shows a 2-dimensional duct with concave sharp corners at *A–A*. Depending on the initial flow deflections the oppositely directed shocks from the corners will intersect as shown. Downstream from this intersection, continuity can be satisfied by another pair of waves (2)–(4), (3)–(5) from the intersection, the conditions being that the static pressures and flow directions must be the same for regions (4) and (5).

Thus, $p_4 = p_5$

and, using δ as the angle of deflection of gas flow at a particular shock (positive upwards),

$$-\delta_{12}+\delta_{24} = \delta_{13}-\delta_{35}$$

If the incident waves are equal, the wave pattern will be symmetrical and flow magnitudes, $|q|$, will be the same for regions (4) and (5). If not then $|q_4| \neq |q_5|$ and a slip plane or vortex sheet occurs, as shown, at the interface of the two regions.

With suitable suffixes,

β is the angle by which the flow direction in a given region is above the *incident flow* direction (1);

θ is the angle of a given shock relative to the flow direction immediately upstream of that shock;

i.e. $\beta_2 = -\delta_{12}+(\beta_1 = 0)$ and, $\beta_4 = \delta_{24}+\beta_2$

Wave Cancellation would occur in region (4) or (5) if the boundary turns through the same angle as the deflection angle of the gas flow caused by the oblique shock wave at that point.

In this problem Figure E7.7a $p_4 = p_5$ across the slip plane and $|q_4| \neq |q_5|$. To find the direction of the slip plane (and hence of the flow directions in (4) and (5)), pressures (ratios) may be determined for slip plane angles of 2° and 4° and interpolated for the necessary condition $p_4 = p_5$.

For $\delta_{12} = -10°$, $\beta_2 = -10°$, by inspection of Table 5 for $\theta_{12} = 36{\cdot}1°$, $M_1 = 2{\cdot}18$

for condition (2) $M_2 = 1{\cdot}8049$, $p_2/p_1 = 1{\cdot}758$,

$T_2/T_1 = 1{\cdot}810$

for condition (3) $M_3 = 1{\cdot}6483$, $p_3/p_1 = 2{\cdot}166$, $T_3/T_1 = 1{\cdot}2637$

$$p_2 = p_a = 1\times10^5 \text{ N/m}^2$$

$$\therefore p_1 = p_2/1{\cdot}758 = 0{\cdot}5688\times10^5 \text{ N/m}^2$$

and $p_3 = p_1\times2{\cdot}166 = 1{\cdot}232\times10^5$ N/m²

assume the slip plane direction $\beta_4 = \beta_5 = +4°$ and, then $\delta_{24} = \beta_4-\beta_2 = 4-(-10) = 14°$

and $\delta_{35} = \beta_5-\beta_3 = 4-14° = -10°$

from Table 5 interpolation proceeds as follows

M_2	T_4/T_2	p_4/p_2	(p_4/p_1)	M_4
1·80	1·2367	2·0295		1·2896
1·8049	1·2369	2·0306	(3·5698)	1·2591
1·82	1.2374	2·0330		1·3106
M_3	T_5/T_3	p_5/p_3	(p_5/p_1)	M_5
1·64	1·1564	1·6426		1·2843
1·6483	1·1565	1·6429	(3·5585)	1·2920
1·66	1·1566	1·643		1·3060

assume the slip plane direction $\beta_4 = \beta_5 = +2°$

then $\delta_{24} = 12°$, $\delta_{35} = -12°$

M_2	T_4/T_2	p_4/p_2	(p_4/p_1)
1·80	1·1970	1·8345	
1·8049	1·1972	1·8365	(3.2286)
1·82	1·1978	1·8386	
M_3	T_5/T_3	p_5/p_3	(p_5/p_1)
1·64	1·1947	1·8234	
1·6483	1·1945	1·8227	(3·9480)
1·66	1·1943	1·8217	

(N. B. change in sign of interpolation)

the solution $\beta_4 = 2°+x$ where for $p_4/p_1 = p_5/p_1$,

interpolation, $$\frac{x}{2-x} = \frac{(3{\cdot}9480-3{\cdot}2286)}{(3{\cdot}5698-3{\cdot}5585)}$$

$$x = 1{\cdot}969$$

$\therefore$ angle of slip plane, $\beta_4 = \beta_5 = 2+1{\cdot}969 = 3{\cdot}97°$

assume $\beta_4 = \beta_5 \approx 4°$

by interpolation, $M_4 = 1{\cdot}2951$ and $M_5 = 1{\cdot}2920$,

now $q_4 = M_4\times a_4$ and $q_5 = M_5\times a_5$

$(a^2 = \gamma RT) \therefore a_4 = 20(T_4)^{0{\cdot}5}$

where, $T_4/T_2 = 1{\cdot}2369$, $T_2/T_1 = 1{\cdot}810$ and $T_1 = 280$ K

$\therefore a_4 = 20(1{\cdot}2369\times1{\cdot}758\times280)^{0{\cdot}5} = 500{\cdot}7$ m/s

$\therefore q_4 = 1{\cdot}2951\times500{\cdot}7 = 648{\cdot}5$ m/s

similarly $a_5 = 20(1{\cdot}1565\times1{\cdot}2637\times280)^{0{\cdot}5} = 404{\cdot}6$ m/s

$\therefore q_5 = 1{\cdot}2920\times404{\cdot}6 = 522{\cdot}7$ m/s

EXAMPLE 7.8 Figure E7.8 shows a two-dimensional supersonic flow with wave A incident on a surface of discontinuity which separates two regions of flow having the same directions and pressures but different velocities. The situation illustrated might serve as a crude model of a shock entering a boundary layer.

Carry out the analysis for the limiting case where the turning angle for wave A is extremely small so that linearised theory for the flow across a wave is valid. Show that,

$$\frac{\theta_5-\theta_2}{\delta} = 2\bigg/\left[1+\frac{M_2^2}{M_1^2}\left(\frac{M_1^2-1}{M_2^2-1}\right)^{0.5}\right]$$

where θ is the angle by which the flow direction in a given region is above the original direction, and that,

$$\frac{p_4-p_1}{p_3-p_1} = 2\Big/\left[1+\frac{M_1^2}{M_2^2}\left(\frac{M_2^2-1}{M_1^2-1}\right)^{0.5}\right]$$

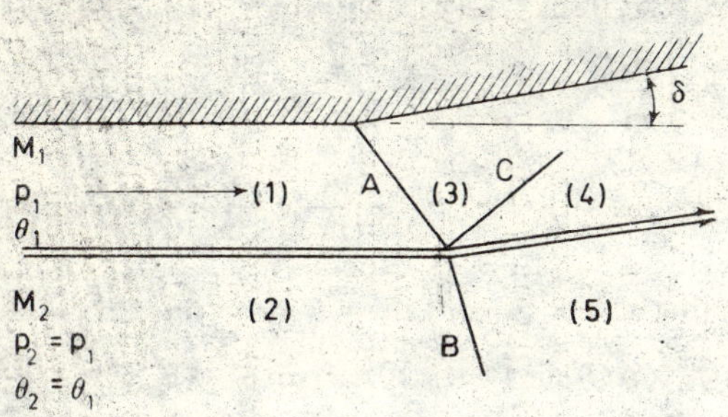

Fig. E7.8

Applying linearised theory to the wave processes in Figure E7·8 and using (3.5), with $\delta = \theta_3 - \theta_1$ for a Type I wave (μ-ve) — Figure 3.3b,

$$\frac{dp}{p_1} = \frac{p_3-p_1}{p_1} = \frac{-\delta\gamma M_1^2}{(M_1^2-1)^{0.5}} \quad \text{and} \quad \frac{p_1-p_3}{p_3} = \frac{+\delta\gamma M_3^2}{(M_3^2-1)^{0.5}} \qquad \text{(A)}$$

whence
$$\frac{p_3}{p_1} = \frac{M_1^2}{M_3^2}\left[\frac{M_3^2-1}{M_1^2-1}\right]^{0.5} \qquad \text{(B)}$$

since, across the discontinuity $p_1 = p_2$ and $p_4 = p_5$

$$\frac{2(p_3-p_1)}{p_1} = \frac{(p_3-p_1)}{p_1} + \frac{(p_5-p_2)}{p_2} - \frac{(p_4-p_3)}{p_1} \qquad \text{(C)}$$

for a Type II wave (μ + ve). — Figure 3.3c,

$$\frac{p_4-p_3}{p_3} = \frac{\gamma M_3^2({}_3\delta_4)}{(M_3^2-1)^{0.5}} = \frac{\gamma M_3^2(\theta_4-\theta_3)}{(M_3^2-1)^{0.5}}$$

from (B), $$\frac{p_4-p_3}{p_1} = \frac{\gamma M_1^2(\theta_4-\theta_3)}{(M_1^2-1)^{0.5}} = \frac{\gamma M_1^2(\theta_5-\delta)}{(M_1^2-1)^{0.5}}$$

(since $\theta_3 = \delta$ and $\theta_5 = \theta_4$)
also, for a Type I wave (μ-ve), Figure 3.3b,

$$\frac{p_5-p_2}{p_2} = -\frac{\gamma M_2^2({}_2\delta_5)}{(M_2^2-1)^{0.5}} = -\frac{\gamma M_2^2(\theta_5)}{(M_2^2-1)^{0.5}} \qquad \text{(D)}$$

substituting the above expressions in (C)

$$-2\delta\frac{\gamma M_1^2}{(M_1^2-1)^{0.5}} = -\delta\frac{\gamma M_1^2}{(M_1^2-1)^{0.5}}$$
$$-\theta_5\frac{\gamma M_2^2}{(M_2^2-1)^{0.5}} - (\theta_5-\delta)\frac{\gamma M_1^2}{(M_1^2-1)^{0.5}}$$

since $\theta_5 = \theta_5 - \theta_2$ i.e., $\theta_2 = 0$ by definition,

then, $$\frac{\theta_5-\theta_2}{\delta} = 2\Big/\left[1+\frac{M_2^2}{M_1^2}\left[\frac{M_1^2-1}{M_2^2-1}\right]^{0.5}\right] \qquad \text{(E)}$$

again, since $p_1 = p_2$ and $p_4 = p_5$ across the discontinuity,

$$\frac{p_4-p_1}{p_3-p_1} = \frac{p_5-p_2}{p_2} \times \frac{p_1}{p_3-p_1}$$

from (A) and (D)

$$\frac{p_4-p_1}{p_3-p_1} = \frac{M_2^2}{M_1^2}\left[\frac{M_1^2-1}{M_2^2-1}\right]^{0.5}\frac{\theta_5}{\delta}$$

from (E), $$\frac{p_4-p_1}{p_3-p_1} = 2\Big/\left[1+\frac{M_1^2}{M_2^2}\left[\frac{M_2^2-1}{M_1^2-1}\right]^{0.5}\right]$$

EXAMPLE 7.9. Figure E7.9 shows air flowing at a Mach number of $M_1 = 2.16$ in a parallel-sided passage of which one side changes direction by 12°. Determine the angles which the shock wave AB and the reflected shock wave BC make with the lower wall.

If the reflected shock wave BC encounters a jet edge at C, describe why the reflection takes the form of an expansion wave. Find the Mach number of the flow behind the expansion wave and the angle through which the jet edge is deflected at C. Determine the total head pressure ratio, p_{04}/p_{01}.

(U. Leeds 1966)

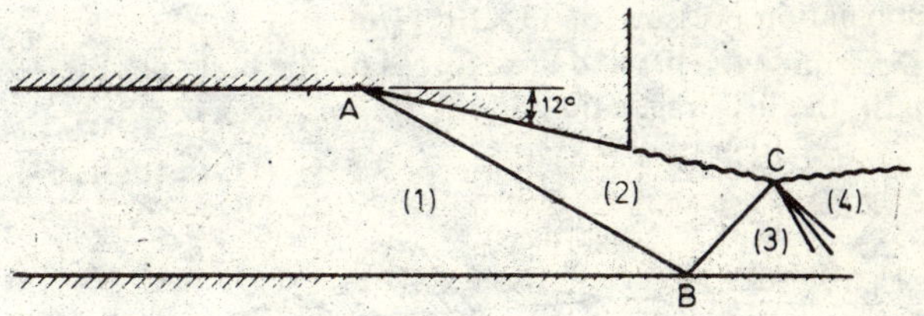

Fig. E7.9

Oblique shock AB

from Table 5, for $M_1 = 2.16$ and ${}_1\delta_2 = 12°$ then $M_2 = 1.71$,

angle of shock wave AB with horizontal, $\underline{\theta_{AB} = 38.53°}$

$p_2/p_1 = 1·9455$ (N.B. $p_2 = p_a = p_4$) and $p_{02}/p_{01} = 0·9707$

Oblique shock BC

from Table 5, for $M_2 = 1·71$ and ${}_2\delta_3 = 12°$ interpolation between $M = 1·70$ and $M = 1·72$ is necessary.

$$\therefore\ M_3 = (1·2673 + 1·2891)/2 = 1·2782$$

and similarly, $p_3/p_2 = 1·8224$, $p_{03}/p_{02} = 0·9784$ and $\theta_{BC} = 49·784°$,

$\therefore$ angle of shock wave BC with horizontal $= 49·784 - 12 = \underline{37·78°}$

Prandtl–Meyer expansion wave

The oblique shock wave process for flow from (2) to (3) causes a change in flow direction as well as an increase in static pressure. Reflection at the free surface of the jet boundary takes the form of an expansion wave to acommodate the resulting pressure gradient and turns the flow in the same direction as did the shock wave BC.

from Table 1, by interpolation for $M_3 = 1·2782$

M	(diff.)	p_{03}/p_3	(diff.)	
1·27		2·6606		
$\underline{M_3 = 1·2782}$	(0·0082)	$\underline{2·6902}$ =		2·6606+0·0361×0·82
1·28	(0.0100)	2·6967	(0·0361)	

similarly,

M	${}_*\delta_3$
1·27	5·36°
$\underline{1·2782}$	$\underline{5·581°}$
1·28	5·63°

for isentropic process (3)→(4), $p_{03} = p_{04}$

$$\therefore\ \frac{p_{04}}{p_4} = \frac{p_{03}}{p_3} \times \frac{p_3}{(p_4 = p_2)} = 2·6902 \times 1·8224 = 4·9026$$

from Table 1, by interpolation for $p_{04}/p_4 = 4·9026$

$\underline{M_4 = 1·6955}$ and ${}_*\delta_4 = 17·679°$

also, for $M_1 = 2{\cdot}16,\ p_{01}/p_1 = 10{\cdot}044$

$${}_3\delta_4 = {}_*\delta_4 - {}_*\delta_3 = 17{\cdot}679 - 5{\cdot}581 = 12{\cdot}098^\circ$$

$\therefore$ total deflection of flow boundary at C due to shock wave BC and expansion wave $= 12 + 12{\cdot}098 = \underline{24{\cdot}1^\circ}$

$$\text{total head pressure ratio } \frac{p_{04}}{p_{01}} = \frac{p_{03}}{p_{01}} = \frac{p_{03}}{p_{02}} \times \frac{p_{02}}{p_{01}}$$

$$\therefore\ p_{04}/p_{01} = 0{\cdot}9784 \times 0{\cdot}9707 = \underline{0{\cdot}9497}$$

EXAMPLE 7.10 The double wedge-shaped body shown in Figure E7.10a is placed with the face AC at an angle of 4° to an air stream of Mach number, $M = 1{\cdot}9$, having a stagnation pressure of 13×10^5 N/m².

Determine the lift and drag forces on the body per metre length, the lift-drag ratio and the lift and drag coefficients.

(U. Leeds 1967)

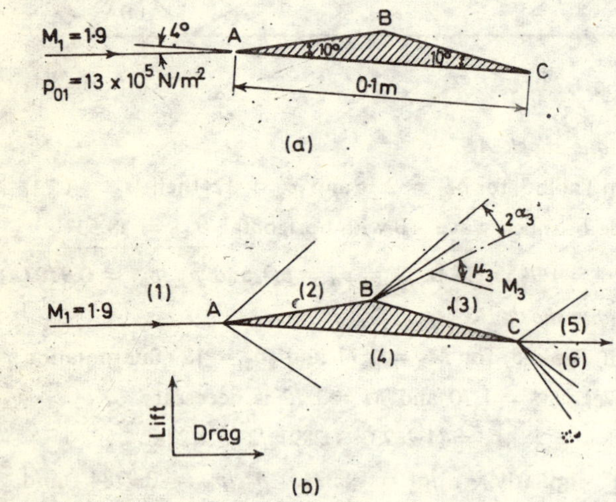

Fig. E7.10

Figure E7.10b, shows the wave distribution at the corners of the section and the regions of uniform conditions between the waves are numbered.

From Table 1 for $M_1 = 1{\cdot}9,\ p_{01}/p_1 = 6{\cdot}7006$

$$\therefore\ p_1 = 13 \times 10^5/6.7006 = 1{\cdot}9401 \times 10^5 \text{ N/m}^2$$

since the angle of incidence is 4°, ${}_1\delta_2 = 6^\circ$ and ${}_1\delta_4 = 4^\circ$

from Table 5 for $M_1 = 1{\cdot}9$ and ${}_1\delta_2 = 6^\circ$,

$$M_2 = 1{\cdot}6901 \approx 1{\cdot}69,\ p_2/p_1 = 1{\cdot}3735,\ p_{02}/p_{01} = 0{\cdot}9968$$

for $M_1 = 3$ and ${}_1\delta_4 = 4^\circ$,

$$M_4 = 1{\cdot}76,\ p_4/p_1 = 1{\cdot}2382$$

from Table 1 for $M_2 = 1{\cdot}69,\ {}_*\delta_2 = 17{\cdot}52$

$$\therefore\ {}_*\delta_3 = {}_*\delta_2 + 20^\circ = 37{\cdot}52^\circ$$

interpolation in Table 1 gives:

$$p_{03}/p_3 = 15{\cdot}3208 + 0{\cdot}2408 \times 5/24 = 15{\cdot}371$$

$$M_3 = 2{\cdot}43 + 0{\cdot}01 \times 5/24 = 2{\cdot}432$$

$$\mu_3 = 24{\cdot}30 - 0{\cdot}11 \times 5/24 = 24{\cdot}28$$

$(p_{02} = p_{03})$,

$$\therefore\ \frac{p_3}{p_1} = \frac{p_3}{p_{03}} \times \frac{p_{01}}{p_1} \times \frac{p_{02}}{p_{01}} = \frac{6{\cdot}7006}{15{\cdot}371} \times 0{\cdot}9968 = 0{\cdot}4345$$

The pressures in regions (2), (3) and (4) are normal to their respective surfaces and the associated forces are obtained by multiplying by the appropriate surface areas. By resolving these forces in the horizontal (X) and vertical (Y) directions the net Lift and Drag forces respectively are obtained:

$$\text{Lift} = (p_4 \times AC)\cos 4^\circ - (p_2 \times AB)\cos 6^\circ - (p_3 \times BC)\cos 14^\circ$$

$$= p_1\left(\frac{p_4}{p_1} \times 0{\cdot}1 \times \cos 4^\circ - \frac{p_2}{p_1} \times \frac{0{\cdot}05}{\cos 10^\circ} \times \cos 6^\circ - \frac{p_3}{p_1} \times \frac{0{\cdot}05}{\cos 10^\circ} \times \cos 14^\circ\right)$$

substituting values gives, Lift $= \underline{6\,344 \text{ N/m run}}$

$$\text{Drag} = (p_4 \times AC)\sin 4^\circ + (p_2 \times AB)\sin 6^\circ - (p_3 \times BC)\sin 14^\circ$$

$$= p_1\left(\frac{p_4}{p_1} \times 0{\cdot}1 \times \sin 4^\circ + \frac{p_2}{p_1} \times \frac{0{\cdot}05}{\cos 10^\circ} \times \sin 6^\circ - \frac{p_3}{p_1} \times \frac{0{\cdot}05}{\cos 10^\circ} \times \sin 14^\circ\right)$$

substituting values gives, Drag $= \underline{2\,053 \text{ N/m run}}$

$$\therefore\ \text{Lift/Drag ratio} = 6\,344/2\,053 = \underline{3{\cdot}09}$$

The Pressure Coefficient is defined as

$$C_p = \frac{p - p_\infty}{\varrho_\infty q_\infty^2/2} \qquad (2.21)$$

or, in this case, $C_p = \dfrac{p - p_1}{\varrho_1 q_1^2/2} = \dfrac{p/p_1 - 1}{\gamma M_1^2/2}$

The Lift and Drag Coefficients are derived from (2.21):

Lift Coefficient,

$$C_L = \frac{(\Delta p \times \text{Area})\uparrow}{\text{Area} \times \rho_\infty q_\infty^2/2} = \frac{\text{Lift}}{\text{Area} \times \gamma p_1 M^2/2}$$

$$\text{Drag Coefficient, } C_D = \frac{\text{Drag}}{\text{Area} \times \gamma p_1 M_1^2/2}$$

where 'Area' is usually the projected area of the section.

In the problem,

$$C_L = 6\,344/(0{\cdot}1 \times 1 \times 0{\cdot}7 \times 1{\cdot}9401 \times 10^5 \times 1.9^2)$$
$$= 0{\cdot}1294$$

and, $C_D = 2\,053/(0{\cdot}1 \times 1 \times 0{\cdot}7 \times 1{\cdot}9401 \times 10^5 \times 1{\cdot}9^2)$

$$= 0{\cdot}0419$$

(and, again Lift/Drag ratio $= C_L/C_D = 3{\cdot}09$)

Note: (1) that the actual pressures on the surfaces of the airfoil need not be calculated and that ratios are usually more convenient.

(2) regions (5) and (6), on either side of the slip plane, need not be analysed for calculations of lift and drag.

(3) the Prandtl–Meyer expansion fan angle ${}_2\alpha_3$ is obtained from μ_2 and μ_3 appropriate to M_2 and M_3,

i.e.

$${}_2\alpha_3 = \mu_2 - (\mu_3 - 20)$$
$$= 36{\cdot}28 - (24{\cdot}28 - 20)$$
$$= 32^\circ$$

alternatively, ${}_2\alpha_3 = {}_*\theta_3 - {}_*\theta_2 = 103{\cdot}18 - 71{\cdot}24 = 31{\cdot}94^\circ$

(the difference in results being due to approximation)

(4) Since $C_p = \dfrac{p/p_1 - 1}{\gamma M_1^2/2}$

then, $C_L = C_4 \cos 4° - C_{p2} \cos 6°/\cos 10°$
$- C_{p3} \cos 14°/\cos 10°$

and, $C_D = C_{p4} \sin 4° + C_{p2} \sin 6°/\cos 10°$
$- C_{p3} \sin 14°/\cos 10°$

Extract from NAVORD REPORT 1488, Two-Dimensional Airfoils, *Handbook of Supersonic Aerodynamics*, 3.

Shock Expansion Method

The shock expansion method utilizes the exact expressions for flow through a plane oblique-compression shock wave and for an expanison flow. It will give exact results for those configurations which do not produce a curved shock wave.

In treating configurations as the double wedge of the preceding subsection, the results will be exact as long as the reflected waves produced by the interaction of the Mach waves originating at the shoulder of the profile and the leading edge shock wave do not strike the after part of the profile. The results are exact for the double wedge, at all Mach numbers and angles of attack shown in the figures. However, at higher angles of attack than shown, the reflected Mach waves will strike the after portion of the double wedge thereby modifying the pressure distribution there. The flattened double wedge will experience this difficulty at the higher Mach numbers at high angles of attack. Although sometimes applied to the biconvex airfoil, the shock expansion method will not give exact results under any condition. The intersection of the Mach waves with the leading edge shock produce a curved shock wave and vorticity. Nevertheless, the shock expansion method gives results sufficiently accurate for most engineering applications. The use of the method of characteristics, which will give exact results in these last cases is rarely justified in practice.

Attached Shock Waves

The fluid flow across the upper and lower airfoil surfaces at the leading edge are independent of each other as long as the shock waves are attached to the leading edge. The shock waves will be attached as long as the greater of the two angles of flow deflection at the leading edge, for a given Mach number, is equal to or less than a certain critical value (i.e. δ_m) for that Mach number.

8 GAS DYNAMICS—WAVE PROPAGATION IN A COMPRESSIBLE FLUID WITHOUT FRICTION IN A CONSTANT AREA DUCT

The content of this section is based entirely on part of the 1958 Akroyd Stuart Memorial Lectures entitled 'Pressure Waves in Gases in Pipes' and given by Professor F. K. Bannister at the University of Nottingham. Accordingly, the reader is recommended to read all the Lectures for a fuller exposition of wave propagation and of wave reflection and interaction.

8.1 PROPAGATION VELOCITY AND PARTICLE VELOCITY

In Section 2.1 the propagation of small pressure waves was discussed as the speed of sound (waves) and a relationship was established between this acoustic velocity, in a stationary fluid, and the properties of a perfect gas, thus:

$$a^2 = \gamma RT = \gamma p/\varrho \tag{2.5}$$

If a small pressure wave travels with speed (a) through a gas which has an initial mean particle velocity (q) then the propagation velocity (c) of the wave is always obtained by taking the algebraic sum of the two velocities,

$$c = a+q = a \pm |q| \tag{8.1}$$

Pressure waves may have pressures greater than or less than the pressure of the undisturbed fluid into which the wave is moving. The former are referred to as compression waves and the latter as rarefaction (expansion) waves.

Figure 8.1 shows the compression and rarefaction waves associated with the movement of a free piston in a uniform duct or pipe.

If the piston moves to the right (positive) from rest the pressure will increase in front of the piston from the initial pressure (p). A compression wave will build up (time t_1) and will move ahead (time t_2) of the fluid particles which move with the same speed (q) as the piston.

Behind the piston the fluid particles must still have the same speed (and direction) as the piston and will move to the right. The pressure will be reduced to the left of the piston from the initial pressure p_1 and a rarefaction or expansion wave will move to the left away from the source of the disturbance of piston movement.

The velocity (q) of the piston corresponding to the particle velocities in front of and behind the piston, becomes uniform after time (t_1), and after time (t_2) the displacement of the piston is (d_2). The maximum pressures, (p_c) and (p_r), of the compression wave and rarefaction respectively are reached at time (t_1) when the piston has reached its maximum velocity.

x_2 ($>d_2$) represents the distance through which the initial disturbance or leading tip of the pressure wave has travelled in time (t_2). If the conditions on both sides of the piston, initially at rest at time ($t_0 = 0$), are assumed to be the same, i.e., p_1, T_1 (a_1), then $x_1 = (a_1 \times t_1)$ and $x_2 = (a_1 \times t_2)$ will be the same for both compression and expansion waves.

8.2 ANALYSIS OF A NON-STEEP FINITE AMPLITUDE WAVE

As the piston in Figure 8.1 accelerates from rest a succession of infinitely small (dp) pressure waves (Figure 8.2a) is generated to give a continuous increase in pressure described as a non-steep finite amplitude wave front (Figure 8.2b). The flat-topped portion of the wave is produced when the piston is moving at uniform velocity. The continuous non-steep increase in pressure of the compression (or expansion) wave process is still isentropic,

$$\frac{a}{a_1} = \left(\frac{T}{T_1}\right)^{0.5} = \left(\frac{p}{p_1}\right)^{\frac{\gamma-1}{2\gamma}} \tag{8.2}$$

For the elemental pressure wave shown in Figure 8.2a, from (2.1) and (2.3),

$$\mathrm{d}q = a\frac{\mathrm{d}\varrho}{\varrho} = a\frac{\mathrm{d}p}{\gamma p} \tag{8.3}$$

Each point on the non-steep wave front moves into fluid which has already been affected by the passage of pressure waves. However, the relative change of velocity caused by each successive part or increment of the wave front is that of a local acoustic wave.

By integrating (8.3) the sum effect of the wave front may be obtained to give the particle velocity (q) for any pressure

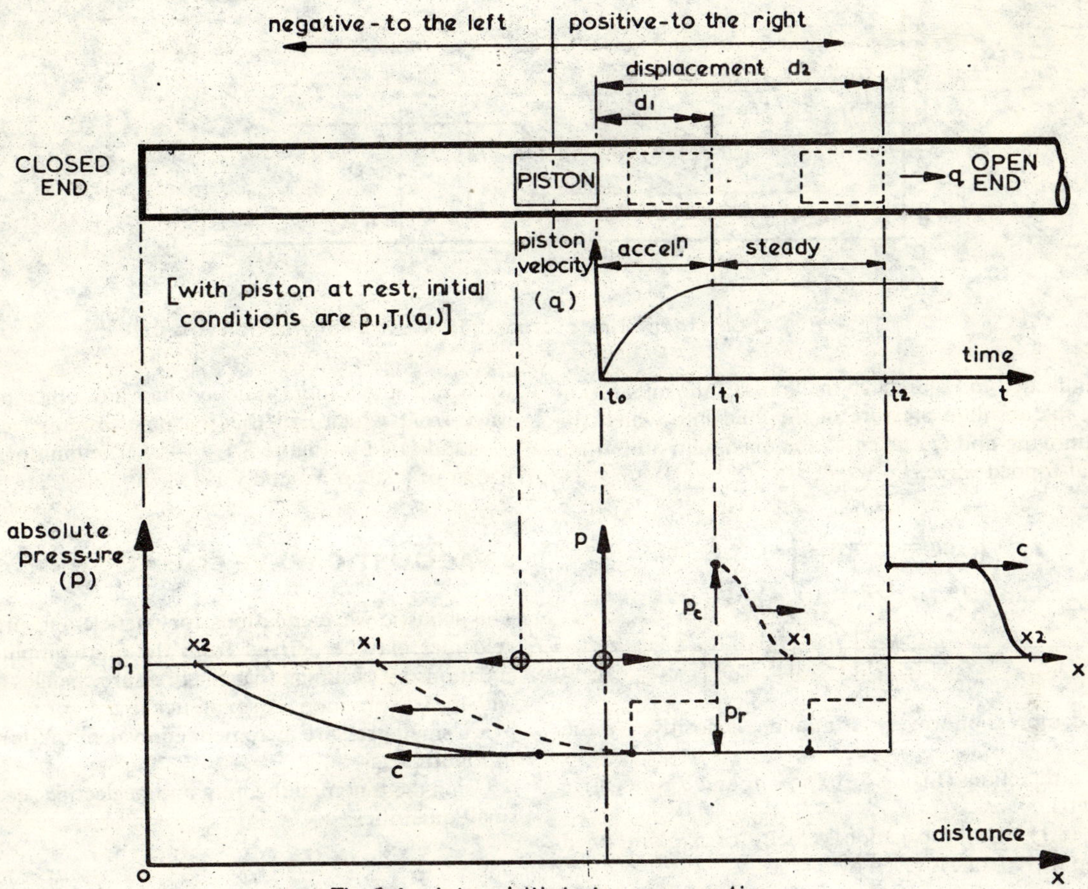

Fig. 8.1 piston initiated wave motion.

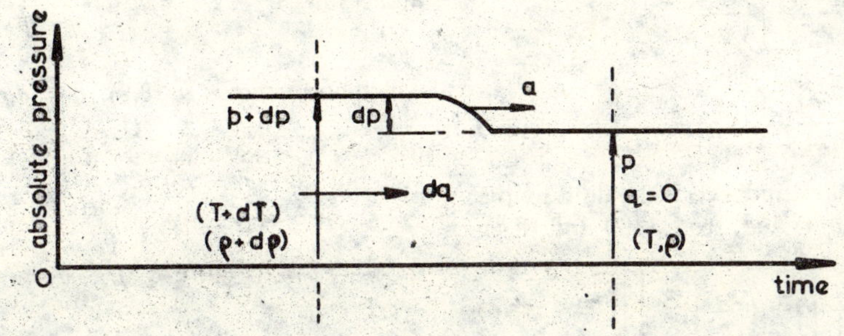

(a) small pressure or sound wave.

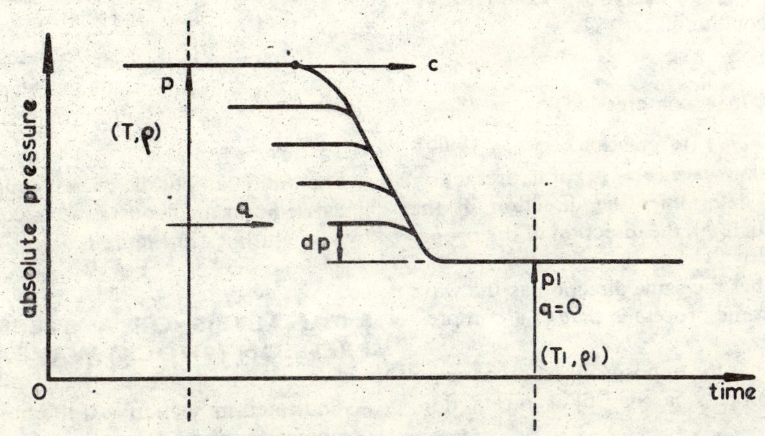

(b) non-steep finite-amplitude wave-compression ($p > p_1$)

Fig. 8.2

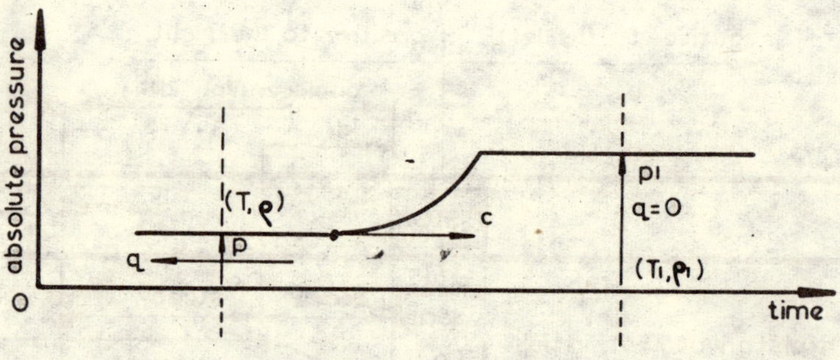

Fig. 8.2 (c) non-steep finite amplitude wave – rarefaction ($p < p_1$)

(p). As indicated in Figure 8.2b the limits of the integration are (p_1), the absolute pressure of the fluid into which the wave is moving and (p) taken as the maximum amplitude of the flat-topped wave.

$$q = \int_0^q dq = \frac{a_1}{(\gamma p_1)^{\frac{\gamma-1}{2\gamma}}} \int_{p_1}^{p} p^{\frac{\gamma-1}{2\gamma}-1}\, dp$$

$$\therefore\ q = \frac{2a_1}{\gamma-1}\left[\left(\frac{p}{p_1}\right)^{\frac{\gamma-1}{2\gamma}} - 1\right] \tag{8.4}$$

For a compression wave, $p > p_1$ and q is positive (to the right).

For a rarefaction, (Figure 8.2c) $p < p_1$ and q is negative (to the left).

From (8.1), the propagation velocity of a point on a non-steep wave front is given by $c = a+q$,

from (8.2) and (8.4),

$$c = a_1\left(\frac{p}{p_1}\right)^{\frac{\gamma-1}{2\gamma}} + \frac{2a_1}{\gamma-1}\left[\left(\frac{p}{p_1}\right)^{\frac{\gamma-1}{2\gamma}} - 1\right]$$

or,

$$c = a_1\left[\frac{\gamma+1}{\gamma-1}\left(\frac{p}{p_1}\right)^{\frac{\gamma-1}{2\gamma}} - \frac{2}{\gamma-1}\right] \tag{8.5}$$

The Mach number of the flow induced in the fluid by the wave propagation is $M = (q/a)$ and, from (8.2) and (8.4),

$$M = \text{abs}\left|\frac{2}{\gamma-1}\left[1-\left(\frac{p_1}{p}\right)^{\frac{\gamma-1}{2\gamma}}\right]\right| \tag{8.6}$$

The modulus is introduced in (8.6) to eliminate direction significance. When $M = 1$ there are two pressure ratios which produce this sonic condition.

For air ($\gamma = 7/5$),

$$p_*/p_1 = 0{\cdot}8^{-7} = 4{\cdot}7684 \text{ (compression)} \tag{8.6a}$$

or

$$p_*/p_1 = 1{\cdot}2^{-7} = 0{\cdot}2791 \text{ (rarefaction)} \tag{8.6b}$$

In (8.4), for a compression wave ($p > p_1$) or a rarefaction ($p < p_1$), the ratio p/p_1 determines the direction of the particle velocity (q) and, in (8.5), the direction of the propagation velocity (c).

If the particle velocity has the same direction as the wave propagation velocity then the process is that of a compression wave, $c = a+|q|$.

If the particle velocity has the opposite direction to wave propagation then the process is that of a rarefaction, $c = a-|q|$.

Since the propagation velocity (c) is a function of the absolute pressure ratio (p/p_1) then the top of the non-steep wave will move more quickly than any other part of the wave front which will thus become progressively steeper —as indicated in Figure 8.1, $t_1 \to t_2$. The limiting condition is that of a steep-fronted wave.

8.3 ACOUSTIC WAVE EQUATIONS

The acoustic wave equations for particle and propagation velocities may be derived from the finite amplitude wave equations by assuming that the pressure changes associated with the wave motion are small such that $p = (p_1+p')$ where p' is a small pressure increment equivalent to (dp) as shown in Figure 8.2a.

Using the binomial theorem and neglecting second order small quantities,

$$\left(\frac{p}{p_1}\right)^{\frac{\gamma-1}{2\gamma}} = \left(1+\frac{p'}{p_1}\right)^{\frac{\gamma-1}{2\gamma}} = 1+\frac{\gamma-1}{2\gamma}\left(\frac{p'}{p_1}\right) \tag{8.7}$$

and, in (8.4),

$$q = \frac{a_1}{\gamma}\frac{p'}{p_1} \tag{8.8a}$$

which corresponds to (8.3), $q = dq = \dfrac{a}{\gamma}\dfrac{dp}{p}$

and putting $p' = p-p_1$,

$$\frac{q}{a_1} = \frac{1}{\gamma}\left(\frac{p}{p_1}-1\right) \tag{8.8b}$$

or $\gamma = 7/5$, (Figure 8.5),

$$\frac{q}{a_1} = -\frac{5}{7}+\frac{5p}{7p_1} \tag{8.8c}$$

with (8.7) in (8.5),

$$\frac{c}{a_1} = 1+\frac{\gamma+1}{2\gamma}\frac{p'}{p_1} \tag{8.9a}$$

giving,

$$\frac{c}{a_1} = 1+\frac{\gamma+1}{2\gamma}\left(\frac{p}{p_1}-1\right) \tag{8.9b}$$

or $\gamma = 7/5$ (Figure 8.5),

$$\frac{c}{a_1} = \frac{1}{7}+\frac{6}{7}\frac{p}{p_1} \tag{8.9c}$$

The finite amplitude wave equations have thus been linearised and the acoustic wave equations are represented by straight lines in Figure 8.5.

8.4 ANALYSIS OF A STEEP-FRONTED COMPRESSION (SHOCK) WAVE

As indicated in Section 8.3, the non-steep wave front of compression wave becomes progressively steeper as the higher pressure points on the wave front overtake those at lower pressures until the wave front becomes vertical as

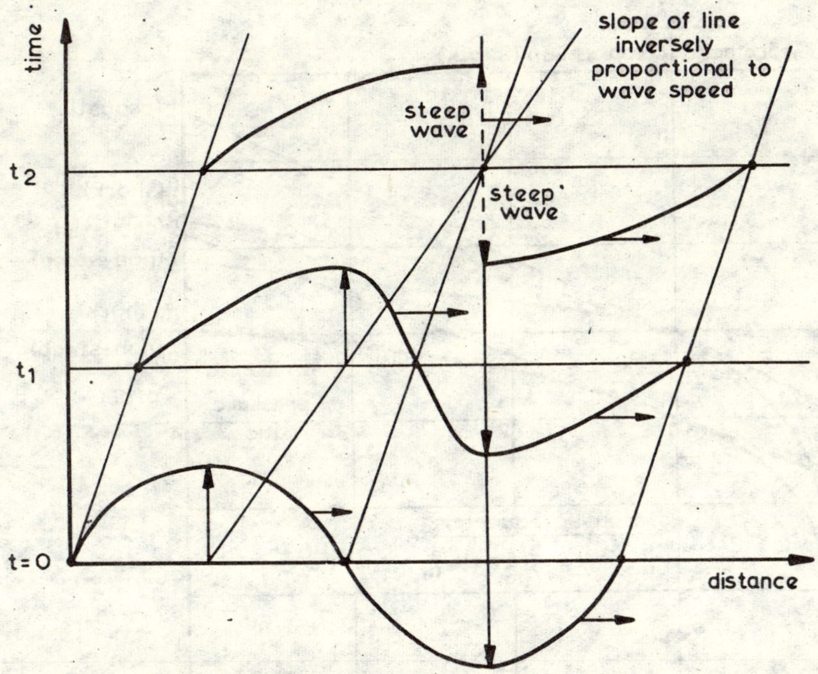

Fig. 8.3. development of steep wave

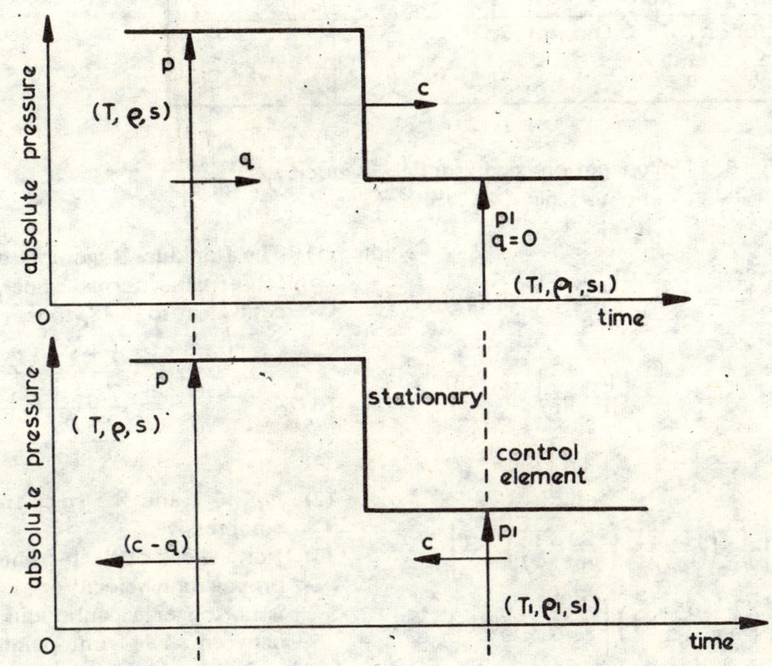

Fig. 8.4. steep-fronted wave (with control element)

shown in Figure 8.3. In a similar manner the non-steep trailing edge or tail of a rarefaction also becomes steeper with the tail point propagating as an acoustic wave.

The abrupt change in pressure of a steep wave can no longer be treated as an isentropic process and is analysed in a manner similar to that of the stationary plane normal shock wave.

For application of the basic equations of fluid mechanics the wave motion is translated into relative fluid (particle) motion by applying an equal and opposite velocity to the fluid thus producing a stationary steep-fronted (shock) wave (Figure 8.4).

For the control element shown, in a constant area duct,

continuity, $$A\varrho(c-q) = A\varrho_1 c \qquad (8.10)$$

momentum, $$A(p_1-p) = A\varrho_1 c[(c-q)-c] \qquad (8.11)$$

energy, $$\frac{\gamma}{\gamma-1}\,\frac{p}{\varrho}+\frac{(c-q)^2}{2} = \frac{\gamma}{\gamma-1}\,\frac{p_1}{\varrho_1}+\frac{c^2}{2} \qquad (8.12)$$

from (8.10) and (8.11), $$\frac{\varrho_1}{\varrho} = 1-\frac{p_1}{\varrho_1}\,\frac{(p/p_1-1)}{c^2} \qquad (8.13a)$$

with $a_1^2 = \gamma p_1/\varrho_1 = \gamma R T_1$,

$$\left(\frac{\varrho_1}{\varrho}-1\right) = -\frac{a_1^2}{\gamma c^2}\left(\frac{p}{p_1}-1\right) \qquad (8.13b)$$

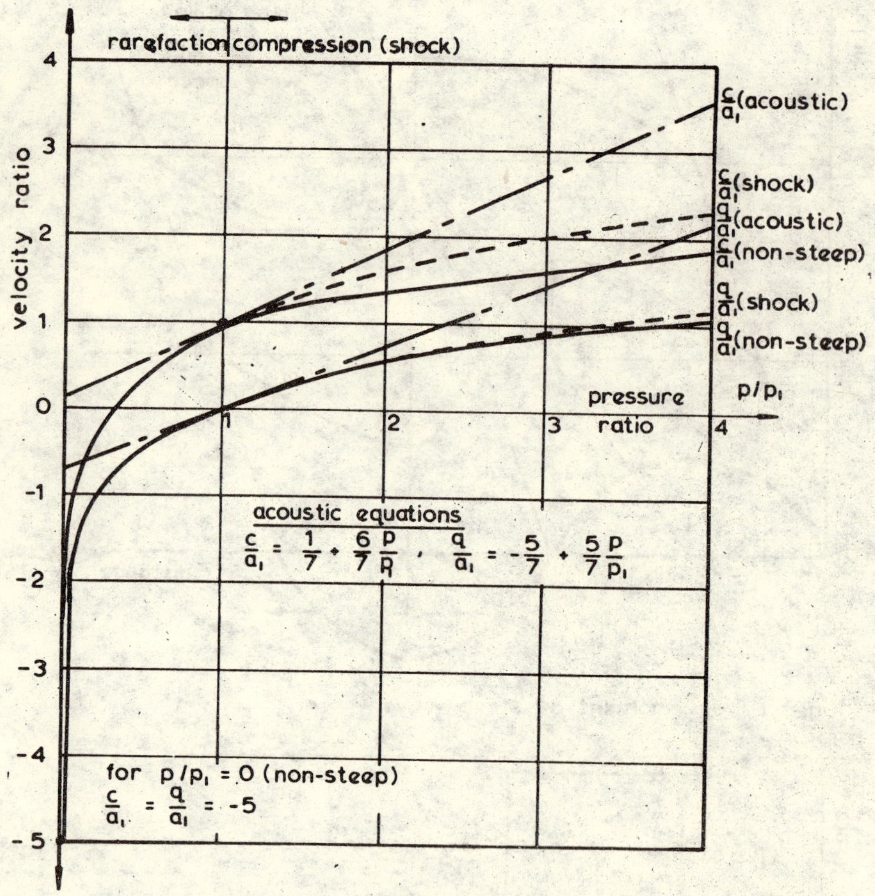

Fig. 8.5 propagation and particle velocity ratio for air ($\gamma = 7/5$)

from (8.10) and (8.12),

$$\frac{a^2}{\gamma-1} + \left(\frac{\varrho_1}{\varrho_2}\right)^2 \frac{c^2}{2} = \frac{a_1^2}{\gamma-1} + \frac{c^2}{2} \qquad (8.14a)$$

or,

$$\left(\frac{\varrho_1}{\varrho}+1\right)\left(\frac{\varrho_1}{\varrho}-1\right) = \frac{2}{\gamma-1}\,\frac{a_1^2}{c^2}\left(1-\frac{a^2}{a_1^2}\right) \qquad (8.14b)$$

from (8.13b),

$$\left(\frac{p}{p_1}-1\right)\left[1-\frac{a_1^2}{2c^2}\left(\frac{p}{p_1}-1\right)+1\right] = \frac{2\gamma}{\gamma-1}\left[\frac{p}{p_1}\left(1-\frac{a_1^2}{\gamma c^2}\left(\frac{p}{p_1}-1\right)\right)-1\right]$$

whence,

$$\frac{c}{a_1} = \left[\frac{\gamma-1}{2\gamma}\left(\frac{\gamma+1}{\gamma-1}\,\frac{p}{p_1}+1\right)\right]^{0.5} \qquad (8.15)$$

with (8.10) and (8.13b),

$$\frac{q}{a_1} = (p/p_1-1)\Bigg/\left[\gamma\left[\frac{\gamma-1}{2\gamma}\left(\frac{\gamma+1}{\gamma-1}\,\frac{p}{p_1}+1\right)\right]^{0.5}\right] \qquad (8.16)$$

also,

$$\frac{a}{a_1} = \left(\frac{T}{T_1}\right)^{0.5} = \left(\frac{p}{p_1}\,\frac{\varrho_1}{\varrho}\right)^{0.5} = \left(\frac{p}{p_1}\right)^{0.5}\left[1-\frac{q}{a_1}\,\frac{a_1}{c}\right]^{0.5} \qquad (8.17)$$

thus, from (8.15) and (8.16),

$$\frac{a}{a_1} = \left[\left(\frac{p}{p_1}\right)\left(\frac{p/p_1+\dfrac{\gamma+1}{\gamma-1}}{\dfrac{\gamma+1}{\gamma-1}\,p/p_1+1}\right)\right]^{0.5} \qquad (8.18)$$

Note: (1) The Rankine–Hugoniot equations (Chapter 4) for a plane normal shock wave give a result equivalent to (8.18) thus,

$$\frac{T}{T_1} = \frac{\dfrac{\gamma+1}{\gamma-1}+\dfrac{p}{p_1}}{\dfrac{\gamma+1}{\gamma-1}+\dfrac{p_1}{p}} \qquad (4.5)$$

(2), $p/p_1 > 1$ and the steep-fronted wave must be compressive.

(3) For a shock or steep-fronted wave moving with propagation velocity (·), $c \neq a \pm |q|$ since there is no longer a continuous relative relationship between wave-front points and the fluid into which the wave is propagating.

(4) Figure 8.5 shows the variation with absolute pressure ratio (p/p_1), of the velocity parameters (c/a_1) and (q/a_1) for acoustic, non-steep and steep waves propagating through air ($\gamma = 7/5$) considered as a perfect gas. (See Section 8.5 for appropriate form of equations.)

The change in entropy caused by the irreversible process of steep-fronted wave propagation may be obtained from (1.20a),

$$s-s_1 = c_p \log_e (T/T_1) - R \log_e (p/p_1)$$

giving,

$$\frac{s-s_1}{R} = \frac{\gamma}{\gamma-1}\log_e\left[\frac{\dfrac{\gamma+1}{\gamma-1}+\dfrac{p}{p_1}}{\dfrac{\gamma+1}{\gamma-1}+\dfrac{p_1}{p}}\right] - \log_e \frac{p}{p_1} \qquad (8.19)$$

Sonic conditions

The Mach number $M = q/a$ of the flow induced by the steep-fronted wave is obtained from (8.16) and (8.17),

$$M = \frac{(p/p_1 - 1)}{\left[\frac{\gamma(\gamma-1)}{2}\left(\frac{p}{p_1}\right)\left(\frac{p}{p_1} + \frac{\gamma+1}{\gamma-1}\right)\right]^{0.5}} \qquad (8.20)$$

for sonic flow, $M = 1$ produces the following quadratic equation with $\gamma = 7/5$ for air,

$$18(p/p_1)^2 - 92(p/p_1) + 25 = 0 \qquad (8.21)$$

whence, with one physically real root since $p/p_1 > 1$,

$$p_*/p_1 = 4{\cdot}8231(5) \qquad (8.21a)$$

8.5 SUMMARY OF EQUATIONS

The following are the equations used in Tables 6.7 for wave propagation in air ($\gamma = 7/5$) without friction in a constant area duct and are plotted in Figures 8.8 and 8.9 (non-steep) wave and Figures 8.10 and 8.11 (steep–fronted wave).

Non-Steep Wave (Figure 8.6), Table 6

isentropic, $$\frac{T}{T_1} = \left(\frac{p}{p_1}\right)^{(\gamma-1)/\gamma} = \left(\frac{p}{p_1}\right)^{2/7} \qquad (8.22a)$$

$$\frac{\varrho}{\varrho_1} = \left(\frac{p}{p_1}\right)^{1/\gamma} = \left(\frac{p}{p_1}\right)^{5/7} \qquad (8.22b)$$

from (8.5), $$\frac{c}{a_1} = \left(\frac{p}{p_1}\right)^{1/7} - - 5 \qquad (8.23)$$

from (8.4), $$\frac{q}{a_1} = 5\left[\left(\frac{p}{p_1}\right)^{1/7} - 1\right] \qquad (8.24)$$

from (8.6), $$M = \frac{q}{a} = \text{abs}\left|\, 5\left[1 - \left(\frac{p}{p_1}\right)^{-1/7}\right]\right| \qquad (8.25)$$

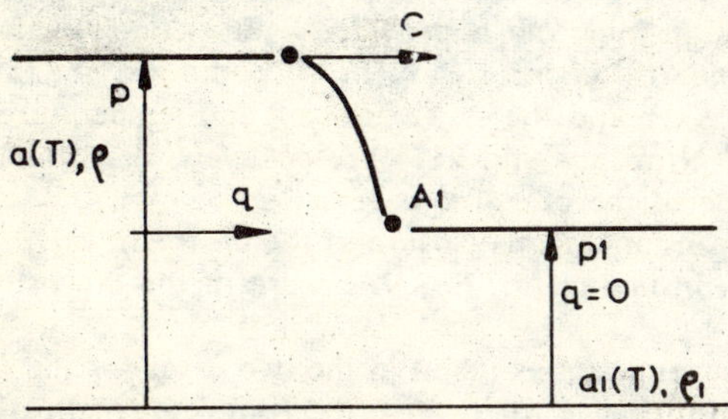

Fig. 8.6. non-steep compression and rarefraction waves, (isentropic)

(A1 propagates at the local acoustic velocity a1)

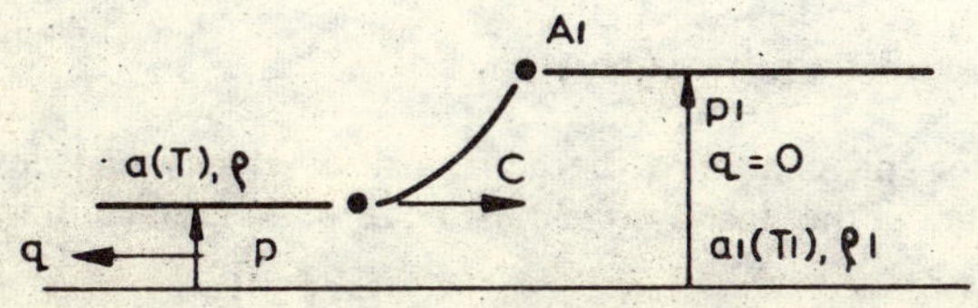

Steep-Fronted Wave (Fig. 8.7), Table 7

shock, $$\frac{T}{T_1} = \frac{p}{p_1}\left[\frac{p/p_1 + 6}{6p/p_1 + 1}\right] \qquad (8.26a)$$

$$\frac{\varrho}{\varrho_1} = \frac{6p/p_1 + 1}{6 + p/p_1} \qquad (8.26b)$$

from (8.15), $$\frac{c}{a_1} = \left[\frac{6p/p_1 + 1}{7}\right]^{0.5} \qquad (8.27)$$

from (8.16), $$\frac{q}{a_1} = \frac{5(p/p_1 - 1)}{[7(6p/p_1 + 1)]^{0.5}} \qquad (8.28)$$

from (8.20), $$M = \frac{q}{a} = \frac{5(p/p_1 - 1)}{[(7p/p_1)(p/p_1 + 6)]^{0.5}} \qquad (8.29)$$

from (8.19), $$\frac{s - s_1}{R} = 3{\cdot}5 \log_e\left[\frac{6 + p/p_1}{6p/p_1 + 1}\right] + 2{\cdot}5 \log_e \frac{p}{p_1} \qquad (8.30)$$

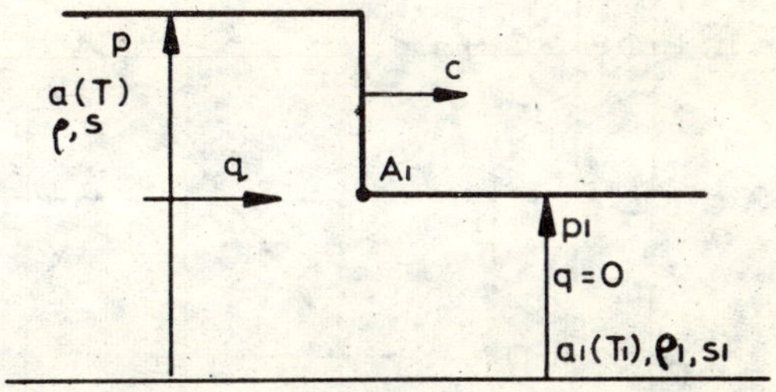

Fig. 8.7. steep-fronted wave, (non-isentropic)

EXAMPLE 8.1 A small plane compression or sound wave travels through a pipe of uniform bore, cross-sectional area 0·1 m², in which the pressure and temperature of the undisturbed air are 30 in Hg abs and 289 K. The amplitude of the wave is 1·5 in Hg gauge and its overall length is 1·5 m; its form is flat-topped with non-steep extremities.

Determine the mass associated with the wave and hence find the local increases in internal and kinetic energies associated with the wave. Determine the displacement and the duration of the displacement, of a gas particle overtaken by the wave. (1 in Hg = 3 386·4 N/m²)

(U. Birmingham)

Figure E8.1 shows the pressure profile of the sound wave where (a) is the propagation velocity of the wave and (q) is the particle velocity induced by the passage of the wave through air with $q = 0$ initially.

Acoustic theory will be used for the analysis but values indicated in brackets have been calculated using non-steep wave theory.

From (8.3), $$q = \mathrm{d}q = a\frac{\mathrm{d}p}{\gamma p} = a\frac{\mathrm{d}\varrho}{\varrho}$$

($a^2 = \gamma RT$), where $a = 20(289)^{0.5} = 340$ m/s

($p = \varrho RT$), $$\varrho = \frac{30 \times 3{\cdot}3864 \times 10^3}{0{\cdot}2871 \times 10^3 \times 289} = 1{\cdot}2244 \text{ kg/m}^3$$

from (8.3), $\mathrm{d}\varrho = 12{\cdot}14 \times 1{\cdot}2244/340$

$= 0{\cdot}0437$ kg/m³ (0·0435)

$\therefore\ \varrho + \mathrm{d}\varrho = 1{\cdot}2681$ kg/m³ (1·2679)

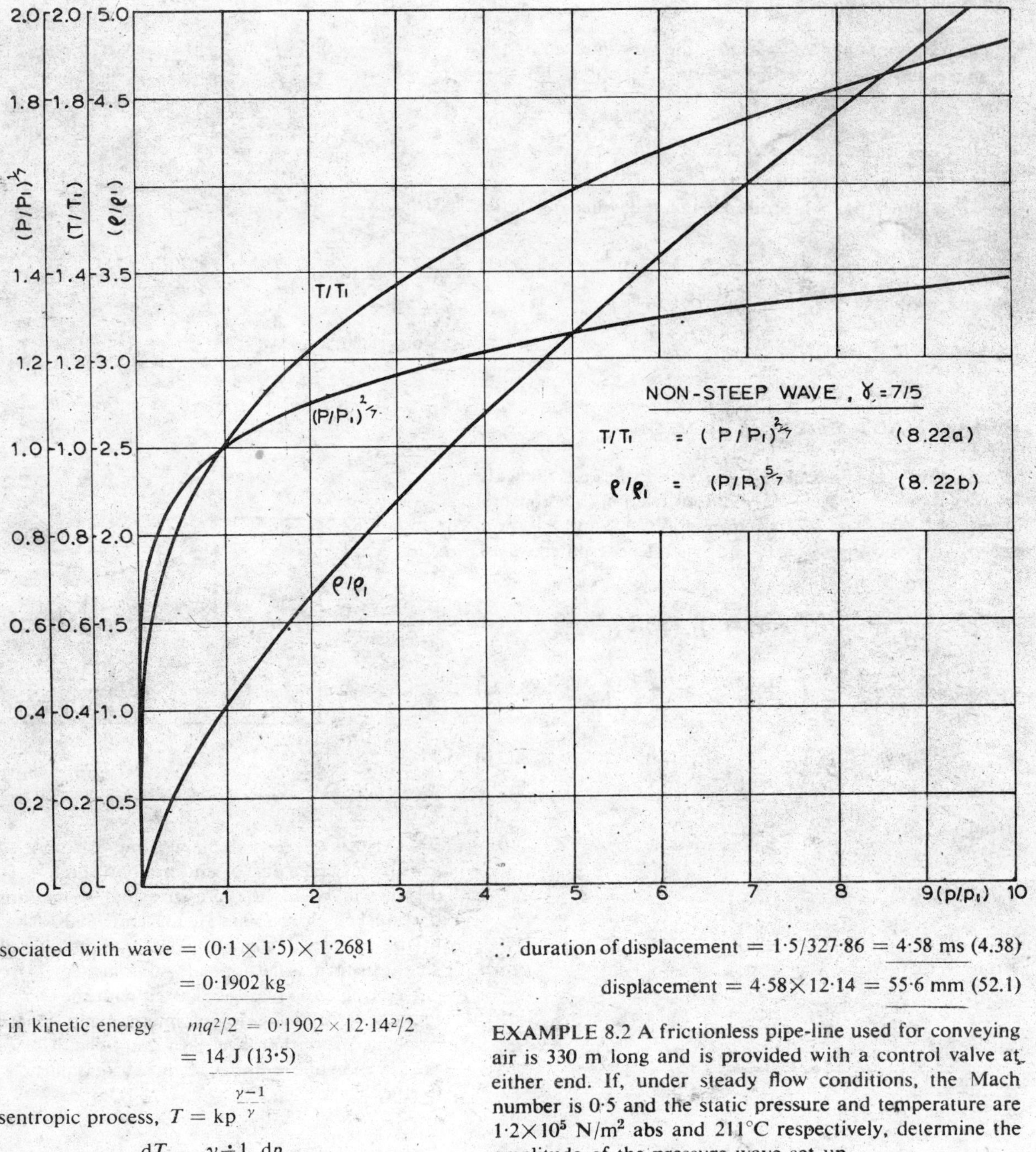

Fig. 8·8

mass associated with wave $= (0{\cdot}1 \times 1{\cdot}5) \times 1{\cdot}2681$

$= 0{\cdot}1902$ kg

increase in kinetic energy $mq^2/2 = 0{\cdot}1902 \times 12{\cdot}14^2/2$

$= 14$ J (13·5)

for an isentropic process, $T = \text{kp}^{\frac{\gamma-1}{\gamma}}$

$$\therefore \frac{dT}{T} = \frac{\gamma-1}{\gamma}\frac{dp}{p}$$

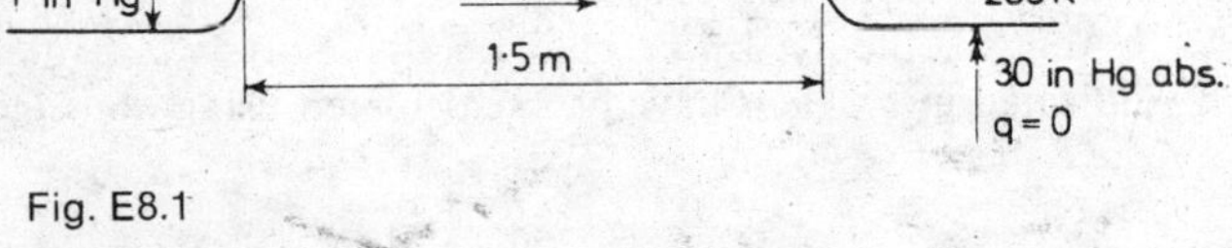

Fig. E8.1

increase in temperature,

$$dT = \frac{2}{7} \times \frac{1{\cdot}5}{30} \times 289 = 4{\cdot}13 \text{ K } (4.05)$$

increase in internal energy, $dU = mc_v dT$

$\therefore dU = 0{\cdot}1902 \times 0{\cdot}718 \times 10^3 \times 4{\cdot}13 = 564$ J (553.1)

relative velocity of wave

$= a - q = 340 - 12{\cdot}14 = 327{\cdot}86$ m/s (342.4)

$\therefore$ duration of displacement $= 1{\cdot}5/327{\cdot}86 = 4{\cdot}58$ ms (4.38)

displacement $= 4{\cdot}58 \times 12{\cdot}14 = 55{\cdot}6$ mm (52.1)

EXAMPLE 8.2 A frictionless pipe-line used for conveying air is 330 m long and is provided with a control valve at either end. If, under steady flow conditions, the Mach number is 0·5 and the static pressure and temperature are $1{\cdot}2 \times 10^5$ N/m² abs and 211°C respectively, determine the amplitude of the pressure wave set up
(a) upon instantaneous closure of the upstream valve,
(b) upon instantaneous (0·3 s) closure of the downstream valve.

Compare the times taken in the two cases for the front of the disturbance to traverse the complete length of the pipe.

(U. Birmingham)

Figure E8.2 represents the details of the problem

$(a^2 = \gamma RT)$, $\therefore a_1 = 20(484)^{0.5} = 440$ m/s

(a) *upstream valve closed*

For this condition the net particle velocity (q_n) downstream of the closed valve must be zero. A rarefaction wave

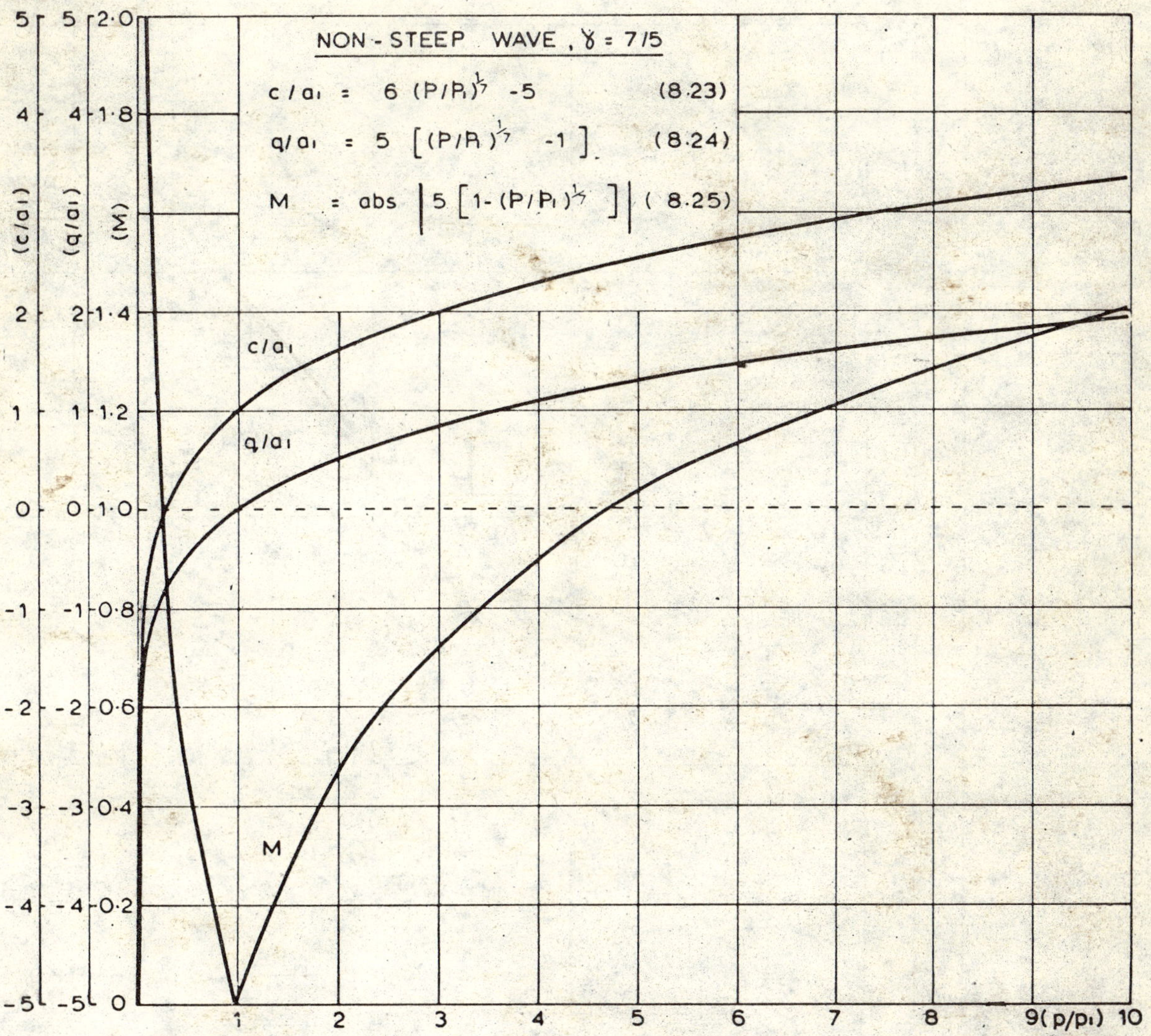

Fig. 8·9

is necessary to produce a negative particle velocity (q_2) to cancel the former steady positive flow speed (q_1),

thus $q_n = 0 = q_2 + q_1$

(b) *downstream valve closed*

In this case a compression wave is necessary to produce a particle velocity to cancel the initial steady flow speed.

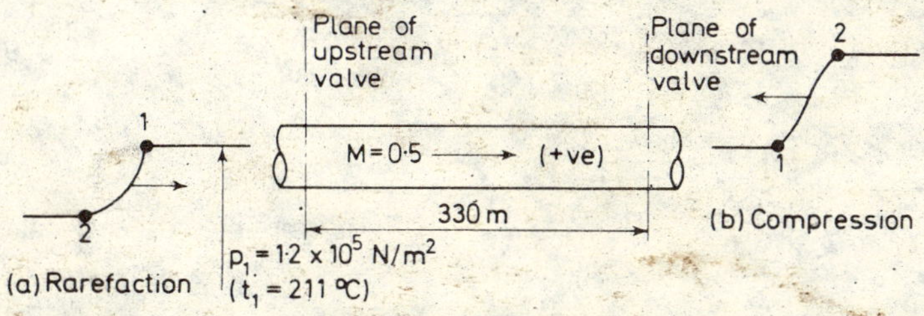

Fig. E8.2

for any non-steep wave (8.24), $\frac{q_2}{a_1} = 5\left[\left(\frac{p_2}{p_1}\right)^{1/7} - 1\right]$

$\therefore\ 0 = 5\times440[(p_2/p_1)^{1/7} - 1] + 0{\cdot}5\times440$

whence, $(p_2/p_1)^{1/7} = 0{\cdot}9$

from Table 6, $p_2/p_1 = 0{\cdot}47 + 0{\cdot}01\times22/27 = 0{\cdot}478$

$\therefore\ p_2 = 0{\cdot}478 + 1{\cdot}2\times10^5 = 0{\cdot}574\times10^5$ N/m²

$\therefore$ amplitude, $\Delta p = p_1 - p_2 = 0{\cdot}626\times10^5$ N/m²

The compression wave moves against the initial flow and is therefore negative,

thus, $q_n = 0 = -5a_1[(p_2/p_1)^{1/7} - 1] + q_1$

whence $(p_2/p_1)^{1/7} = 1{\cdot}1$

from Table 6, $p_2/p_1 = 1{\cdot}95 + 0{\cdot}05\times1/40 = 1{\cdot}951$

$\therefore\ p_2 = 1{\cdot}951\times1{\cdot}2\times10^5 = 2{\cdot}341\times10^5$ N/m²

$\therefore$ amplitude, $\Delta p = p_2 - p_1 = 1{\cdot}141\times10^5$ N/m²

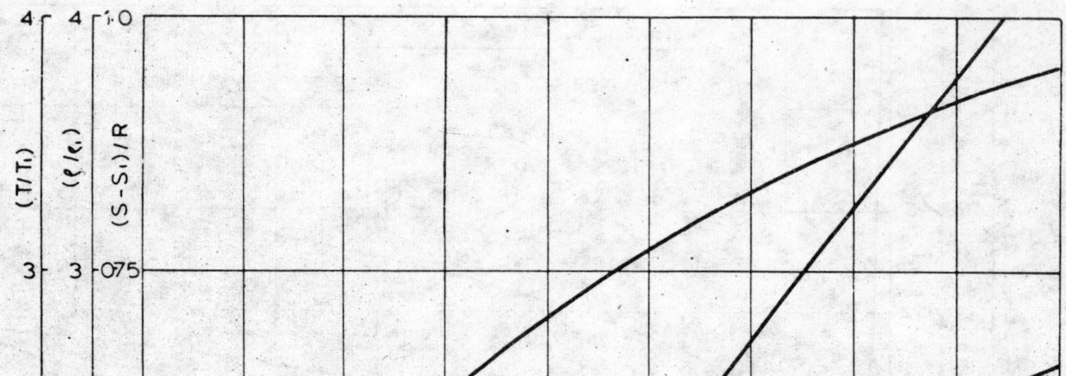

Fig. 8·10

(a) the leading tip (1) of the rarefaction will move at the local speed of sound relative to the steady flow, i.e., the absolute propagation velocity, $c_1 = a_1 + q_1$

$\therefore$ traverse time, $t_1 = 330/c_1 = 330/(440+220) = 0{\cdot}5$ s

alternatively, considering point (2), the absolute propagation velocity $c_2 = (a_2+q_2)+q_1$

where $q_2 = -q_1$ and $\dfrac{a_2}{a_1} = \left(\dfrac{T_2}{T_1}\right)^{0.5} = \left(\dfrac{p_2}{p_1}\right)^{1/7} = 0{\cdot}9$

$\therefore c_2 = 0{\cdot}9a_1$

$\therefore$ traverse time, $t_2 = 330/c_2 = 330/(440 \times 0{\cdot}9) = 0.833$ s

(b) the leading tip (1) of the compression wave will move with the local speed of sound (a_1) in the opposite direction to the initial steady flow (q_1).

$\therefore$ absolute propagation velocity, $c_1 = -a_1 + q_1$

$\therefore$ traverse time, $t_1 = 330/c_1 = 330/(440-220) = 1{\cdot}5$ s

as before, considering point (2), the absolute propagation velocity, $c_2 = a_2 = a_1(p_2/p_1)^{1/7} = 1.1a_1$

$\therefore$ traverse time, $t_2 = 330/c_2 = 330/(440 \times 1{\cdot}1) = 0{\cdot}682$ s

Even allowing for a closing time of 0·3 s the above result indicates that point (2) will arrive at the other end before point (1)!. Assuming that the non-steep wave becomes a steep-fronted wave when point (2) overtakes point (1) a different approach must be used:
for a valve opening time of 0·3 s, point (1) will travel a distance, $x_1 = 0{\cdot}3(440-220) = 66$ m before (it is assumed) point (2) leaves the valve;

the velocity of (2) relative to (1) is $c_2 - c_1 = 484 - 220 = 264$ m/s

$\therefore$ time for (2) to overtake (1) is $66/264 = 0{\cdot}25$ s

$\therefore$ remaining distance $= 330 - 66 - 0{\cdot}25 \times 220 = 210$ m

from Table 7, for $p_2/p_1 = 1{\cdot}951$ (for a steep wave)

$c_2/a_1 = 1{\cdot}3470 + 0{\cdot}0158 \times 1/50 = 1{\cdot}3473$

$\therefore$ absolute propagation velocity,
$c_2 = 1{\cdot}3473a_1 - 220 = 392{\cdot}8$ m/s

$\therefore$ remaining time $= 210/392{\cdot}8 = 0{\cdot}563$ s

$\therefore$ total traverse time $= 0{\cdot}3 + 0{\cdot}25 + 0{\cdot}563 = 1{\cdot}113$ s

EXAMPLE 8.3 Two non-steep, flat-topped waves of amplitude p_2 and p_3 (absolute) are generated in air contained in a uniform frictionless duct such that they approach each

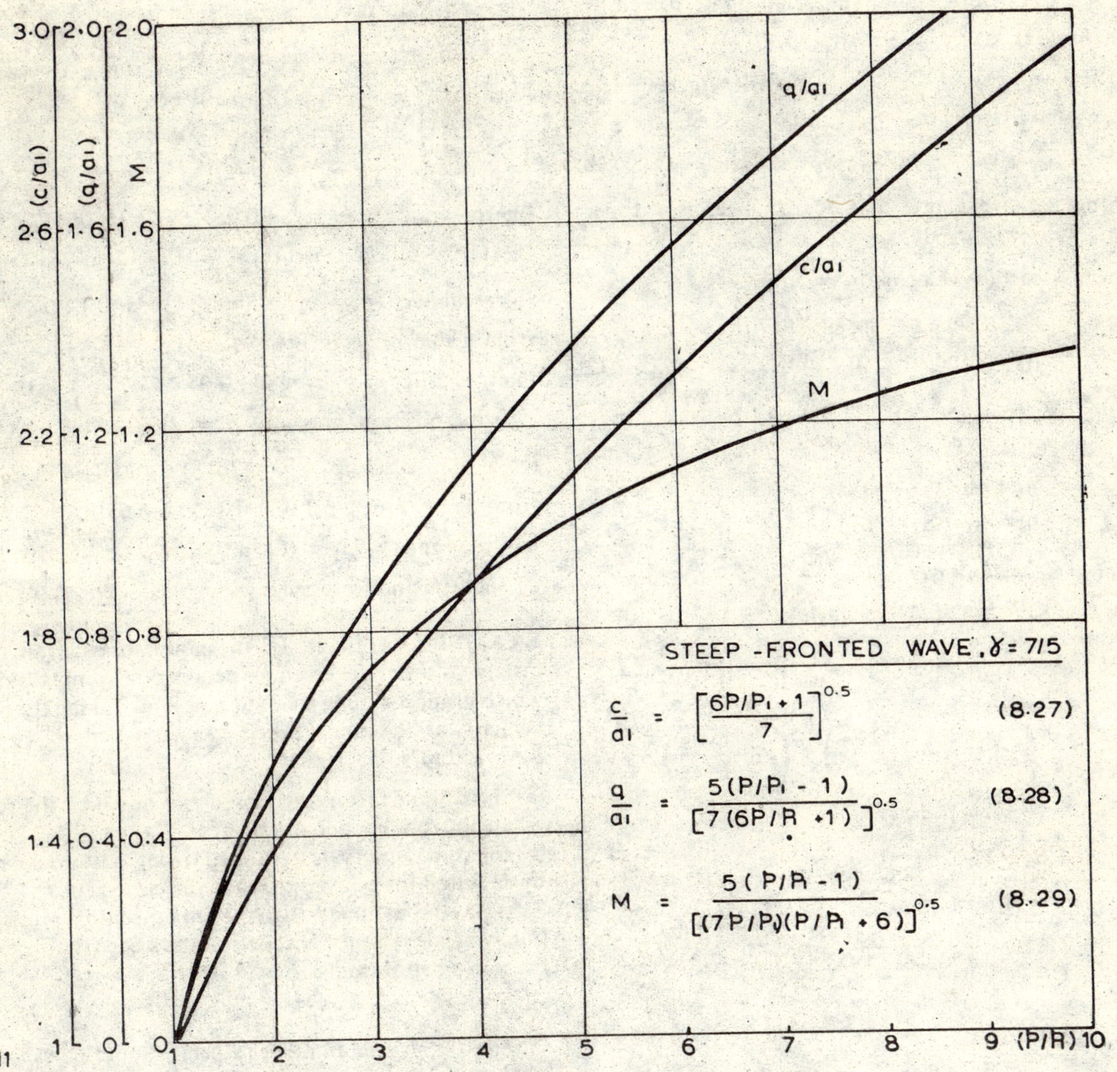

Fig. 8·11

other from opposite directions. If the pressure of the air in the undisturbed condition is p_1, show that the resultant static pressure, p_n, during superposition of the two waves is given by:

$$x_n = x_2 + x_3 - 1$$

where $x = (p/p_1)^{1/7}$

Two non-steep, flat-topped waves (A) and (B) approach each other through undisturbed air of pressure 0·8 bar (1 bar $= 1 \times 10^5$ N/m^2). Determine the pressures of superposition for the following absolute incident amplitudes:

(a) $p_A = 0{\cdot}88$ bar, $p_B = 0{\cdot}76$ bar (cf. acoustic theory)

(b) $p_A = 1{\cdot}6$ bar, $p_B = 2{\cdot}12$ bar

(c) $p_A = 2{\cdot}0$ bar, $p_B = 0{\cdot}04$ bar

(U. Birmingham)

Figure E8.3a shows the pressure profile of the incident waves of amplitude p_2 and p_3 and the superposition condition p_n.

Consider wave-point B to establish the superposition particle velocity q_n with directions positive to the right.

In general (8.24) gives, $q = 5a_1[(p/p_1)^{1/7} - 1]$

$\because$ q_n relative to condition (2) is $q_{n2} = +5a_2\,[(p_n/p_2)^{1/7} - 1]$

particle velocity of (2) is $q_2 = -5a_1[(p_2/p_1)^{1/7} - 1]$

$\therefore$ absolute superposition velocity $q_n = q_{n2} + q_2$

$\therefore$ $q_n = 5a_2[(p_n/p_2)^{1/7} - 1] - 5a_1[(p_2/p_1)^{1/7} - 1]$

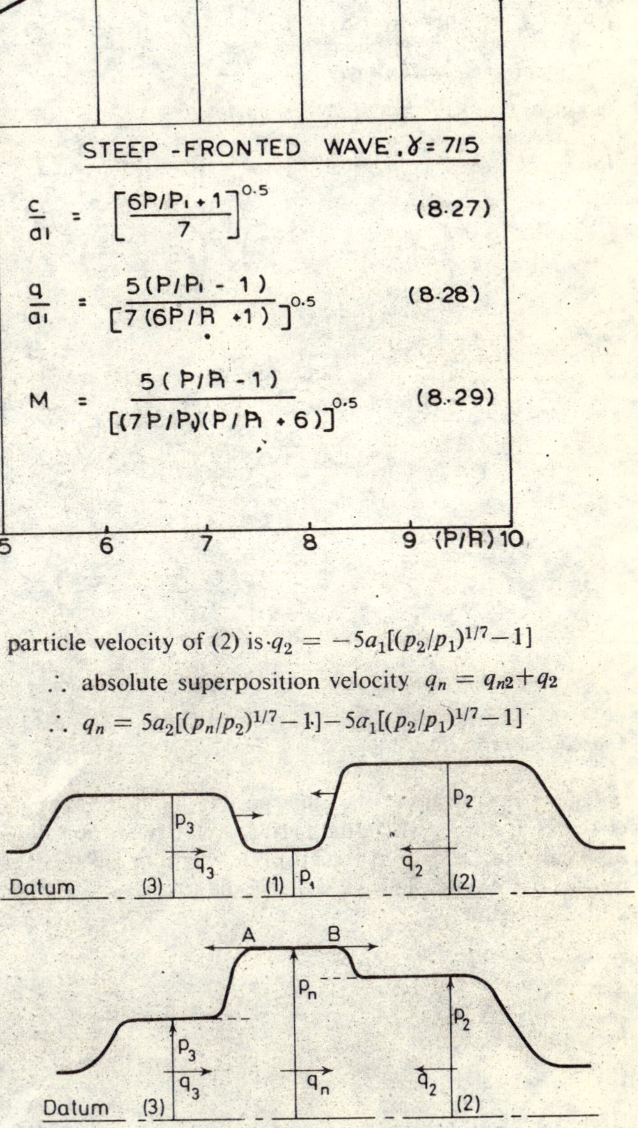

Fig. E8.3a

now, for isentropic processes $a/a_1 = (p/p_1)^{1/7} = x$

$\therefore$ $a_2 = x_2 a_1$

$\therefore$ $q_n = 5a_1(x_n - x_2) - 5a_1(x_2 - 1)$

$\therefore$ $q_n = 5a_1(x_n - 2x_2 + 1)$ (A)

If wave-point A is now considered, Figure E48, q_n relative to condition (3) is $q_{n3} = -5a_3[(p_n/p_3)^{1/7}-1]$

and particle velocity of (3) is $q_3 = +5a_1[(p_3/p_1)^{1/7}-1]$

whence, as above, $q_n = q_{n3}+q_3$

giving, $\quad q_n = 5a_1(2x_3-x_n-1)$ (B)

Comparing the two expressions for q_n gives (for superposition), $x_n = x_2+x_3-1$ (C)

superposition particle velocity, $q_n = 5a_1(x_3-x_2)$

or $\quad q_n = 5a_1[(p_3/p_1)^{1/7}-1]-5a_1[(p_2/p_1)^{1/7}-1]$

i.e. $\quad q_n = q_1+q_2$ (positive to right) (D)

propagation velocity of wave point B, $c_B = a_n+q_n$ ($\xrightarrow{+}$)

$$\therefore c_B = a_1(p_n/p_1)^{17} - (p_2/p_1)^{17} - (p_2/p_1)^{17}]$$

$$= a_1(x_n+5x_3-5x_2)$$

from (C), $c_B = a_1(6x_3-4x_2-1)$ (E)

For small or acoustic waves

(C) gives $(p_n/p_1)^{1/7} = (p_2/p_1)^{1/7}+(p_3/p_1)^{1/7}-1$

for small waves, $(p/p_1)^{1/7} = (1+p'/p_1)^{1/7} = \left(1+\frac{1}{7}\frac{p_1}{p_1}\right)$

$$\therefore \left(1+\frac{1}{7}\frac{p_n'}{p_1}\right) = \left(1+\frac{1}{7}\frac{p_2'}{p_1}\right)+\left(1+\frac{1}{7}\frac{p_3'}{p_1}\right)-1$$

or $\quad p_n' = p_2'+p_3'$ (F)

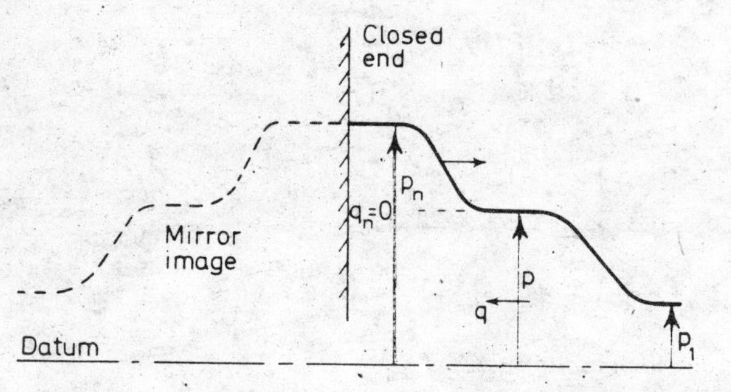

Fig. E8.3b

Closed end reflection

Figure E8.3b shows the superposition condition at a closed end, during which the particle velocity at the solid boundary must be zero. Alternatively a 'mirror image' model may be used so that in Figure E8.3a and in (c), $p_3 = p_2 = p$ and (C) becomes,

$$x_n = 2x-1$$

or $\quad (p_n/p_1)^{1/7} = (p/p_1)^{1/7}-1$ (T)

for *acoustic* waves, as above,

$$p_n' = 2p'$$ (H)

from (E), $c = a_1(2x-1)$

or, from (G), $c = a_n+(q_n = 0)$ (J)

(a) for small waves (acoustic theory),

from (F), $p_n' = p_A'+p_B'$

$= (0{\cdot}88-0{\cdot}8)+(0{\cdot}76-0{\cdot}8)$

$= 0{\cdot}04$ bar

superposition pressure, $p_n = 0{\cdot}8+0{\cdot}04 = 0{\cdot}84$ bar

for finite amplitude, non-steep waves,

from (C), $(p_n/p_1)^{1/7} = (p_A/p_1)^{1/7}+(p_B/p_1)^{1/7}-1$

$= (0{\cdot}88/0{\cdot}8)^{1/7}+(0{\cdot}76/0{\cdot}8)^{1/7}-1$

from Table 6, $(p_n/p_1)^{1/7} = 1{\cdot}0137+0{\cdot}9927-1 = 1{\cdot}0064$

interpolating, $p_n/p_1 = 1+0{\cdot}05\times64/70 = 1{\cdot}0457$

$\therefore$ superposition pressure, $p_n = 1{\cdot}0457+0{\cdot}8 = 0{\cdot}8366$ **bar**

(b) as above, $(p_n/p_1)^{1/7} = (1{\cdot}6/0{\cdot}8)^{1/7}+(2{\cdot}12/0{\cdot}8)^{1/7}-1$

from Table 6, $(p_n/p_1)^{1/7} = 1{\cdot}1041+1{\cdot}1494-1 = 1{\cdot}2535$

interpolating, $p_n/p_1 = 4{\cdot}85+0{\cdot}05\times5/19 = 4{\cdot}8632$

$\therefore$ superposition pressure, $p_n = 4{\cdot}8632\times0{\cdot}8 = 3{\cdot}891$ **bar**

(c) as above, $(p_n/p_1)^{1/7} = (2/0{\cdot}8)^{1/7}+(0{\cdot}04/0{\cdot}8)^{1/7}-1$

from Table 6, $(p_n/p_1)^{1/7} = 1{\cdot}1399+0{\cdot}6518-1 = 0{\cdot}7917$

interpolating, $p_n/p_1 = 0{\cdot}19+0{\cdot}01\times29/58 = 0{\cdot}195$

$\therefore$ superposition pressure, $p_n = 0{\cdot}195\times0{\cdot}8 = 0{\cdot}156$ **bar**

EXAMPLE 8.4 Figure E8.4a shows, for a certain (datum) instant of time, the essential details of a compression wave which is approaching a closed pipe end. Taking the acoustic velocity of the undisturbed gas (assumed to be air) as $a_1 = 300$ m/s, determine

(a) the distance from the closed end at which the reflected steep front encounters point B,
(b) the time interval, reckoned from datum conditions, required for point B to reach the closed end.

It may be assumed that the superposition pressure, p_n, following closed end reflection of the steep front is given by the non-steep formula

$$(p_n/p_1)^{1/7} = 2(p/p_1)^{1/7}-1$$

but the analysis in all other respects should be exact.

(U. Birmingham 1955)

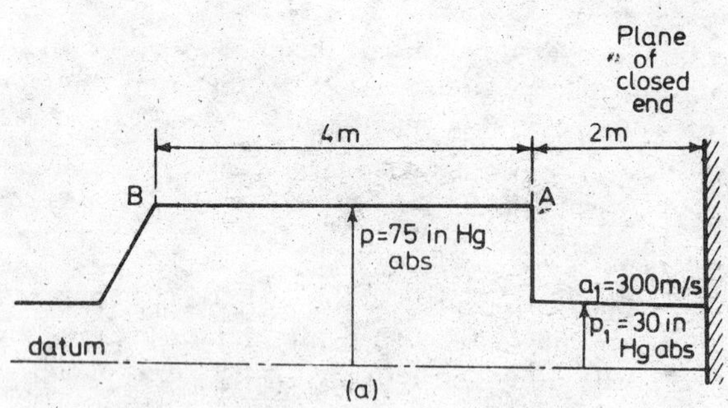

Fig. E8.4a

For the pressure profile shown, $p/p_1 = 75/30 = 25$

from Table 7, $\quad c_A/a_1 = +1{\cdot}5119$ ($\xrightarrow{+}$)

$$c_B = q+a$$

or $\quad c_B/a_1 = q/a_1+a/a_1$

from Table 7, $q/a_1 = +0{\cdot}7087$, $T/T_1 = 1{\cdot}3281$

$(a^2 = \gamma RT)$, $\therefore c_B/a_1 = 0{\cdot}7087+(1{\cdot}3281)^{0{\cdot}5} = 1{\cdot}8611$

Point B is therefore moving faster than point A. If the wave were non-steep B would move with the same speed as A.

time for A to reach closed end, t_A, is given by

$$t_A(a_1) = 2/1{\cdot}5119 = \underline{1{\cdot}3228\ \text{m}}$$

distance travelled by B in time t_A

$$= 1{\cdot}3228 \times 1{\cdot}8611 = \underline{2{\cdot}462\ \text{m}}$$

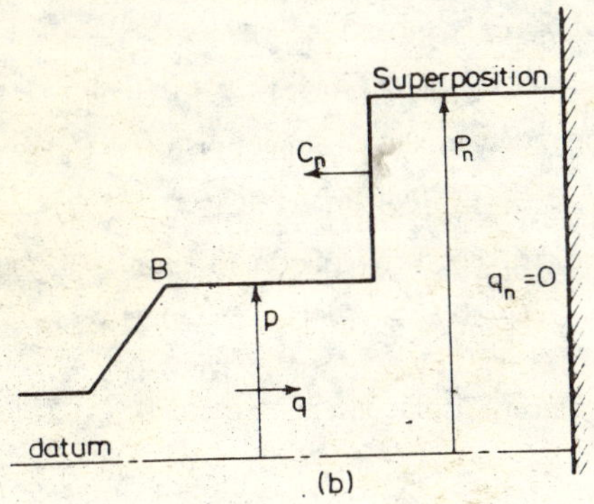

Fig. E8.4b

Superposition (Figure E8.4b)

$$(p_n/p_1)^{1/7} = 2(p/p_1)^{1/7} - 1 = 2(2{\cdot}5)^{1/7} - 1$$

from Table 6, $\quad = 2 \times 1{\cdot}1399 - 1 = 1{\cdot}2798$

interpolating, $p_n/p_1 = 5{\cdot}60 + 0{\cdot}1 \times 8/33 = 5{\cdot}625$

$\therefore\ p_n/p = 5{\cdot}625/2{\cdot}5 = 2{\cdot}25$

from Table 7, $c_n/a = 1{\cdot}4392$ (relative to q)

or $\ c_n/a_1 = 1{\cdot}4392 \times a/a_1 = 1{\cdot}4392 \times 1{\cdot}1524 = 1{\cdot}6585$

$\therefore$ time for B to meet reflected wave,

$$t_B = \frac{(6 - 2{\cdot}462)}{(c_n - q) + (q + a)}$$

or, $\quad t_B(a_1) = 3{\cdot}538/(1{\cdot}6585 + 1{\cdot}1524) = \underline{1{\cdot}2587\ \text{m}}$

distance travelled by B in time t_B

$$= 1{\cdot}2587 \times 1{\cdot}8611 = \underline{2{\cdot}343\ \text{m}}$$

(a) Distance from closed end at which the reflected steep front encounters point B is

$$6 - (2{\cdot}343 + 2{\cdot}462) = \underline{1{\cdot}195\ \text{m}}$$

(b) When B meets the reflected steep wave the propagation velocity becomes

$$c_B = a_n + q - q = a_n$$

from Table 7, for $p_n/p_1 = 5{\cdot}625$,

$$T_n/T_1 = 1{\cdot}8775 + 0{\cdot}25 \times 0{\cdot}0171 = 1{\cdot}8818$$

$$\therefore\ a_n/a_1 = (1{\cdot}8818)^{0{\cdot}5} = 1{\cdot}3718$$

$\therefore$ time for B to move through the superposition condition is

$$t_n(a_1) = 1{\cdot}195/1{\cdot}3718 = 0{\cdot}8711\ \text{m}$$

$\therefore$ total time for B to reach closed end is

$$t = t_A + t_B + t_n = (1{\cdot}3228 + 1{\cdot}2587 + 0{\cdot}8711)/300 = \underline{0{\cdot}0115\text{s}}$$

EXAMPLE 8.5 Deduce an exact expression for the superposition pressure arising during the process of closed end reflection of a steep-fronted, flat topped wave of absolute amplitude, p, propagated through a column of air, of which pressure and the acoustic velocity in the undisturbed state, are denoted by p_1 and a_1 respectively.

A frictionless pipe of uniform bore contains air at 16°C and 1×10^5 N/m² absolute pressure, which is confined between a piston at one end and a fixed closure at the other. Assuming for present purposes that the piston is accelerated instantaneously to such a uniform speed that the gauge amplitude of the resultant pressure wave in the air column is 1×10^5 N/m² determine:

(a) the pressure and temperature of the air within the superposition region resulting from the reflection of the initial wave-front at the closed end of the pipe,
(b) the piston displacement, expressed as a proportion of the initial pipe length, at the instant when the wave reflected from the closed end returns to the piston face,
(c) the comparative state of the air column (i.e. the pressure and temperature) resulting from an isentropic compression for the same piston displacement as in (*b*).

(U. Birmingham 1956)

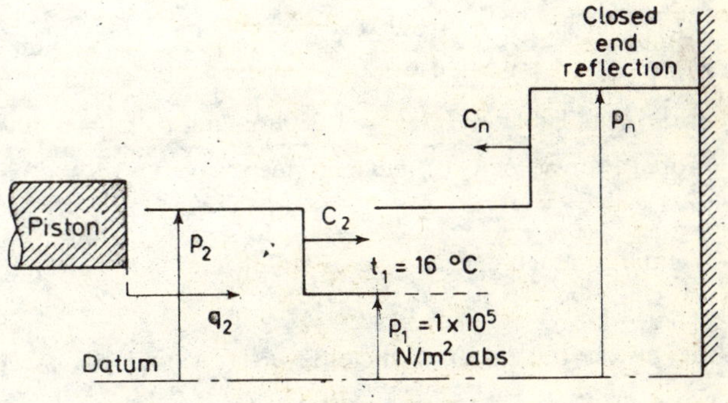

Fig. E8.5

Figure E8.5 shows the pressure profile associated with the steep wave propagation and reflection at the closed end of the pipe. At this solid surface the superposition particle velocity is zero and is the algebraic sum of the incident and reflected particle velocities:

from (8.26a), $\quad \dfrac{a}{a_1} = \left(\dfrac{T}{T_1}\right)^{0{\cdot}5} = \left[\dfrac{p}{p_1}\left[\dfrac{p/p_1 + 6}{6p/p_1 + 1}\right]\right]^{0{\cdot}5}$

if $x = p/p_1$, then $\quad \dfrac{a}{a_1} = \left[\dfrac{x(x+6)}{6x+1}\right]^{0{\cdot}5} \qquad$ (A)

from (8.27), $\quad \dfrac{c}{a_1} = [(6x+1)/7]^{0{\cdot}5} \qquad$ (B)

from (8.28), $\quad \dfrac{q}{a_1} = \dfrac{5(x-1)}{[7(6x+1)]^{0{\cdot}5}}$

q_{n2} relative to condition (2) $\quad = -\dfrac{5a_2(p_n/p_2 - 1)}{[7(6p_n/p_2 + 1)]^{0{\cdot}5}}$

$\therefore$ absolute velocity, $q_n = q_{n2} + q_2 = 0$

from (A)

$$\therefore 0 = +\frac{5a_1(x_2 - 1)}{[7(6x_2+1)]^{0{\cdot}5}} - 5a_1\left[\frac{x_2(x_2+6)}{6x_2+1}\right]^{0{\cdot}5}\left[\frac{y_n - 1}{7(6y_n+1)^{0{\cdot}5}}\right]$$

where $y_n = p_n/p_2$

$$\therefore\ x_2 - 1 = [x_2(x_2+6)]^{0{\cdot}5}\frac{(y_n - 1)}{(6y_n+1)^{0{\cdot}5}}$$

or $\qquad (x_2 - 1)^2(6y_n + 1) = (x_2^2 + 6x_2)(y_n - 1)^2$

whence, $y_n^2 - y_n\left[\dfrac{8x_2^2+6}{6x_2+x_2^2}\right]+\left[\dfrac{8x_2-1}{6x_2+x_2^2}\right] = 0$

solving the quadratic gives,

$$y_n = \frac{8x_2^2-x_2}{6x_2+x_2^2} \quad \text{or} \quad \frac{x_2+6}{6x_2+x_2^2}$$

for $x_2 > 1$, $y_n > 1$ and the 1st root is appropriate,

$$\therefore \quad \frac{p_n}{p_1} = \frac{p_n}{p_2}\times\frac{p_2}{p_1} = \frac{p_2}{p_1}\left[\frac{8p_2/p_1-1}{6+p_2/p_1}\right]$$

For the incident wave $p_2/p_1 = 2$

$\therefore\ p_n/p_1 = 2\left[\dfrac{8\times2-1}{2+6}\right] = 3{\cdot}75 \qquad \therefore\ p_n/p_2 = 1{\cdot}875$

(a) $\qquad p_n = 3{\cdot}75\times10^5\ \text{N/m}^2$

from Table 7, $\qquad T_n/T_1 = 1{\cdot}4836$

$$\therefore\ T_n = 1{\cdot}4836\times289 = 428{\cdot}8\ \text{K}$$

$(a^2 = \gamma RT)$, $\therefore\ a_1 = 20(289)^{0{\cdot}5} = 340$ m/s and

$a_n/a_1 = (1{\cdot}4836)^{0{\cdot}5} = 1{\cdot}218$

also, from Table 7, $T_2/T_1 = 1{\cdot}2308$ giving $a_2/a_1 = 1{\cdot}1094$

propagation velocity of incident steep wave ($p_2/p_1 = 2$) is

$$c_2/a_1 = +1{\cdot}3628$$

$\therefore$ time to reach end of pipe of length L m is (t_2) given by

$$t_2(a_1) = L/1{\cdot}3628\ \text{m}$$

the piston travels at particle velocity behind the steep wave, $\qquad q_2/a_1 = +0{\cdot}5241$

$\therefore$ distance travelled by piston in time (t_2)

$$= 0{\cdot}5241\times L/1{\cdot}3628 = 0{\cdot}3846L\ \text{m}$$

during superposition the reflected wave travels with a propagation velocity of

$$c_n/a_2 = 1{\cdot}3148+0{\cdot}5\times0{\cdot}0162 = 1{\cdot}3229$$

(from Table 7, for $p_n/p_2 = 1{\cdot}875$)

c_n is the propagation velocity relative to the piston moving with speed q_2 and

$$\dot{c}_n/a_1 = c_n/a_2\times a_2/a_1 = 1{\cdot}3229\times1{\cdot}1094 = 1{\cdot}4676$$

$\therefore$ time for reflected wave to meet piston is (t_n) given by

$$t_n(a_1) = (L-0{\cdot}3846L)/1{\cdot}4676 = 0{\cdot}4193L\ \text{m}$$

$\therefore$ distance travelled by piston in time

$$(t_n) = 0{\cdot}5241\times0{\cdot}4193L = 0{\cdot}2198L\text{m}$$

(b) total piston displacement $= 0{\cdot}3956L+0{\cdot}2198L =$ $0{\cdot}6044L$m

(c) isentropic compression due to piston displacement corresponds to a volume ratio of $(L-0{\cdot}6044L)/L = 0{\cdot}3956$ and the density change is therefore

$$\varrho/\varrho_1 = 1/0{\cdot}3956 = 2{\cdot}5278$$

from Table 6 (interpolating),

$$p/p_1 = 3{\cdot}65+0{\cdot}05\times64/246 = 3{\cdot}663$$

and $\qquad T/T_1 = 1{\cdot}4476+0{\cdot}0057\times64/246 = 1{\cdot}4491$

$\therefore$ for the same piston displacement (steep wave results in brackets),

pressure $= 3{\cdot}663\times10^5\ \text{N/m}^2$ $(3{\cdot}75\times10^5\ \text{N/m}^2)$

temperature $= 1{\cdot}4491\times289 = 418{\cdot}8$ K (428·8K)

TABLES FOR THE COMPRESSIBLE FLOW OF AIR ($\gamma = 1.4$, $R = 0.2871$ kJ/kg K)

TABLE 1 STEADY ISENTROPIC FLOW OF AIR WITH PRANDTL–MEYER DATA

Flow and fluid parameters are calculated with the local Mach number (M) as the independent variable for the range, $M = 0 \rightarrow (0{\cdot}01) \rightarrow 5$.
Summaries of equations used are given in Sections 2·8 and 3·4 (Prandtl–Meyer flow for $M > 1$ only). The values of the following parameters are tabulated:

T_o/T ratio of total head temperature to static temperature,
p_o/p ratio of total head pressure to static pressure,
A/A_* ratio of local cross-sectional area to cross-sectional area at which $M = 1$,
q/a_* ratio of local flow velocity to the speed of sound at the sonic condition, $M = 1$,

$MN = ma_o/Ap_o$, mass flow number

(Prandtl–Meyer parameters - Figure 3·8)
μ Mach angle (degrees),
$_*\theta$ angular (degrees) displacement of local Mach line from the reference sonic condition, $M = 1$,
$_*\delta$ angular (degrees) change in flow direction from the reference sonic condition

M	T_o/T	p_o/p	A/A_*	q/a_*	MN	μ	$_*\theta$ (for supersonic flow only)	$_*\delta$
0	1	1	∞	0	0			
0.01	1.0000	1.0001	57.8738	0.0110	0.0140			
0.02	1.0001	1.0003	28.9421	0.0219	0.0280			
0.03	1.0002	1.0006	19.3005	0.0329	0.0420			
0.04	1.0003	1.0011	14.4815	0.0438	0.0559			
0.05	1.0005	1.0018	11.5914	0.0548	0.0699			
0.06	1.0007	1.0025	9.6659	0.0657	0.0838			
0.07	1.0010	1.0034	8.2915	0.0766	0.0977			
0.08	1.0013	1.0045	7.2616	0.0876	0.1116			
0.09	1.0016	1.0057	6.4613	0.0985	0.1254			
0.10	1.0020	1.0070	5.8218	0.1094	0.1392			
0.11	1.0024	1.0085	5.2992	0.1204	0.1529			
0.12	1.0029	1.0101	4.8643	0.1313	0.1666			
0.13	1.0034	1.0119	4.4969	0.1422	0.1802			
0.14	1.0039	1.0138	4.1824	0.1531	0.1937			
0.15	1.0045	1.0158	3.9103	0.1639	0.2072			
0.16	1.0051	1.0180	3.6727	0.1748	0.2206			
0.17	1.0058	1.0204	3.4635	0.1857	0.2339			
0.18	1.0065	1.0229	3.2779	0.1965	0.2472			
0.19	1.0072	1.0255	3.1123	0.2074	0.2603			
0.20	1.0080	1.0283	2.9635	0.2182	0.2734			
0.21	1.0088	1.0312	2.8293	0.2290	0.2864			
0.22	1.0097	1.0343	2.7076	0.2398	0.2992			
0.23	1.0106	1.0375	2.5968	0.2506	0.3120			
0.24	1.0115	1.0409	2.4956	0.2614	0.3247			
0.25	1.0125	1.0444	2.4027	0.2722	0.3372			
0.26	1.0135	1.0481	2.3173	0.2829	0.3496			
0.27	1.0146	1.0520	2.2385	0.2936	0.3619			
0.28	1.0157	1.0560	2.1656	0.3043	0.3741			
0.29	1.0168	1.0601	2.0979	0.3150	0.3862			
0.30	1.0180	1.0644	2.0351	0.3257	0.3981			
0.31	1.0192	1.0689	1.9765	0.3364	0.4099			
0.32	1.0205	1.0735	1.9219	0.3470	0.4216			
0.33	1.0218	1.0783	1.8707	0.3576	0.4331			
0.34	1.0231	1.0833	1.8229	0.3682	0.4445			
0.35	1.0245	1.0884	1.7780	0.3788	0.4557			
0.36	1.0259	1.0937	1.7358	0.3893	0.4668			
0.37	1.0274	1.0992	1.6961	0.3999	0.4777			
0.38	1.0289	1.1048	1.6587	0.4104	0.4884			
0.39	1.0304	1.1106	1.6234	0.4209	0.4991			

M	T_o/T	p_o/p	A/A_*	q/a_*	MN	μ	$_*\theta$ (for supersonic flow only)	$_*\delta$
0.40	1.0320	1.1166	1.5901	0.4313	0.5095			
0.41	1.0336	1.1227	1.5587	0.4418	0.5198			
0.42	1.0353	1.1290	1.5289	0.4522	0.5299			
0.43	1.0370	1.1355	1.5007	0.4626	0.5399			
0.44	1.0387	1.1422	1.4740	0.4729	0.5496			
0.45	1.0405	1.1491	1.4487	0.4833	0.5593			
0.46	1.0423	1.1561	1.4246	0.4936	0.5687			
0.47	1.0442	1.1634	1.4018	0.5038	0.5780			
0.48	1.0461	1.1708	1.3801	0.5141	0.5870			
0.49	1.0480	1.1784	1.3595	0.5243	0.5960			
0.50	1.0500	1.1862	1.3398	0.5345	0.6047			
0.51	1.0520	1.1942	1.3212	0.5447	0.6132			
0.52	1.0541	1.2024	1.3034	0.5548	0.6216			
0.53	1.0562	1.2108	1.2865	0.5649	0.6298			
0.54	1.0583	1.2194	1.2703	0.5750	0.6378			
0.55	1.0605	1.2283	1.2549	0.5851	0.6456			
0.56	1.0627	1.2373	1.2403	0.5951	0.6532			
0.57	1.0650	1.2465	1.2263	0.6051	0.6607			
0.58	1.0673	1.2560	1.2130	0.6150	0.6679			
0.59	1.0696	1.2656	1.2003	0.6249	0.6750			
0.60	1.0720	1.2755	1.1882	0.6348	0.6819			
0.61	1.0744	1.2856	1.1767	0.6447	0.6886			
0.62	1.0769	1.2959	1.1656	0.6545	0.6951			
0.63	1.0794	1.3065	1.1552	0.6643	0.7014			
0.64	1.0819	1.3173	1.1451	0.6740	0.7075			
0.65	1.0845	1.3283	1.1356	0.6837	0.7134			
0.66	1.0871	1.3396	1.1265	0.6934	0.7192			
0.67	1.0898	1.3511	1.1179	0.7031	0.7247			
0.68	1.0925	1.3628	1.1097	0.7127	0.7301			
0.69	1.0952	1.3748	1.1018	0.7223	0.7353			
0.70	1.0980	1.3871	1.0944	0.7318	0.7403			
0.71	1.1008	1.3996	1.0873	0.7413	0.7451			
0.72	1.1037	1.4124	1.0806	0.7508	0.7498			
0.73	1.1066	1.4254	1.0742	0.7602	0.7542			
0.74	1.1095	1.4387	1.0681	0.7696	0.7585			
0.75	1.1125	1.4523	1.0624	0.7789	0.7626			
0.76	1.1155	1.4661	1.0570	0.7883	0.7665			
0.77	1.1186	1.4802	1.0519	0.7975	0.7702			
0.78	1.1217	1.4947	1.0471	0.8068	0.7738			
0.79	1.1248	1.5094	1.0425	0.8160	0.7772			

M	T_o/T	p_o/p	A/A_*	q/a_*	MN	μ (for supersonic flow only)	θ	δ
0.80	1.1280	1.5243	1.0382	0.8251	0.7804			
0.81	1.1312	1.5396	1.0342	0.8343	0.7834			
0.82	1.1345	1.5552	1.0305	0.8433	0.7862			
0.83	1.1378	1.5711	1.0270	0.8524	0.7889			
0.84	1.1411	1.5873	1.0237	0.8614	0.7914			
0.85	1.1445	1.6038	1.0207	0.8704	0.7938			
0.86	1.1479	1.6207	1.0179	0.8793	0.7960			
0.87	1.1514	1.6378	1.0153	0.8882	0.7980			
0.88	1.1549	1.6553	1.0129	0.8970	0.7998			
0.89	1.1584	1.6731	1.0108	0.9058	0.8015			
0.90	1.1620	1.6913	1.0089	0.9146	0.8031			
0.91	1.1656	1.7098	1.0071	0.9233	0.8044			
0.92	1.1693	1.7287	1.0056	0.9320	0.8057			
0.93	1.1730	1.7479	1.0043	0.9407	0.8068			
0.94	1.1767	1.7675	1.0031	0.9493	0.8077			
0.95	1.1805	1.7874	1.0021	0.9578	0.8085			
0.96	1.1843	1.8078	1.0014	0.9663	0.8091			
0.97	1.1882	1.8285	1.0008	0.9748	0.8096			
0.98	1.1921	1.8496	1.0003	0.9833	0.8099			
0.99	1.1960	1.8710	1.0001	0.9916	0.8101			
1.00	1.2000	1.8929	1.0000	1.0000	0.8102	90	0	0
1.01	1.2040	1.9152	1.0001	1.0083	0.8101	81.93	8.11	0.04
1.02	1.2081	1.9379	1.0003	1.0166	0.8099	78.64	11.49	0.13
1.03	1.2122	1.9610	1.0007	1.0248	0.8096	76.14	14.09	0.23
1.04	1.2163	1.9846	1.0013	1.0330	0.8091	74.06	16.29	0.35
1.05	1.2205	2.0085	1.0020	1.0411	0.8085	72.25	18.24	0.49
1.06	1.2247	2.0330	1.0029	1.0492	0.8078	70.63	20.01	0.64
1.07	1.2290	2.0578	1.0039	1.0573	0.8070	69.16	21.64	0.80
1.08	1.2333	2.0831	1.0051	1.0653	0.8061	67.81	23.16	0.97
1.09	1.2376	2.1089	1.0064	1.0733	0.8050	66.55	24.59	1.15
1.10	1.2420	2.1351	1.0079	1.0812	0.8038	65.38	25.96	1.34
1.11	1.2464	2.1618	1.0095	1.0891	0.8025	64.28	27.26	1.53
1.12	1.2509	2.1890	1.0113	1.0970	0.8011	63.23	28.50	1.74
1.13	1.2554	2.2167	1.0132	1.1048	0.7996	62.25	29.70	1.94
1.14	1.2599	2.2449	1.0153	1.1126	0.7980	61.31	30.85	2.16
1.15	1.2645	2.2736	1.0175	1.1203	0.7963	60.41	31.97	2.38
1.16	1.2691	2.3028	1.0198	1.1280	0.7945	59.55	33.06	2.61
1.17	1.2738	2.3325	1.0222	1.1356	0.7926	58.73	34.11	2.84
1.18	1.2785	2.3628	1.0248	1.1432	0.7905	57.94	35.14	3.07
1.19	1.2832	2.3936	1.0276	1.1508	0.7884	57.18	36.14	3.31
1.20	1.2880	2.4250	1.0304	1.1583	0.7863	56.44	37.12	3.56
1.21	1.2928	2.4569	1.0334	1.1658	0.7840	55.74	38.07	3.81
1.22	1.2977	2.4894	1.0366	1.1732	0.7816	55.05	39.01	4.06
1.23	1.3026	2.5224	1.0398	1.1806	0.7791	54.39	39.92	4.31
1.24	1.3075	2.5560	1.0432	1.1879	0.7766	53.75	40.82	4.57
1.25	1.3125	2.5903	1.0468	1.1952	0.7740	53.13	41.70	4.83
1.26	1.3175	2.6251	1.0504	1.2025	0.7713	52.53	42.57	5.09
1.27	1.3226	2.6606	1.0542	1.2097	0.7685	51.94	43.42	5.36
1.28	1.3277	2.6967	1.0581	1.2169	0.7657	51.38	44.25	5.63
1.29	1.3328	2.7334	1.0621	1.2240	0.7628	50.82	45.08	5.90
1.30	1.3380	2.7707	1.0663	1.2311	0.7598	50.28	45.89	6.17
1.31	1.3432	2.8088	1.0706	1.2382	0.7568	49.76	46.68	6.44
1.32	1.3485	2.8474	1.0750	1.2452	0.7536	49.25	47.47	6.72
1.33	1.3538	2.8868	1.0796	1.2522	0.7505	48.75	48.25	7.00
1.34	1.3591	2.9269	1.0842	1.2591	0.7472	48.27	49.01	7.28

M	T_o/T	p_o/p	A/A_*	q/a_*	MN	μ (for supersonic flow only)	θ	δ
1.35	1.3645	2.9676	1.0890	1.2660	0.7439	47.79	49.77	7.56
1.36	1.3699	3.0091	1.0940	1.2729	0.7406	47.33	50.51	7.84
1.37	1.3754	3.0513	1.0990	1.2797	0.7372	46.88	51.25	8.13
1.38	1.3809	3.0942	1.1042	1.2864	0.7337	46.44	51.97	8.41
1.39	1.3864	3.1378	1.1095	1.2932	0.7302	46.01	52.69	8.70
1.40	1.3920	3.1823	1.1149	1.2999	0.7267	45.59	53.40	8.99
1.41	1.3976	3.2275	1.1205	1.3065	0.7231	45.17	54.10	9.28
1.42	1.4033	3.2734	1.1262	1.3131	0.7194	44.77	54.80	9.57
1.43	1.4090	3.3202	1.1320	1.3197	0.7157	44.37	55.48	9.86
1.44	1.4147	3.3678	1.1379	1.3262	0.7120	43.98	56.16	10.15
1.45	1.4205	3.4162	1.1440	1.3327	0.7082	43.60	56.84	10.44
1.46	1.4263	3.4654	1.1501	1.3392	0.7044	43.23	57.50	10.73
1.47	1.4322	3.5155	1.1565	1.3456	0.7006	42.86	58.16	11.02
1.48	1.4381	3.5665	1.1629	1.3520	0.6967	42.51	58.81	11.32
1.49	1.4440	3.6183	1.1695	1.3583	0.6928	42.16	59.46	11.61
1.50	1.4500	3.6710	1.1762	1.3646	0.6888	41.81	60.09	11.91
1.51	1.4560	3.7247	1.1830	1.3708	0.6849	41.47	60.73	12.20
1.52	1.4621	3.7792	1.1899	1.3770	0.6809	41.14	61.36	12.49
1.53	1.4682	3.8347	1.1970	1.3832	0.6768	40.81	61.98	12.79
1.54	1.4743	3.8911	1.2042	1.3894	0.6728	40.49	62.59	13.09
1.55	1.4805	3.9485	1.2116	1.3955	0.6687	40.18	63.20	13.38
1.56	1.4867	4.0068	1.2190	1.4015	0.6646	39.87	63.81	13.68
1.57	1.4930	4.0662	1.2266	1.4075	0.6605	39.56	64.41	13.97
1.58	1.4993	4.1266	1.2344	1.4135	0.6564	39.27	65.00	14.27
1.59	1.5056	4.1880	1.2422	1.4195	0.6522	38.97	65.59	14.56
1.60	1.5120	4.2504	1.2502	1.4254	0.6480	38.68	66.18	14.86
1.61	1.5184	4.3139	1.2584	1.4313	0.6438	38.40	66.76	15.16
1.62	1.5249	4.3785	1.2666	1.4371	0.6396	38.12	67.33	15.45
1.63	1.5314	4.4442	1.2750	1.4429	0.6354	37.84	67.90	15.75
1.64	1.5379	4.5110	1.2836	1.4487	0.6312	37.57	68.47	16.04
1.65	1.5445	4.5789	1.2922	1.4544	0.6270	37.31	69.03	16.34
1.66	1.5511	4.6479	1.3010	1.4601	0.6227	37.04	69.59	16.63
1.67	1.5578	4.7181	1.3100	1.4657	0.6185	36.78	70.14	16.93
1.68	1.5645	4.7896	1.3190	1.4713	0.6142	36.53	70.69	17.22
1.69	1.5712	4.8622	1.3283	1.4769	0.6100	36.28	71.24	17.52
1.70	1.5780	4.9360	1.3376	1.4825	0.6057	36.03	71.78	17.81
1.71	1.5848	5.0111	1.3471	1.4880	0.6014	35.79	72.31	18.10
1.72	1.5917	5.0874	1.3567	1.4935	0.5972	35.55	72.85	18.40
1.73	1.5986	5.1650	1.3665	1.4989	0.5929	35.31	73.38	18.69
1.74	1.6055	5.2439	1.3764	1.5043	0.5886	35.08	73.90	18.98
1.75	1.6125	5.3241	1.3865	1.5097	0.5843	34.85	74.42	19.27
1.76	1.6195	5.4057	1.3967	1.5150	0.5801	34.62	74.94	19.56
1.77	1.6266	5.4886	1.4070	1.5203	0.5758	34.40	75.46	19.86
1.78	1.6337	5.5729	1.4175	1.5256	0.5715	34.18	75.97	20.15
1.79	1.6408	5.6587	1.4282	1.5308	0.5673	33.96	76.47	20.44
1.80	1.6480	5.7458	1.4390	1.5360	0.5630	33.75	76.98	20.73
1.81	1.6552	5.8344	1.4499	1.5411	0.5588	33.54	77.48	21.01
1.82	1.6625	5.9244	1.4610	1.5463	0.5545	33.33	77.97	21.30
1.83	1.6698	6.0160	1.4723	1.5514	0.5503	33.12	78.47	21.59
1.84	1.6771	6.1091	1.4836	1.5564	0.5461	32.92	78.96	21.88
1.85	1.6845	6.2037	1.4952	1.5614	0.5419	32.72	79.44	22.16
1.86	1.6919	6.2998	1.5069	1.5664	0.5377	32.52	79.93	22.45
1.87	1.6994	6.3976	1.5187	1.5714	0.5335	32.33	80.41	22.73
1.88	1.7069	6.4970	1.5308	1.5763	0.5293	32.13	80.88	23.02
1.89	1.7144	6.5980	1.5429	1.5812	0.5251	31.94	81.36	23.30

M	T_o/T	p_o/p	A/A_*	q/a_*	MN	μ (for supersonic flow only)	θ (for supersonic flow only)	δ (for supersonic flow only)
1.90	1.7220	6.7006	1.5553	1.5861	0.5209	31.76	81.83	23.59
1.91	1.7296	6.8050	1.5677	1.5909	0.5168	31.57	82.30	23.87
1.92	1.7373	6.9111	1.5804	1.5957	0.5126	31.39	82.76	24.15
1.93	1.7450	7.0189	1.5932	1.6005	0.5085	31.21	83.22	24.43
1.94	1.7527	7.1284	1.6062	1.6052	0.5044	31.03	83.68	24.71
1.95	1.7605	7.2398	1.6193	1.6099	0.5003	30.85	84.14	24.99
1.96	1.7683	7.3530	1.6326	1.6146	0.4963	30.68	84.59	25.27
1.97	1.7762	7.4680	1.6461	1.6192	0.4922	30.51	85.04	25.55
1.98	1.7841	7.5849	1.6597	1.6239	0.4881	30.33	85.49	25.83
1.99	1.7920	7.7037	1.6735	1.6284	0.4841	30.17	85.94	26.10
2.00	1.8000	7.8244	1.6875	1.6330	0.4801	30.00	86.38	26.38
2.01	1.8080	7.9471	1.7016	1.6375	0.4761	29.84	86.82	26.66
2.02	1.8161	8.0718	1.7160	1.6420	0.4721	29.67	87.26	26.93
2.03	1.8242	8.1985	1.7305	1.6465	0.4682	29.51	87.69	27.20
2.04	1.8323	8.3273	1.7451	1.6509	0.4643	29.35	88.12	27.48
2.05	1.8405	8.4581	1.7600	1.6553	0.4603	29.20	88.55	27.75
2.06	1.8487	8.5911	1.7750	1.6597	0.4564	29.04	88.98	28.02
2.07	1.8570	8.7262	1.7902	1.6640	0.4526	28.89	89.40	28.29
2.08	1.8653	8.8635	1.8056	1.6683	0.4487	28.74	89.82	28.56
2.09	1.8736	9.0030	1.8212	1.6726	0.4449	28.59	90.24	28.83
2.10	1.8820	9.1447	1.8369	1.6769	0.4411	28.44	90.66	29.10
2.11	1.8904	9.2887	1.8529	1.6811	0.4373	28.29	91.07	29.36
2.12	1.8989	9.4350	1.8690	1.6853	0.4335	28.14	91.49	29.63
2.13	1.9074	9.5836	1.8853	1.6895	0.4297	28.00	91.90	29.90
2.14	1.9159	9.7347	1.9018	1.6936	0.4260	27.86	92.30	30.16
2.15	1.9245	9.8881	1.9185	1.6977	0.4223	27.72	92.71	30.43
2.16	1.9331	10.0440	1.9354	1.7018	0.4186	27.58	93.11	30.69
2.17	1.9418	10.2023	1.9525	1.7059	0.4149	27.44	93.51	30.95
2.18	1.9505	10.3632	1.9698	1.7099	0.4113	27.30	93.91	31.21
2.19	1.9592	10.5267	1.9873	1.7139	0.4077	27.17	94.30	31.47
2.20	1.9680	10.6927	2.0050	1.7179	0.4041	27.04	94.70	31.73
2.21	1.9768	10.8614	2.0229	1.7219	0.4005	26.90	95.09	31.99
2.22	1.9857	11.0327	2.0409	1.7258	0.3970	26.77	95.48	32.25
2.23	1.9946	11.2068	2.0592	1.7297	0.3934	26.64	95.86	32.51
2.24	2.0035	11.3836	2.0777	1.7336	0.3899	26.51	96.25	32.76
2.25	2.0125	11.5631	2.0964	1.7374	0.3865	26.39	96.63	33.02
2.26	2.0215	11.7455	2.1153	1.7412	0.3830	26.26	97.01	33.27
2.27	2.0306	11.9308	2.1345	1.7450	0.3796	26.14	97.39	33.53
2.28	2.0397	12.1190	2.1538	1.7488	0.3762	26.01	97.77	33.78
2.29	2.0488	12.3101	2.1734	1.7526	0.3728	25.89	98.14	34.03
2.30	2.0580	12.5043	2.1931	1.7563	0.3694	25.77	98.51	34.28
2.31	2.0672	12.7015	2.2131	1.7600	0.3661	25.65	98.88	34.53
2.32	2.0765	12.9017	2.2333	1.7637	0.3628	25.53	99.25	34.78
2.33	2.0858	13.1051	2.2538	1.7673	0.3595	25.42	99.62	35.03
2.34	2.0951	13.3116	2.2744	1.7709	0.3562	25.30	99.98	35.28
2.35	2.1045	13.5214	2.2953	1.7745	0.3530	25.18	100.34	35.53
2.36	2.1139	13.7344	2.3164	1.7781	0.3498	25.07	100.70	35.77
2.37	2.1234	13.9507	2.3377	1.7817	0.3466	24.96	101.06	36.02
2.38	2.1329	14.1704	2.3593	1.7852	0.3434	24.85	101.42	36.26
2.39	2.1424	14.3935	2.3811	1.7887	0.3403	24.73	101.77	36.50
2.40	2.1520	14.6200	2.4031	1.7922	0.3371	24.62	102.12	36.75
2.41	2.1616	14.8500	2.4254	1.7956	0.3340	24.52	102.47	36.99
2.42	2.1713	15.0836	2.4479	1.7991	0.3310	24.41	102.82	37.23
2.43	2.1810	15.3208	2.4706	1.8025	0.3279	24.30	103.17	37.47
2.44	2.1907	15.5616	2.4936	1.8059	0.3249	24.19	103.51	37.71

M	T_o/T	p_o/p	A/A_*	q/a_*	MN	μ (for supersonic flow only)	θ (for supersonic flow only)	δ (for supersonic flow only)
2.45	2.2005	15.8061	2.5168	1.8092	0.3219	24.09	103.86	37.95
2.46	2.2103	16.0544	2.5403	1.8126	0.3189	23.99	104.20	38.18
2.47	2.2202	16.3064	2.5640	1.8159	0.3160	23.88	104.54	38.42
2.48	2.2301	16.5623	2.5880	1.8192	0.3131	23.78	104.88	38.66
2.49	2.2400	16.8222	2.6122	1.8225	0.3101	23.68	105.21	38.89
2.50	2.2500	17.086	2.6367	1.8257	0.3073	23.58	105.55	39.12
2.51	2.2600	17.354	2.6615	1.8290	0.3044	23.48	105.88	39.36
2.52	2.2701	17.626	2.6865	1.8322	0.3016	23.38	106.21	39.59
2.53	2.2802	17.902	2.7117	1.8354	0.2988	23.28	106.54	39.82
2.54	2.2903	18.182	2.7372	1.8386	0.2960	23.18	106.87	40.05
2.55	2.3005	18.466	2.7630	1.8417	0.2932	23.09	107.19	40.28
2.56	2.3107	18.755	2.7891	1.8448	0.2905	22.99	107.52	40.51
2.57	2.3210	19.048	2.8154	1.8479	0.2878	22.90	107.84	40.74
2.58	2.3313	19.346	2.8420	1.8510	0.2851	22.81	108.16	40.96
2.59	2.3416	19.648	2.8688	1.8541	0.2824	22.71	108.48	41.19
2.60	2.3520	19.954	2.8960	1.8571	0.2798	22.62	108.79	41.41
2.61	2.3624	20.265	2.9234	1.8602	0.2771	22.53	109.11	41.64
2.62	2.3729	20.581	2.9511	1.8632	0.2745	22.44	109.43	41.86
2.63	2.3834	20.901	2.9791	1.8662	0.2720	22.35	109.74	42.09
2.64	2.3939	21.227	3.0073	1.8691	0.2694	22.26	110.05	42.31
2.65	2.4045	21.557	3.0359	1.8721	0.2669	22.17	110.36	42.53
2.66	2.4151	21.892	3.0647	1.8750	0.2644	22.08	110.67	42.75
2.67	2.4258	22.232	3.0938	1.8779	0.2619	22.00	110.97	42.97
2.68	2.4365	22.577	3.1233	1.8808	0.2594	21.91	111.28	43.19
2.69	2.4472	22.927	3.1530	1.8837	0.2570	21.82	111.58	43.40
2.70	2.4580	23.283	3.1830	1.8865	0.2545	21.74	111.88	43.62
2.71	2.4688	23.644	3.2133	1.8894	0.2521	21.65	112.18	43.84
2.72	2.4797	24.010	3.2440	1.8922	0.2498	21.57	112.48	44.05
2.73	2.4906	24.381	3.2749	1.8950	0.2474	21.49	112.78	44.27
2.74	2.5015	24.758	3.3061	1.8978	0.2451	21.41	113.08	44.48
2.75	2.5125	25.140	3.3377	1.9005	0.2427	21.32	113.37	44.69
2.76	2.5235	25.528	3.3695	1.9033	0.2404	21.24	113.66	44.91
2.77	2.5346	25.922	3.4017	1.9060	0.2382	21.16	113.95	45.12
2.78	2.5457	26.322	3.4342	1.9087	0.2359	21.08	114.24	45.33
2.79	2.5568	26.727	3.4670	1.9114	0.2337	21.00	114.53	45.54
2.80	2.5680	27.138	3.5001	1.9140	0.2315	20.92	114.82	45.75
2.81	2.5792	27.556	3.5336	1.9167	0.2293	20.85	115.11	45.95
2.82	2.5905	27.979	3.5674	1.9193	0.2271	20.77	115.39	46.16
2.83	2.6018	28.408	3.6015	1.9219	0.2250	20.69	115.67	46.37
2.84	2.6131	28.844	3.6359	1.9246	0.2228	20.62	115.96	46.57
2.85	2.6245	29.286	3.6707	1.9271	0.2207	20.54	116.24	46.78
2.86	2.6359	29.735	3.7058	1.9297	0.2186	20.47	116.52	46.98
2.87	2.6474	30.190	3.7413	1.9323	0.2166	20.39	116.79	47.19
2.88	2.6589	30.651	3.7771	1.9348	0.2145	20.32	117.07	47.39
2.89	2.6704	31.119	3.8133	1.9373	0.2125	20.24	117.35	47.59
2.90	2.6820	31.594	3.8498	1.9398	0.2105	20.17	117.62	47.79
2.91	2.6936	32.076	3.8866	1.9423	0.2085	20.10	117.89	47.99
2.92	2.7053	32.564	3.9238	1.9448	0.2065	20.03	118.16	48.19
2.93	2.7170	33.060	3.9614	1.9472	0.2045	19.96	118.43	48.39
2.94	2.7287	33.563	3.9993	1.9497	0.2026	19.89	118.70	48.59
2.95	2.7405	34.073	4.0376	1.9521	0.2007	19.81	118.97	48.78
2.96	2.7523	34.590	4.0763	1.9545	0.1988	19.75	119.23	48.98
2.97	2.7642	35.114	4.1153	1.9569	0.1969	19.68	119.50	49.18
2.98	2.7761	35.646	4.1547	1.9593	0.1950	19.61	119.76	49.37
2.99	2.7880	36.186	4.1944	1.9616	0.1932	19.54	120.03	49.56

M	T_o/T	P_o/p	A/A_*	q/a_*	MN	μ (for supersonic flow only)	θ	δ
3.00	2.8000	36.733	4.2346	1.9640	0.1913	19.47	120.29	49.76
3.01	2.8120	37.288	4.2751	1.9663	0.1895	19.40	120.55	49.95
3.02	2.8241	37.850	4.3160	1.9686	0.1877	19.34	120.80	50.14
3.03	2.8362	38.421	4.3573	1.9709	0.1859	19.27	121.06	50.33
3.04	2.8483	39.000	4.3989	1.9732	0.1842	19.20	121.32	50.52
3.05	2.8605	39.586	4.4410	1.9755	0.1824	19.14	121.57	50.71
3.06	2.8727	40.182	4.4835	1.9777	0.1807	19.07	121.83	50.90
3.07	2.8850	40.785	4.5263	1.9800	0.1790	19.01	122.08	51.09
3.08	2.8973	41.397	4.5696	1.9822	0.1773	18.95	122.33	51.28
3.09	2.9096	42.017	4.6132	1.9844	0.1756	18.88	122.58	51.46
3.10	2.9220	42.646	4.6573	1.9866	0.1740	18.82	122.83	51.65
3.11	2.9344	43.284	4.7018	1.9888	0.1723	18.76	123.08	51.83
3.12	2.9469	43.931	4.7467	1.9910	0.1707	18.69	123.33	52.02
3.13	2.9594	44.586	4.7920	1.9931	0.1691	18.63	123.57	52.20
3.14	2.9719	45.251	4.8377	1.9953	0.1675	18.57	123.82	52.39
3.15	2.9845	45.925	4.8838	1.9974	0.1659	18.51	124.06	52.57
3.16	2.9971	46.608	4.9304	1.9995	0.1643	18.45	124.30	52.75
3.17	3.0098	47.301	4.9774	2.0016	0.1628	18.39	124.54	52.93
3.18	3.0225	48.003	5.0248	2.0037	0.1612	18.33	124.78	53.11
3.19	3.0352	48.715	5.0727	2.0058	0.1597	18.27	125.02	53.29
3.20	3.0480	49.437	5.1210	2.0079	0.1582	18.21	125.26	53.47
3.21	3.0608	50.169	5.1697	2.0099	0.1567	18.15	125.50	53.65
3.22	3.0737	50.910	5.2189	2.0119	0.1552	18.09	125.73	53.83
3.23	3.0866	51.662	5.2685	2.0140	0.1538	18.03	125.97	54.00
3.24	3.0995	52.424	5.3186	2.0160	0.1523	17.98	126.20	54.18
3.25	3.1125	53.196	5.3691	2.0180	0.1509	17.92	126.43	54.35
3.26	3.1255	53.979	5.4201	2.0200	0.1495	17.86	126.67	54.53
3.27	3.1386	54.773	5.4715	2.0220	0.1481	17.81	126.90	54.70
3.28	3.1517	55.577	5.5234	2.0239	0.1467	17.75	127.13	54.88
3.29	3.1648	56.393	5.5758	2.0259	0.1453	17.70	127.35	55.05
3.30	3.1780	57.219	5.6286	2.0278	0.1439	17.64	127.58	55.22
3.31	3.1912	58.056	5.6820	2.0297	0.1426	17.58	127.81	55.39
3.32	3.2045	58.905	5.7358	2.0317	0.1413	17.53	128.03	55.56
3.33	3.2178	59.765	5.7900	2.0336	0.1399	17.48	128.26	55.73
3.34	3.2311	60.637	5.8448	2.0355	0.1386	17.42	128.48	55.90
3.35	3.2445	61.520	5.9000	2.0373	0.1373	17.37	128.70	56.07
3.36	3.2579	62.415	5.9558	2.0392	0.1360	17.31	128.93	56.24
3.37	3.2714	63.323	6.0120	2.0411	0.1348	17.26	129.15	56.41
3.38	3.2849	64.242	6.0687	2.0429	0.1335	17.21	129.37	56.58
3.39	3.2984	65.173	6.1260	2.0447	0.1323	17.16	129.59	56.74
3.40	3.3120	66.117	6.1837	2.0466	0.1310	17.10	129.80	56.91
3.41	3.3256	67.074	6.2419	2.0484	0.1298	17.05	130.02	57.07
3.42	3.3393	68.043	6.3007	2.0502	0.1286	17.00	130.24	57.24
3.43	3.3530	69.025	6.3600	2.0520	0.1274	16.95	130.45	57.40
3.44	3.3667	70.020	6.4198	2.0537	0.1262	16.90	130.66	57.56
3.45	3.3805	71.029	6.4801	2.0555	0.1250	16.85	130.88	57.73
3.46	3.3943	72.050	6.5409	2.0573	0.1239	16.80	131.09	57.89
3.47	3.4082	73.085	6.6023	2.0590	0.1227	16.75	131.30	58.05
3.48	3.4221	74.134	6.6642	2.0607	0.1216	16.70	131.51	58.21
3.49	3.4360	75.196	6.7266	2.0625	0.1204	16.65	131.72	58.37
3.50	3.4500	76.272	6.7896	2.0642	0.1193	16.60	131.93	58.53
3.51	3.4640	77.363	6.8532	2.0659	0.1182	16.55	132.14	58.69
3.52	3.4781	78.467	6.9172	2.0676	0.1171	16.50	132.34	58.85
3.53	3.4922	79.586	6.9819	2.0693	0.1160	16.46	132.55	59.00
3.54	3.5063	80.720	7.0471	2.0709	0.1150	16.41	132.75	59.16

M	T_o/T	P_o/p	A/A_*	q/a_*	MN	μ (for supersonic flow only)	θ	δ
3.55	3.5205	81.868	7.1128	2.0726	0.1139	16.36	132.96	59.32
3.56	3.5347	83.031	7.1791	2.0743	0.1129	16.31	133.16	59.47
3.57	3.5490	84.210	7.2460	2.0759	0.1118	16.27	133.36	59.63
3.58	3.5633	85.403	7.3135	2.0775	0.1108	16.22	133.56	59.78
3.59	3.5776	86.612	7.3815	2.0792	0.1098	16.17	133.76	59.94
3.60	3.5920	87.837	7.4501	2.0808	0.1087	16.13	133.96	60.09
3.61	3.6064	89.077	7.5193	2.0824	0.1077	16.08	134.16	60.24
3.62	3.6209	90.334	7.5891	2.0840	0.1068	16.04	134.36	60.40
3.63	3.6354	91.606	7.6595	2.0856	0.1058	15.99	134.56	60.55
3.64	3.6499	92.895	7.7305	2.0871	0.1048	15.95	134.75	60.70
3.65	3.6645	94.200	7.8020	2.0887	0.1038	15.90	134.95	60.85
3.66	3.6791	95.522	7.8742	2.0903	0.1029	15.86	135.14	61.00
3.67	3.6938	96.861	7.9470	2.0918	0.1019	15.81	135.34	61.15
3.68	3.7085	98.217	8.0204	2.0933	0.1010	15.77	135.53	61.30
3.69	3.7232	99.590	8.0944	2.0949	0.1001	15.72	135.72	61.45
3.70	3.7380	100.981	8.1691	2.0964	0.0992	15.68	135.91	61.60
3.71	3.7528	102.389	8.2443	2.0979	0.0983	15.64	136.11	61.74
3.72	3.7677	103.815	8.3202	2.0994	0.0974	15.59	136.30	61.89
3.73	3.7826	105.259	8.3968	2.1009	0.0965	15.55	136.48	62.04
3.74	3.7975	106.721	8.4739	2.1024	0.0956	15.51	136.67	62.18
3.75	3.8125	108.202	8.5517	2.1039	0.0947	15.47	136.86	62.33
3.76	3.8275	109.701	8.6302	2.1053	0.0939	15.42	137.05	62.47
3.77	3.8426	111.219	8.7093	2.1068	0.0930	15.38	137.23	62.61
3.78	3.8577	112.757	8.7891	2.1082	0.0922	15.34	137.42	62.76
3.79	3.8728	114.313	8.8695	2.1097	0.0913	15.30	137.60	62.90
3.80	3.8880	115.889	8.9506	2.1111	0.0905	15.26	137.79	63.04
3.81	3.9032	117.485	9.0323	2.1125	0.0897	15.22	137.97	63.19
3.82	3.9185	119.100	9.1148	2.1140	0.0889	15.18	138.15	63.33
3.83	3.9338	120.736	9.1979	2.1154	0.0881	15.14	138.33	63.47
3.84	3.9491	122.391	9.2817	2.1168	0.0873	15.09	138.51	63.61
3.85	3.9645	124.068	9.3661	2.1182	0.0865	15.05	138.69	63.75
3.86	3.9799	125.765	9.4513	2.1195	0.0857	15.01	138.87	63.89
3.87	3.9954	127.483	9.5372	2.1209	0.0850	14.98	139.05	64.03
3.88	4.0109	129.223	9.6237	2.1223	0.0842	14.94	139.23	64.16
3.89	4.0264	130.984	9.7110	2.1236	0.0834	14.90	139.41	64.30
3.90	4.0420	132.766	9.7990	2.1250	0.0827	14.86	139.58	64.44
3.91	4.0576	134.570	9.8877	2.1263	0.0819	14.82	139.76	64.58
3.92	4.0733	136.397	9.9771	2.1277	0.0812	14.78	139.93	64.71
3.93	4.0890	138.246	10.0672	2.1290	0.0805	14.74	140.11	64.85
3.94	4.1047	140.118	10.1581	2.1303	0.0798	14.70	140.28	64.98
3.95	4.1205	142.012	10.2496	2.1316	0.0790	14.66	140.45	65.12
3.96	4.1363	143.929	10.3420	2.1329	0.0783	14.63	140.63	65.25
3.97	4.1522	145.870	10.4350	2.1342	0.0776	14.59	140.80	65.39
3.98	4.1681	147.835	10.5289	2.1355	0.0769	14.55	140.97	65.52
3.99	4.1840	149.823	10.6234	2.1368	0.0763	14.51	141.14	65.65
4.00	4.2000	151.835	10.719	2.1381	0.0756	14.48	141.31	65.78
4.01	4.2160	153.872	10.815	2.1394	0.0749	14.44	141.48	65.92
4.02	4.2321	155.933	10.912	2.1406	0.0742	14.40	141.64	66.05
4.03	4.2482	158.019	11.009	2.1419	0.0736	14.37	141.81	66.18
4.04	4.2643	160.131	11.108	2.1431	0.0729	14.33	141.98	66.31
4.05	4.2805	162.267	11.207	2.1444	0.0723	14.29	142.14	66.44
4.06	4.2967	164.429	11.307	2.1456	0.0717	14.26	142.31	66.57
4.07	4.3130	166.618	11.408	2.1468	0.0710	14.22	142.47	66.70
4.08	4.3293	168.832	11.509	2.1480	0.0704	14.19	142.64	66.83
4.09	4.3456	171.073	11.611	2.1493	0.0698	14.15	142.80	66.95

M	T_o/T	p_o/p	A/A_*	q/a_*	MN	μ (for supersonic flow only)	θ	δ
4.10	4.3620	173.340	11.715	2.1505	0.0692	14.12	142.97	67.08
4.11	4.3784	175.635	11.819	2.1517	0.0686	14.08	143.13	67.21
4.12	4.3949	177.957	11.923	2.1529	0.0679	14.05	143.29	67.34
4.13	4.4114	180.306	12.029	2.1540	0.0674	14.01	143.45	67.46
4.14	4.4279	182.683	12.135	2.1552	0.0668	13.98	143.61	67.59
4.15	4.4445	185.089	12.243	2.1564	0.0662	13.94	143.77	67.71
4.16	4.4611	187.523	12.351	2.1576	0.0656	13.91	143.93	67.84
4.17	4.4778	189.985	12.460	2.1587	0.0650	13.88	144.09	67.96
4.18	4.4945	192.477	12.570	2.1599	0.0645	13.84	144.25	68.09
4.19	4.5112	194.997	12.680	2.1610	0.0639	13.81	144.40	68.21
4.20	4.5280	197.548	12.792	2.1622	0.0633	13.77	144.56	68.33
4.21	4.5448	200.128	12.904	2.1633	0.0628	13.74	144.72	68.46
4.22	4.5617	202.739	13.017	2.1644	0.0622	13.71	144.87	68.58
4.23	4.5786	205.380	13.131	2.1655	0.0617	13.67	145.03	68.70
4.24	4.5955	208.052	13.246	2.1667	0.0612	13.64	145.18	68.82
4.25	4.6125	210.755	13.362	2.1678	0.0606	13.61	145.33	68.94
4.26	4.6295	213.489	13.479	2.1689	0.0601	13.58	145.49	69.06
4.27	4.6466	216.255	13.597	2.1700	0.0596	13.54	145.64	69.18
4.28	4.6637	219.054	13.715	2.1711	0.0591	13.51	145.79	69.30
4.29	4.6808	221.884	13.835	2.1721	0.0586	13.48	145.94	69.42
4.30	4.6980	224.748	13.955	2.1732	0.0581	13.45	146.09	69.54
4.31	4.7152	227.644	14.076	2.1743	0.0576	13.42	146.24	69.66
4.32	4.7325	230.574	14.198	2.1754	0.0571	13.38	146.39	69.78
4.33	4.7498	233.538	14.322	2.1764	0.0566	13.35	146.54	69.89
4.34	4.7671	236.535	14.446	2.1775	0.0561	13.32	146.69	70.01
4.35	4.7845	239.568	14.571	2.1785	0.0556	13.29	146.84	70.13
4.36	4.8019	242.634	14.697	2.1796	0.0551	13.26	146.99	70.24
4.37	4.8194	245.736	14.823	2.1806	0.0547	13.23	147.13	70.36
4.38	4.8369	248.873	14.951	2.1816	0.0542	13.20	147.28	70.48
4.39	4.8544	252.047	15.080	2.1827	0.0537	13.17	147.42	70.59
4.40	4.8720	255.256	15.210	2.1837	0.0533	13.14	147.57	70.71
4.41	4.8896	258.501	15.341	2.1847	0.0528	13.11	147.71	70.82
4.42	4.9073	261.784	15.472	2.1857	0.0524	13.08	147.86	70.93
4.43	4.9250	265.104	15.605	2.1867	0.0519	13.05	148.00	71.05
4.44	4.9427	268.461	15.739	2.1877	0.0515	13.02	148.14	71.16
4.45	4.9605	271.856	15.873	2.1887	0.0510	12.99	148.29	71.27
4.46	4.9783	275.290	16.009	2.1897	0.0506	12.96	148.43	71.39
4.47	4.9962	278.762	16.146	2.1907	0.0502	12.93	148.57	71.50
4.48	5.0141	282.273	16.284	2.1917	0.0498	12.90	148.71	71.61
4.49	5.0320	285.824	16.422	2.1926	0.0493	12.87	148.85	71.72
4.50	5.0500	289.414	16.562	2.1936	0.0489	12.84	148.99	71.83
4.51	5.0680	293.045	16.703	2.1946	0.0485	12.81	149.13	71.94
4.52	5.0861	296.716	16.845	2.1955	0.0481	12.78	149.27	72.05
4.53	5.1042	300.428	16.988	2.1965	0.0477	12.75	149.41	72.16
4.54	5.1223	304.182	17.132	2.1974	0.0473	12.72	149.55	72.27

M	T_o/T	p_o/p	A/A_*	q/a_*	MN	μ (for supersonic flow only)	θ	δ
4.55	5.1405	307.977	17.277	2.1984	0.0469	12.70	149.68	72.38
4.56	5.1587	311.815	17.423	2.1993	0.0465	12.67	149.82	72.49
4.57	5.1770	315.695	17.570	2.2002	0.0461	12.64	149.96	72.60
4.58	5.1953	319.618	17.718	2.2012	0.0457	12.61	150.09	72.70
4.59	5.2136	323.585	17.867	2.2021	0.0453	12.58	150.23	72.81
4.60	5.2320	327.595	18.018	2.2030	0.0450	12.56	150.36	72.92
4.61	5.2504	331.649	18.169	2.2039	0.0446	12.53	150.50	73.03
4.62	5.2689	335.748	18.322	2.2048	0.0442	12.50	150.63	73.13
4.63	5.2874	339.893	18.476	2.2057	0.0439	12.47	150.77	73.24
4.64	5.3059	344.082	18.630	2.2066	0.0435	12.45	150.90	73.34
4.65	5.3245	348.318	18.786	2.2075	0.0431	12.42	151.03	73.45
4.66	5.3431	352.600	18.943	2.2084	0.0428	12.39	151.16	73.55
4.67	5.3618	356.929	19.101	2.2093	0.0424	12.36	151.29	73.66
4.68	5.3805	361.305	19.261	2.2102	0.0421	12.34	151.42	73.76
4.69	5.3992	365.728	19.421	2.2110	0.0417	12.31	151.56	73.87
4.70	5.4180	370.200	19.583	2.2119	0.0414	12.28	151.69	73.97
4.71	5.4368	374.720	19.746	2.2128	0.0410	12.26	151.82	74.07
4.72	5.4557	379.290	19.910	2.2136	0.0407	12.23	151.94	74.18
4.73	5.4746	383.909	20.075	2.2145	0.0404	12.21	152.07	74.28
4.74	5.4935	388.577	20.241	2.2154	0.0400	12.18	152.20	74.38
4.75	5.5125	393.297	20.408	2.2162	0.0397	12.15	152.33	74.48
4.76	5.5315	398.067	20.577	2.2170	0.0394	12.13	152.46	74.58
4.77	5.5506	402.888	20.747	2.2179	0.0391	12.10	152.58	74.68
4.78	5.5697	407.761	20.918	2.2187	0.0387	12.08	152.71	74.79
4.79	5.5888	412.687	21.090	2.2196	0.0384	12.05	152.84	74.89
4.80	5.6080	417.665	21.264	2.2204	0.0381	12.02	152.96	74.99
4.81	5.6272	422.696	21.438	2.2212	0.0378	12.00	153.09	75.09
4.82	5.6465	427.782	21.614	2.2220	0.0375	11.97	153.21	75.19
4.83	5.6658	432.921	21.792	2.2228	0.0372	11.95	153.34	75.28
4.84	5.6851	438.116	21.970	2.2236	0.0369	11.92	153.46	75.38
4.85	5.7045	443.365	22.150	2.2245	0.0366	11.90	153.58	75.48
4.86	5.7239	448.670	22.331	2.2253	0.0363	11.87	153.71	75.58
4.87	5.7434	454.032	22.513	2.2261	0.0360	11.85	153.83	75.68
4.88	5.7629	459.450	22.696	2.2268	0.0357	11.82	153.95	75.78
4.89	5.7824	464.926	22.881	2.2276	0.0354	11.80	154.07	75.87
4.90	5.8020	470.459	23.067	2.2284	0.0351	11.78	154.19	75.97
4.91	5.8216	476.051	23.254	2.2292	0.0348	11.75	154.31	76.07
4.92	5.8413	481.702	23.443	2.2300	0.0346	11.73	154.43	76.16
4.93	5.8610	487.412	23.633	2.2308	0.0343	11.70	154.55	76.26
4.94	5.8807	493.181	23.824	2.2315	0.0340	11.68	154.67	76.35
4.95	5.9005	499.012	24.017	2.2323	0.0337	11.66	154.79	76.45
4.96	5.9203	504.903	24.211	2.2331	0.0335	11.63	154.91	76.54
4.97	5.9402	510.856	24.406	2.2338	0.0332	11.61	155.03	76.64
4.98	5.9601	516.871	24.603	2.2346	0.0329	11.58	155.15	76.73
4.99	5.9800	522.949	24.801	2.2353	0.0327	11.56	155.27	76.83
5.00	6.0000	529.090	25.000	2.2361	0.0324	11.54	155.38	76.92

TABLE 2 STEADY ADIABATIC FLOW OF AIR THROUGH A PLANE NORMAL SHOCK WAVE

Flow and fluid parameters are calculated with the incident Mach number (M_1) as the independent variable for the range, $M_1 = 1{\cdot}0 \rightarrow (0{\cdot}01) \rightarrow 5$. The summary of equations used is given in Section 4·2. The values of the following parameters are tabulated:

T_2/T_1 static temperature ratio across the shock
p_2/p_1 static pressure ratio across the shock
$q_2/q_1 = \rho_1/\rho_2$, velocity ratio
M_2 Mach number downstream of shock
p_{o2}/p_{o1} total head pressure ratio across the shock
$(s_2 - s_1)/R$ parameter for entropy change across the shock
(Note: (4·14) gives the simple logarithmic relation between the last two parameters)

M_1	T_2/T_1	p_2/p_1	$q_2/q_1 = \rho_1/\rho_2$	M_2	p_{o2}/p_{o1}	$(s_2 - s_1)/R$
1.00	1.0000	1.0000	1.0000	1.0000	1.0000	0.0000
1.01	1.0066	1.0234	0.9836	0.9901	1.0000	+1.2708$_{10}$ -6
1.02	1.0132	1.0471	0.9676	0.9805	1.0000	+9.9682$_{10}$ -6
1.03	1.0198	1.0710	0.9522	0.9712	1.0000	+3.2995$_{10}$ -5
1.04	1.0263	1.0952	0.9371	0.9620	0.9999	+7.6720$_{10}$ -5
1.05	1.0328	1.1196	0.9225	0.9531	0.9999	+1.4702$_{10}$ -4
1.06	1.0393	1.1442	0.9083	0.9444	0.9998	+2.4931$_{10}$ -4
1.07	1.0458	1.1690	0.8945	0.9360	0.9996	+3.8859$_{10}$ -4
1.08	1.0522	1.1941	0.8811	0.9277	0.9994	+5.6946$_{10}$ -4
1.09	1.0586	1.2194	0.8681	0.9196	0.9992	+7.9615$_{10}$ -4
1.10	1.0649	1.2450	0.8554	0.9118	0.9989	+1.0726$_{10}$ -3
1.11	1.0713	1.2708	0.8430	0.9041	0.9986	+1.4023$_{10}$ -3
1.12	1.0776	1.2968	0.8310	0.8966	0.9982	+1.7885$_{10}$ -3
1.13	1.0840	1.3230	0.8193	0.8892	0.9978	+2.2344$_{10}$ -3
1.14	1.0903	1.3495	0.8079	0.8820	0.9973	+2.7426$_{10}$ -3
1.15	1.0966	1.3762	0.7968	0.8750	0.9967	+3.3157$_{10}$ -3
1.16	1.1029	1.4032	0.7860	0.8682	0.9961	+3.9560$_{10}$ -3
1.17	1.1092	1.4304	0.7754	0.8615	0.9953	+4.6655$_{10}$ -3
1.18	1.1154	1.4578	0.7652	0.8549	0.9946	+5.4462$_{10}$ -3
1.19	1.1217	1.4854	0.7551	0.8485	0.9937	+6.2997$_{10}$ -3
1.20	1.1280	1.5133	0.7454	0.8422	0.9928	+7.2277$_{10}$ -3
1.21	1.1343	1.5414	0.7358	0.8360	0.9918	+8.2314$_{10}$ -3
1.22	1.1405	1.5698	0.7266	0.8300	0.9907	+9.3123$_{10}$ -3
1.23	1.1468	1.5984	0.7175	0.8241	0.9896	+1.0471$_{10}$ -2
1.24	1.1531	1.6272	0.7086	0.8183	0.9884	+1.1710$_{10}$ -2
1.25	1.1594	1.6562	0.7000	0.8126	0.9871	+1.3028$_{10}$ -2
1.26	1.1657	1.6855	0.6916	0.8071	0.9857	+1.4427$_{10}$ -2
1.27	1.1720	1.7150	0.6833	0.8016	0.9842	+1.5907$_{10}$ -2
1.28	1.1783	1.7448	0.6753	0.7963	0.9827	+1.7470$_{10}$ -2
1.29	1.1846	1.7748	0.6674	0.7911	0.9811	+1.9114$_{10}$ -2
1.30	1.1909	1.8050	0.6598	0.7860	0.9794	+2.0842$_{10}$ -2
1.31	1.1972	1.8354	0.6523	0.7809	0.9776	+2.2653$_{10}$ -2
1.32	1.2035	1.8661	0.6449	0.7760	0.9758	+2.4547$_{10}$ -2
1.33	1.2099	1.8970	0.6378	0.7712	0.9738	+2.6524$_{10}$ -2
1.34	1.2162	1.9282	0.6308	0.7664	0.9718	+2.8586$_{10}$ -2
1.35	1.2226	1.9596	0.6239	0.7618	0.9697	+3.0730$_{10}$ -2
1.36	1.2290	1.9912	0.6172	0.7572	0.9676	+3.2959$_{10}$ -2
1.37	1.2354	2.0230	0.6107	0.7527	0.9653	+3.5270$_{10}$ -2
1.38	1.2418	2.0551	0.6042	0.7483	0.9630	+3.7665$_{10}$ -2
1.39	1.2482	2.0874	0.5980	0.7440	0.9607	+4.0143$_{10}$ -2

M_1	T_2/T_1	p_2/p_1	$q_2/q_1 = \rho_1/\rho_2$	M_2	p_{o2}/p_{o1}	$(s_2 - s_1)/R$
1.40	1.2547	2.1200	0.5918	0.7397	0.9582	+4.2705$_{10}$ -2
1.41	1.2612	2.1528	0.5858	0.7355	0.9557	+4.5348$_{10}$ -2
1.42	1.2676	2.1858	0.5799	0.7314	0.9531	+4.8074$_{10}$ -2
1.43	1.2741	2.2190	0.5742	0.7274	0.9504	+5.0883$_{10}$ -2
1.44	1.2807	2.2525	0.5685	0.7235	0.9476	+5.3772$_{10}$ -2
1.45	1.2872	2.2862	0.5630	0.7196	0.9448	+5.6743$_{10}$ -2
1.46	1.2938	2.3202	0.5576	0.7157	0.9420	+5.9795$_{10}$ -2
1.47	1.3003	2.3544	0.5523	0.7120	0.9390	+6.2927$_{10}$ -2
1.48	1.3069	2.3888	0.5471	0.7083	0.9360	+6.6139$_{10}$ -2
1.49	1.3136	2.4234	0.5420	0.7047	0.9329	+6.9430$_{10}$ -2
1.50	1.3202	2.4583	0.5370	0.7011	0.9298	+7.2800$_{10}$ -2
1.51	1.3269	2.4934	0.5321	0.6976	0.9266	+7.6249$_{10}$ -2
1.52	1.3336	2.5288	0.5274	0.6941	0.9233	+7.9775$_{10}$ -2
1.53	1.3403	2.5644	0.5227	0.6907	0.9200	+8.3378$_{10}$ -2
1.54	1.3470	2.6002	0.5180	0.6874	0.9166	+8.7058$_{10}$ -2
1.55	1.3538	2.6362	0.5135	0.6841	0.9132	+9.0813$_{10}$ -2
1.56	1.3606	2.6725	0.5091	0.6809	0.9097	+9.4644$_{10}$ -
1.57	1.3674	2.7090	0.5047	0.6777	0.9062	+9.8550$_{10}$ -2
1.58	1.3742	2.7458	0.5005	0.6746	0.9026	+1.0253$_{10}$ -1
1.59	1.3811	2.7828	0.4963	0.6715	0.8989	+1.0658$_{10}$ -1
1.60	1.3880	2.8200	0.4922	0.6684	0.8952	+1.1071$_{10}$ -1
1.61	1.3949	2.8574	0.4882	0.6655	0.8915	+1.1491$_{10}$ -1
1.62	1.4018	2.8951	0.4842	0.6625	0.8877	+1.1917$_{10}$ -1
1.63	1.4088	2.9330	0.4803	0.6596	0.8838	+1.2351$_{10}$ -1
1.64	1.4158	2.9712	0.4765	0.6568	0.8799	+1.2792$_{10}$ -1
1.65	1.4228	3.0096	0.4728	0.6540	0.8760	+1.3240$_{10}$ -1
1.66	1.4299	3.0482	0.4691	0.6512	0.8720	+1.3695$_{10}$ -1
1.67	1.4369	3.0870	0.4655	0.6485	0.8680	+1.4157$_{10}$ -1
1.68	1.4440	3.1261	0.4619	0.6458	0.8639	+1.4625$_{10}$ -1
1.69	1.4512	3.1654	0.4584	0.6431	0.8599	+1.5100$_{10}$ -1
1.70	1.4583	3.2050	0.4550	0.6405	0.8557	+1.5581$_{10}$ -1
1.71	1.4655	3.2448	0.4517	0.6380	0.8516	+1.6069$_{10}$ -1
1.72	1.4727	3.2848	0.4484	0.6355	0.8474	+1.6563$_{10}$ -1
1.73	1.4800	3.3250	0.4451	0.6330	0.8431	+1.7064$_{10}$ -1
1.74	1.4873	3.3655	0.4419	0.6305	0.8389	+1.7571$_{10}$ -1
1.75	1.4946	3.4062	0.4388	0.6281	0.8346	+1.8084$_{10}$ -1
1.76	1.5019	3.4472	0.4357	0.6257	0.8302	+1.8604$_{10}$ -1
1.77	1.5093	3.4884	0.4327	0.6234	0.8259	+1.9129$_{10}$ -1
1.78	1.5167	3.5298	0.4297	0.6210	0.8215	+1.9661$_{10}$ -1
1.79	1.5241	3.5714	0.4268	0.6188	0.8171	+2.0198$_{10}$ -1

M_1	T_2/T_1	p_2/p_1	$q_2/q_1 = \rho_1/\rho_2$	M_2	p_{o2}/p_{o1}	$(s_2 - s_1)/R$
1.80	1.5316	3.6133	0.4239	0.6165	0.8127	+2.0741$_{10}$ -1
1.81	1.5391	3.6554	0.4210	0.6143	0.8082	+2.1290$_{10}$ -1
1.82	1.5466	3.6978	0.4182	0.6121	0.8038	+2.1845$_{10}$ -1
1.83	1.5541	3.7404	0.4155	0.6099	0.7993	+2.2405$_{10}$ -1
1.84	1.5617	3.7832	0.4128	0.6078	0.7948	+2.2971$_{10}$ -1
1.85	1.5693	3.8262	0.4102	0.6057	0.7902	+2.3543$_{10}$ -1
1.86	1.5770	3.8695	0.4075	0.6036	0.7857	+2.4120$_{10}$ -1
1.87	1.5847	3.9130	0.4050	0.6016	0.7811	+2.4702$_{10}$ -1
1.88	1.5924	3.9568	0.4024	0.5996	0.7765	+2.5290$_{10}$ -1
1.89	1.6001	4.0008	0.4000	0.5976	0.7720	+2.5882$_{10}$ -1
1.90	1.6079	4.0450	0.3975	0.5956	0.7674	+2.6480$_{10}$ -1
1.91	1.6157	4.0894	0.3951	0.5937	0.7627	+2.7083$_{10}$ -1
1.92	1.6236	4.1341	0.3927	0.5918	0.7581	+2.7691$_{10}$ -1
1.93	1.6314	4.1790	0.3904	0.5899	0.7535	+2.8305$_{10}$ -1
1.94	1.6394	4.2242	0.3881	0.5880	0.7488	+2.8922$_{10}$ -1
1.95	1.6473	4.2696	0.3858	0.5862	0.7442	+2.9545$_{10}$ -1
1.96	1.6553	4.3152	0.3836	0.5844	0.7395	+3.0173$_{10}$ -1
1.97	1.6633	4.3610	0.3814	0.5826	0.7349	+3.0805$_{10}$ -1
1.98	1.6713	4.4071	0.3792	0.5808	0.7302	+3.1442$_{10}$ -1
1.99	1.6794	4.4534	0.3771	0.5791	0.7255	+3.2083$_{10}$ -1
2.00	1.6875	4.5000	0.3750	0.5774	0.7209	+3.2729$_{10}$ -1
2.01	1.6956	4.5468	0.3729	0.5757	0.7162	+3.3379$_{10}$ -1
2.02	1.7038	4.5938	0.3709	0.5740	0.7115	+3.4034$_{10}$ -1
2.03	1.7120	4.6410	0.3689	0.5723	0.7069	+3.4693$_{10}$ -1
2.04	1.7203	4.6885	0.3669	0.5707	0.7022	+3.5357$_{10}$ -1
2.05	1.7285	4.7362	0.3650	0.5691	0.6975	+3.6024$_{10}$ -1
2.06	1.7369	4.7842	0.3630	0.5675	0.6928	+3.6696$_{10}$ -1
2.07	1.7452	4.8324	0.3611	0.5659	0.6882	+3.7371$_{10}$ -1
2.08	1.7536	4.8808	0.3593	0.5643	0.6835	+3.8051$_{10}$ -1
2.09	1.7620	4.9294	0.3574	0.5628	0.6789	+3.8735$_{10}$ -1
2.10	1.7705	4.9783	0.3556	0.5613	0.6742	+3.9422$_{10}$ -1
2.11	1.7789	5.0274	0.3538	0.5598	0.6696	+4.0114$_{10}$ -1
2.12	1.7875	5.0768	0.3521	0.5583	0.6649	+4.0809$_{10}$ -1
2.13	1.7960	5.1264	0.3503	0.5568	0.6603	+4.1508$_{10}$ -1
2.14	1.8046	5.1762	0.3486	0.5554	0.6557	+4.2210$_{10}$ -1
2.15	1.8132	5.2262	0.3469	0.5540	0.6511	+4.2917$_{10}$ -1
2.16	1.8219	5.2765	0.3453	0.5525	0.6464	+4.3626$_{10}$ -1
2.17	1.8306	5.3270	0.3436	0.5511	0.6419	+4.4340$_{10}$ -1
2.18	1.8393	5.3778	0.3420	0.5498	0.6373	+4.5056$_{10}$ -1
2.19	1.8481	5.4288	0.3404	0.5484	0.6327	+4.5776$_{10}$ -1
2.20	1.8569	5.4800	0.3388	0.5471	0.6281	+4.6500$_{10}$ -1
2.21	1.8657	5.5314	0.3373	0.5457	0.6236	+4.7227$_{10}$ -1
2.22	1.8746	5.5831	0.3358	0.5444	0.6191	+4.7956$_{10}$ -1
2.23	1.8835	5.6350	0.3342	0.5431	0.6145	+4.8690$_{10}$ -1
2.24	1.8924	5.6872	0.3327	0.5418	0.6100	+4.9426$_{10}$ -1
2.25	1.9014	5.7396	0.3313	0.5406	0.6055	+5.0165$_{10}$ -1
2.26	1.9104	5.7922	0.3298	0.5393	0.6011	+5.0908$_{10}$ -1
2.27	1.9194	5.8450	0.3284	0.5381	0.5966	+5.1653$_{10}$ -1
2.28	1.9285	5.8981	0.3270	0.5368	0.5921	+5.2401$_{10}$ -1
2.29	1.9376	5.9514	0.3256	0.5356	0.5877	+5.3152$_{10}$ -1
2.30	1.9468	6.0050	0.3242	0.5344	0.5833	+5.3906$_{10}$ -1
2.31	1.9560	6.0588	0.3228	0.5332	0.5789	+5.4663$_{10}$ -1
2.32	1.9652	6.1128	0.3215	0.5321	0.5745	+5.5423$_{10}$ -1
2.33	1.9745	6.1670	0.3202	0.5309	0.5702	+5.6185$_{10}$ -1
2.34	1.9838	6.2215	0.3189	0.5297	0.5658	+5.6950$_{10}$ -1
2.35	1.9931	6.2762	0.3176	0.5286	0.5615	+5.7717$_{10}$ -1
2.36	2.0025	6.3312	0.3163	0.5275	0.5572	+5.8487$_{10}$ -1
2.37	2.0119	6.3864	0.3150	0.5264	0.5529	+5.9260$_{10}$ -1
2.38	2.0213	6.4418	0.3138	0.5253	0.5486	+6.0035$_{10}$ -1
2.39	2.0308	6.4974	0.3126	0.5242	0.5444	+6.0812$_{10}$ -1

M_1	T_2/T_1	p_2/p_1	$q_2/q_1 = \rho_1/\rho_2$	M_2	p_{o2}/p_{o1}	$(s_2 - s_1)/R$
2.40	2.0403	6.5533	0.3113	0.5231	0.5401	+6.1592$_{10}$ -1
2.41	2.0499	6.6094	0.3101	0.5221	0.5359	+6.2374$_{10}$ -1
2.42	2.0595	6.6658	0.3090	0.5210	0.5317	+6.3159$_{10}$ -1
2.43	2.0691	6.7224	0.3078	0.5200	0.5276	+6.3945$_{10}$ -1
2.44	2.0788	6.7792	0.3066	0.5189	0.5234	+6.4734$_{10}$ -1
2.45	2.0885	6.8362	0.3055	0.5179	0.5193	+6.5525$_{10}$ -1
2.46	2.0982	6.8935	0.3044	0.5169	0.5152	+6.6319$_{10}$ -1
2.47	2.1080	6.9510	0.3033	0.5159	0.5111	+6.7114$_{10}$ -1
2.48	2.1178	7.0088	0.3022	0.5149	0.5071	+6.7911$_{10}$ -1
2.49	2.1276	7.0668	0.3011	0.5140	0.5030	+6.8711$_{10}$ -1
2.50	2.1375	7.1250	0.3000	0.5130	0.4990	+6.9512$_{10}$ -1
2.51	2.1474	7.1834	0.2989	0.5120	0.4950	+7.0315$_{10}$ -1
2.52	2.1574	7.2421	0.2979	0.5111	0.4911	+7.1120$_{10}$ -1
2.53	2.1674	7.3010	0.2969	0.5102	0.4871	+7.1928$_{10}$ -1
2.54	2.1774	7.3602	0.2958	0.5092	0.4832	+7.2736$_{10}$ -1
2.55	2.1875	7.4196	0.2948	0.5083	0.4793	+7.3547$_{10}$ -1
2.56	2.1976	7.4792	0.2938	0.5074	0.4754	+7.4360$_{10}$ -1
2.57	2.2077	7.5390	0.2928	0.5065	0.4715	+7.5174$_{10}$ -1
2.58	2.2179	7.5991	0.2919	0.5056	0.4677	+7.5990$_{10}$ -1
2.59	2.2281	7.6594	0.2909	0.5047	0.4639	+7.6807$_{10}$ -1
2.60	2.2383	7.7200	0.2899	0.5039	0.4601	+7.7626$_{10}$ -1
2.61	2.2486	7.7808	0.2890	0.5030	0.4564	+7.8447$_{10}$ -1
2.62	2.2590	7.8418	0.2881	0.5022	0.4526	+7.9269$_{10}$ -1
2.63	2.2693	7.9030	0.2871	0.5013	0.4489	+8.0093$_{10}$ -1
2.64	2.2797	7.9645	0.2862	0.5005	0.4452	+8.0918$_{10}$ -1
2.65	2.2902	8.0262	0.2853	0.4996	0.4416	+8.1745$_{10}$ -1
2.66	2.3006	8.0882	0.2844	0.4988	0.4379	+8.2573$_{10}$ -1
2.67	2.3111	8.1504	0.2836	0.4980	0.4343	+8.3402$_{10}$ -1
2.68	2.3217	8.2128	0.2827	0.4972	0.4307	+8.4233$_{10}$ -1
2.69	2.3323	8.2754	0.2818	0.4964	0.4271	+8.5065$_{10}$ -1
2.70	2.3429	8.3383	0.2810	0.4956	0.4236	+8.5899$_{10}$ -1
2.71	2.3536	8.4014	0.2801	0.4949	0.4201	+8.6734$_{10}$ -1
2.72	2.3642	8.4648	0.2793	0.4941	0.4166	+8.7570$_{10}$ -1
2.73	2.3750	8.5284	0.2785	0.4933	0.4131	+8.8407$_{10}$ -1
2.74	2.3858	8.5922	0.2777	0.4926	0.4097	+8.9245$_{10}$ -1
2.75	2.3966	8.6562	0.2769	0.4918	0.4062	+9.0085$_{10}$ -1
2.76	2.4074	8.7205	0.2761	0.4911	0.4028	+9.0925$_{10}$ -1
2.77	2.4183	8.7850	0.2753	0.4903	0.3994	+9.1767$_{10}$ -1
2.78	2.4292	8.8498	0.2745	0.4896	0.3961	+9.2610$_{10}$ -1
2.79	2.4402	8.9148	0.2737	0.4889	0.3928	+9.3454$_{10}$ -1
2.80	2.4512	8.9800	0.2730	0.4882	0.3895	+9.4298$_{10}$ -1
2.81	2.4622	9.0455	0.2722	0.4875	0.3862	+9.5144$_{10}$ -1
2.82	2.4733	9.1111	0.2715	0.4868	0.3829	+9.5991$_{10}$ -1
2.83	2.4844	9.1771	0.2707	0.4861	0.3797	+9.6839$_{10}$ -1
2.84	2.4955	9.2432	0.2700	0.4854	0.3765	+9.7687$_{10}$ -1
2.85	2.5067	9.3096	0.2693	0.4847	0.3733	+9.8537$_{10}$ -1
2.86	2.5179	9.3762	0.2685	0.4840	0.3701	+9.9387$_{10}$ -1
2.87	2.5292	9.4431	0.2678	0.4833	0.3670	+1.0024$_{10}$ 0
2.88	2.5405	9.5101	0.2671	0.4827	0.3639	+1.0109$_{10}$ 0
2.89	2.5518	9.5775	0.2664	0.4820	0.3608	+1.0194$_{10}$ 0
2.90	2.5632	9.6450	0.2658	0.4814	0.3577	+1.0280$_{10}$ 0
2.91	2.5746	9.7128	0.2651	0.4807	0.3547	+1.0365$_{10}$ 0
2.92	2.5861	9.7808	0.2644	0.4801	0.3517	+1.0451$_{10}$ 0
2.93	2.5976	9.8491	0.2637	0.4795	0.3487	+1.0536$_{10}$ 0
2.94	2.6091	9.9175	0.2631	0.4788	0.3457	+1.0622$_{10}$ 0
2.95	2.6206	9.9863	0.2624	0.4782	0.3428	+1.0708$_{10}$ 0
2.96	2.6322	10.0552	0.2618	0.4776	0.3398	+1.0793$_{10}$ 0
2.97	2.6439	10.1244	0.2611	0.4770	0.3369	+1.0879$_{10}$ 0
2.98	2.6555	10.1938	0.2605	0.4764	0.3340	+1.0965$_{10}$ 0
2.99	2.6673	10.2635	0.2599	0.4758	0.3312	+1.1051$_{10}$ 0

M_1	T_2/T_1	p_2/p_1	$= q_2/q_1$ ρ_1/ρ_2	M_2	p_{o2}/p_{o1}	$(s_2 - s_1)/R$	M_1	T_2/T_1	p_2/p_1	$= q_2/q_1$ ρ_1/ρ_2	M_2	p_{o2}/p_{o1}	$(s_2 - s_1)/R$
3.00	2.6790	10.3333	0.2593	0.4752	0.3283	+1.1137$_{10}$ 0	3.60	3.4537	14.9533	0.2310	0.4474	0.1953	+1.6332$_{10}$ 0
3.01	2.6908	10.4035	0.2586	0.4746	0.3255	+1.1223$_{10}$ 0	3.61	3.4678	15.0375	0.2306	0.4471	0.1936	+1.6418$_{10}$ 0
3.02	2.7026	10.4738	0.2580	0.4740	0.3227	+1.1309$_{10}$ 0	3.62	3.4819	15.1218	0.2303	0.4467	0.1920	+1.6504$_{10}$ 0
3.03	2.7145	10.5444	0.2574	0.4734	0.3200	+1.1395$_{10}$ 0	3.63	3.4961	15.2064	0.2299	0.4463	0.1903	+1.6590$_{10}$ 0
3.04	2.7264	10.6152	0.2568	0.4729	0.3172	+1.1481$_{10}$ 0	3.64	3.5103	15.2912	0.2296	0.4460	0.1887	+1.6677$_{10}$ 0
3.05	2.7383	10.6863	0.2562	0.4723	0.3145	+1.1568$_{10}$ 0	3.65	3.5245	15.3763	0.2292	0.4456	0.1871	+1.6763$_{10}$ 0
3.06	2.7503	10.7575	0.2557	0.4717	0.3118	+1.1654$_{10}$ 0	3.66	3.5388	15.4615	0.2289	0.4453	0.1855	+1.6849$_{10}$ 0
3.07	2.7623	10.8291	0.2551	0.4712	0.3091	+1.1740$_{10}$ 0	3.67	3.5531	15.5471	0.2285	0.4450	0.1839	+1.6935$_{10}$ 0
3.08	2.7744	10.9008	0.2545	0.4706	0.3065	+1.1827$_{10}$ 0	3.68	3.5674	15.6328	0.2282	0.4446	0.1823	+1.7021$_{10}$ 0
3.09	2.7865	10.9728	0.2539	0.4701	0.3038	+1.1913$_{10}$ 0	3.69	3.5818	15.7188	0.2279	0.4443	0.1807	+1.7107$_{10}$ 0
3.10	2.7986	11.0450	0.2534	0.4695	0.3012	+1.1999$_{10}$ 0	3.70	3.5962	15.8050	0.2275	0.4439	0.1792	+1.7193$_{10}$ 0
3.11	2.8108	11.1175	0.2528	0.4690	0.2986	+1.2086$_{10}$ 0	3.71	3.6107	15.8915	0.2272	0.4436	0.1777	+1.7279$_{10}$ 0
3.12	2.8230	11.1901	0.2523	0.4685	0.2960	+1.2172$_{10}$ 0	3.72	3.6252	15.9781	0.2269	0.4433	0.1761	+1.7365$_{10}$ 0
3.13	2.8352	11.2631	0.2517	0.4679	0.2935	+1.2259$_{10}$ 0	3.73	3.6397	16.0651	0.2266	0.4430	0.1746	+1.7451$_{10}$ 0
3.14	2.8475	11.3362	0.2512	0.4674	0.2910	+1.2345$_{10}$ 0	3.74	3.6543	16.1522	0.2262	0.4426	0.1731	+1.7536$_{10}$ 0
3.15	2.8598	11.4096	0.2507	0.4669	0.2885	+1.2432$_{10}$ 0	3.75	3.6689	16.2396	0.2259	0.4423	0.1717	+1.7622$_{10}$ 0
3.16	2.8722	11.4832	0.2501	0.4664	0.2860	+1.2519$_{10}$ 0	3.76	3.6836	16.3272	0.2256	0.4420	0.1702	+1.7708$_{10}$ 0
3.17	2.8846	11.5571	0.2496	0.4659	0.2835	+1.2605$_{10}$ 0	3.77	3.6983	16.4151	0.2253	0.4417	0.1687	+1.7794$_{10}$ 0
3.18	2.8970	11.6311	0.2491	0.4654	0.2811	+1.2692$_{10}$ 0	3.78	3.7130	16.5031	0.2250	0.4414	0.1673	+1.7879$_{10}$ 0
3.19	2.9095	11.7055	0.2486	0.4648	0.2786	+1.2779$_{10}$ 0	3.79	3.7278	16.5915	0.2247	0.4410	0.1659	+1.7965$_{10}$ 0
3.20	2.9220	11.7800	0.2480	0.4643	0.2762	+1.2865$_{10}$ 0	3.80	3.7426	16.6800	0.2244	0.4407	0.1645	+1.8050$_{10}$ 0
3.21	2.9345	11.8548	0.2475	0.4639	0.2738	+1.2952$_{10}$ 0	3.81	3.7575	16.7688	0.2241	0.4404	0.1631	+1.8136$_{10}$ 0
3.22	2.9471	11.9298	0.2470	0.4634	0.2715	+1.3039$_{10}$ 0	3.82	3.7723	16.8578	0.2238	0.4401	0.1617	+1.8221$_{10}$ 0
3.23	2.9598	12.0051	0.2465	0.4629	0.2691	+1.3125$_{10}$ 0	3.83	3.7873	16.9471	0.2235	0.4398	0.1603	+1.8307$_{10}$ 0
3.24	2.9724	12.0805	0.2460	0.4624	0.2668	+1.3212$_{10}$ 0	3.84	3.8022	17.0365	0.2232	0.4395	0.1589	+1.8392$_{10}$ 0
3.25	2.9851	12.1563	0.2456	0.4619	0.2645	+1.3299$_{10}$ 0	3.85	3.8172	17.1263	0.2229	0.4392	0.1576	+1.8477$_{10}$ 0
3.26	2.9979	12.2322	0.2451	0.4614	0.2622	+1.3386$_{10}$ 0	3.86	3.8323	17.2162	0.2226	0.4389	0.1563	+1.8562$_{10}$ 0
3.27	3.0106	12.3084	0.2446	0.4610	0.2600	+1.3472$_{10}$ 0	3.87	3.8473	17.3064	0.2223	0.4386	0.1549	+1.8648$_{10}$ 0
3.28	3.0234	12.3848	0.2441	0.4605	0.2577	+1.3559$_{10}$ 0	3.88	3.8625	17.3968	0.2220	0.4383	0.1536	+1.8733$_{10}$ 0
3.29	3.0363	12.4615	0.2437	0.4600	0.2555	+1.3646$_{10}$ 0	3.89	3.8776	17.4875	0.2217	0.4380	0.1523	+1.8818$_{10}$ 0
3.30	3.0492	12.5383	0.2432	0.4596	0.2533	+1.3733$_{10}$ 0	3.90	3.8928	17.5783	0.2215	0.4377	0.1510	+1.8903$_{10}$ 0
3.31	3.0621	12.6155	0.2427	0.4591	0.2511	+1.3820$_{10}$ 0	3.91	3.9080	17.6695	0.2212	0.4375	0.1497	+1.8988$_{10}$ 0
3.32	3.0751	12.6928	0.2423	0.4587	0.2489	+1.3906$_{10}$ 0	3.92	3.9233	17.7608	0.2209	0.4372	0.1485	+1.9073$_{10}$ 0
3.33	3.0881	12.7704	0.2418	0.4582	0.2468	+1.3993$_{10}$ 0	3.93	3.9386	17.8524	0.2206	0.4369	0.1472	+1.9158$_{10}$ 0
3.34	3.1011	12.8482	0.2414	0.4578	0.2446	+1.4080$_{10}$ 0	3.94	3.9540	17.9442	0.2203	0.4366	0.1460	+1.9243$_{10}$ 0
3.35	3.1142	12.9263	0.2409	0.4573	0.2425	+1.4167$_{10}$ 0	3.95	3.9694	18.0363	0.2201	0.4363	0.1448	+1.9327$_{10}$ 0
3.36	3.1273	13.0045	0.2405	0.4569	0.2404	+1.4253$_{10}$ 0	3.96	3.9848	18.1285	0.2198	0.4360	0.1435	+1.9412$_{10}$ 0
3.37	3.1405	13.0831	0.2400	0.4565	0.2383	+1.4340$_{10}$ 0	3.97	4.0003	18.2211	0.2195	0.4358	0.1423	+1.9497$_{10}$ 0
3.38	3.1537	13.1618	0.2396	0.4560	0.2363	+1.4427$_{10}$ 0	3.98	4.0158	18.3138	0.2193	0.4355	0.1411	+1.9581$_{10}$ 0
3.39	3.1669	13.2408	0.2392	0.4556	0.2342	+1.4514$_{10}$ 0	3.99	4.0313	18.4068	0.2190	0.4352	0.1399	+1.9666$_{10}$ 0
3.40	3.1802	13.3200	0.2388	0.4552	0.2322	+1.4600$_{10}$ 0	4.00	4.0469	18.5000	0.2187	0.4350	0.1388	+1.9750$_{10}$ 0
3.41	3.1935	13.3995	0.2383	0.4548	0.2302	+1.4687$_{10}$ 0	4.01	4.0625	18.5935	0.2185	0.4347	0.1376	+1.9835$_{10}$ 0
3.42	3.2069	13.4791	0.2379	0.4544	0.2282	+1.4774$_{10}$ 0	4.02	4.0782	18.6871	0.2182	0.4344	0.1364	+1.9919$_{10}$ 0
3.43	3.2203	13.5591	0.2375	0.4540	0.2263	+1.4861$_{10}$ 0	4.03	4.0938	18.7811	0.2180	0.4342	0.1353	+2.0003$_{10}$ 0
3.44	3.2337	13.6392	0.2371	0.4535	0.2243	+1.4947$_{10}$ 0	4.04	4.1096	18.8752	0.2177	0.4339	0.1342	+2.0088$_{10}$ 0
3.45	3.2472	13.7196	0.2367	0.4531	0.2224	+1.5034$_{10}$ 0	4.05	4.1254	18.9696	0.2175	0.4336	0.1330	+2.0172$_{10}$ 0
3.46	3.2607	13.8002	0.2363	0.4527	0.2205	+1.5121$_{10}$ 0	4.06	4.1412	19.0642	0.2172	0.4334	0.1319	+2.0256$_{10}$ 0
3.47	3.2742	13.8811	0.2359	0.4523	0.2186	+1.5207$_{10}$ 0	4.07	4.1570	19.1591	0.2170	0.4331	0.1308	+2.0340$_{10}$ 0
3.48	3.2878	13.9621	0.2355	0.4519	0.2167	+1.5294$_{10}$ 0	4.08	4.1729	19.2541	0.2167	0.4329	0.1297	+2.0424$_{10}$ 0
3.49	3.3014	14.0435	0.2351	0.4515	0.2148	+1.5381$_{10}$ 0	4.09	4.1888	19.3495	0.2165	0.4326	0.1286	+2.0508$_{10}$ 0
3.50	3.3151	14.1250	0.2347	0.4512	0.2129	+1.5467$_{10}$ 0	4.10	4.2048	19.4450	0.2162	0.4324	0.1276	+2.0592$_{10}$ 0
3.51	3.3287	14.2068	0.2343	0.4508	0.2111	+1.5554$_{10}$ 0	4.11	4.2208	19.5408	0.2160	0.4321	0.1265	+2.0676$_{10}$ 0
3.52	3.3425	14.2888	0.2339	0.4504	0.2093	+1.5640$_{10}$ 0	4.12	4.2368	19.6368	0.2158	0.4319	0.1254	+2.0760$_{10}$ 0
3.53	3.3563	14.3711	0.2335	0.4500	0.2075	+1.5727$_{10}$ 0	4.13	4.2529	19.7331	0.2155	0.4316	0.1244	+2.0843$_{10}$ 0
3.54	3.3701	14.4535	0.2332	0.4496	0.2057	+1.5813$_{10}$ 0	4.14	4.2690	19.8295	0.2153	0.4314	0.1234	+2.0927$_{10}$ 0
3.55	3.3839	14.5363	0.2328	0.4492	0.2039	+1.5900$_{10}$ 0	4.15	4.2852	19.9263	0.2151	0.4311	0.1223	+2.1010$_{10}$ 0
3.56	3.3978	14.6192	0.2324	0.4489	0.2022	+1.5986$_{10}$ 0	4.16	4.3014	20.0232	0.2148	0.4309	0.1213	+2.1094$_{10}$ 0
3.57	3.4117	14.7024	0.2321	0.4485	0.2004	+1.6073$_{10}$ 0	4.17	4.3176	20.1204	0.2146	0.4306	0.1203	+2.1177$_{10}$ 0
3.58	3.4257	14.7858	0.2317	0.4481	0.1987	+1.6159$_{10}$ 0	4.18	4.3339	20.2178	0.2144	0.4304	0.1193	+2.1261$_{10}$ 0
3.59	3.4397	14.8695	0.2313	0.4478	0.1970	+1.6245$_{10}$ 0	4.19	4.3502	20.3155	0.2141	0.4302	0.1183	+2.1344$_{10}$ 0

TABLE 2

M_1	T_2/T_1	p_2/p_1	$q_2/q_1 = \rho_1/\rho_2$	M_2	p_{o2}/p_{o1}	$(s_2 - s_1)/R$
4.20	4.3666	20.4133	0.2139	0.4299	0.1173	+2.1427$_{10}$ 0
4.21	4.3830	20.5115	0.2137	0.4297	0.1164	+2.1511$_{10}$ 0
4.22	4.3994	20.6098	0.2135	0.4295	0.1154	+2.1594$_{10}$ 0
4.23	4.4159	20.7084	0.2132	0.4292	0.1144	+2.1677$_{10}$ 0
4.24	4.4324	20.8072	0.2130	0.4290	0.1135	+2.1760$_{10}$ 0
4.25	4.4489	20.9063	0.2128	0.4288	0.1126	+2.1843$_{10}$ 0
4.26	4.4655	21.0055	0.2126	0.4286	0.1116	+2.1925$_{10}$ 0
4.27	4.4821	21.1051	0.2124	0.4283	0.1107	+2.2008$_{10}$ 0
4.28	4.4988	21.2048	0.2122	0.4281	0.1098	+2.2091$_{10}$ 0
4.29	4.5155	21.3048	0.2119	0.4279	0.1089	+2.2174$_{10}$ 0
4.30	4.5322	21.4050	0.2117	0.4277	0.1080	+2.2256$_{10}$ 0
4.31	4.5490	21.5055	0.2115	0.4275	0.1071	+2.2339$_{10}$ 0
4.32	4.5658	21.6061	0.2113	0.4272	0.1062	+2.2421$_{10}$ 0
4.33	4.5827	21.7071	0.2111	0.4270	0.1054	+2.2503$_{10}$ 0
4.34	4.5995	21.8082	0.2109	0.4268	0.1045	+2.2586$_{10}$ 0
4.35	4.6165	21.9096	0.2107	0.4266	0.1036	+2.2668$_{10}$ 0
4.36	4.6334	22.0112	0.2105	0.4264	0.1028	+2.2750$_{10}$ 0
4.37	4.6505	22.1131	0.2103	0.4262	0.1020	+2.2832$_{10}$ 0
4.38	4.6675	22.2151	0.2101	0.4260	0.1011	+2.2914$_{10}$ 0
4.39	4.6846	22.3175	0.2099	0.4258	0.1003	+2.2996$_{10}$ 0
4.40	4.7017	22.4200	0.2097	0.4255	0.0995	+2.3078$_{10}$ 0
4.41	4.7189	22.5228	0.2095	0.4253	0.0987	+2.3160$_{10}$ 0
4.42	4.7361	22.6258	0.2093	0.4251	0.0979	+2.3241$_{10}$ 0
4.43	4.7533	22.7291	0.2091	0.4249	0.0971	+2.3323$_{10}$ 0
4.44	4.7706	22.8325	0.2089	0.4247	0.0963	+2.3405$_{10}$ 0
4.45	4.7879	22.9363	0.2087	0.4245	0.0955	+2.3486$_{10}$ 0
4.46	4.8053	23.0402	0.2086	0.4243	0.0947	+2.3568$_{10}$ 0
4.47	4.8227	23.1444	0.2084	0.4241	0.0940	+2.3649$_{10}$ 0
4.48	4.8401	23.2488	0.2082	0.4239	0.0932	+2.3730$_{10}$ 0
4.49	4.8576	23.3534	0.2080	0.4237	0.0924	+2.3811$_{10}$ 0
4.50	4.8751	23.4583	0.2078	0.4236	0.0917	+2.3893$_{10}$ 0
4.51	4.8926	23.5634	0.2076	0.4234	0.0910	+2.3974$_{10}$ 0
4.52	4.9102	23.6688	0.2075	0.4232	0.0902	+2.4055$_{10}$ 0
4.53	4.9279	23.7744	0.2073	0.4230	0.0895	+2.4136$_{10}$ 0
4.54	4.9455	23.8802	0.2071	0.4228	0.0888	+2.4216$_{10}$ 0
4.55	4.9632	23.9862	0.2069	0.4226	0.0881	+2.4297$_{10}$ 0
4.56	4.9810	24.0925	0.2067	0.4224	0.0874	+2.4378$_{10}$ 0
4.57	4.9987	24.1990	0.2066	0.4222	0.0867	+2.4458$_{10}$ 0
4.58	5.0166	24.3058	0.2064	0.4220	0.0860	+2.4539$_{10}$ 0
4.59	5.0344	24.4128	0.2062	0.4219	0.0853	+2.4619$_{10}$ 0
4.60	5.0523	24.5200	0.2060	0.4217	0.0846	+2.4700$_{10}$ 0
4.61	5.0703	24.6274	0.2059	0.4215	0.0839	+2.4780$_{10}$ 0
4.62	5.0882	24.7351	0.2057	0.4213	0.0832	+2.4860$_{10}$ 0
4.63	5.1063	24.8430	0.2055	0.4211	0.0826	+2.4941$_{10}$ 0
4.64	5.1243	24.9512	0.2054	0.4210	0.0819	+2.5021$_{10}$ 0
4.65	5.1424	25.0596	0.2052	0.4208	0.0813	+2.5101$_{10}$ 0
4.66	5.1605	25.1682	0.2050	0.4206	0.0806	+2.5181$_{10}$ 0
4.67	5.1787	25.2770	0.2049	0.4204	0.0800	+2.5260$_{10}$ 0
4.68	5.1969	25.3861	0.2047	0.4203	0.0793	+2.5340$_{10}$ 0
4.69	5.2151	25.4954	0.2046	0.4201	0.0787	+2.5420$_{10}$ 0
4.70	5.2334	25.6050	0.2044	0.4199	0.0781	+2.5499$_{10}$ 0
4.71	5.2518	25.7148	0.2042	0.4197	0.0775	+2.5579$_{10}$ 0
4.72	5.2701	25.8248	0.2041	0.4196	0.0769	+2.5659$_{10}$ 0
4.73	5.2885	25.9350	0.2039	0.4194	0.0762	+2.5738$_{10}$ 0
4.74	5.3070	26.0455	0.2038	0.4192	0.0756	+2.5817$_{10}$ 0
4.75	5.3254	26.1562	0.2036	0.4191	0.0750	+2.5896$_{10}$ 0
4.76	5.3440	26.2672	0.2034	0.4189	0.0745	+2.5976$_{10}$ 0
4.77	5.3625	26.3784	0.2033	0.4187	0.0739	+2.6055$_{10}$ 0
4.78	5.3811	26.4898	0.2031	0.4186	0.0733	+2.6134$_{10}$ 0
4.79	5.3997	26.6014	0.2030	0.4184	0.0727	+2.6213$_{10}$ 0
4.80	5.4184	26.7133	0.2028	0.4183	0.0721	+2.6291$_{10}$ 0
4.81	5.4371	26.8254	0.2027	0.4181	0.0716	+2.6370$_{10}$ 0
4.82	5.4559	26.9378	0.2025	0.4179	0.0710	+2.6449$_{10}$ 0
4.83	5.4747	27.0504	0.2024	0.4178	0.0705	+2.6528$_{10}$ 0
4.84	5.4935	27.1632	0.2022	0.4176	0.0699	+2.6606$_{10}$ 0
4.85	5.5124	27.2762	0.2021	0.4175	0.0694	+2.6685$_{10}$ 0
4.86	5.5313	27.3895	0.2019	0.4173	0.0688	+2.6763$_{10}$ 0
4.87	5.5502	27.5030	0.2018	0.4172	0.0683	+2.6841$_{10}$ 0
4.88	5.5692	27.6168	0.2017	0.4170	0.0677	+2.6919$_{10}$ 0
4.89	5.5882	27.7308	0.2015	0.4169	0.0672	+2.6998$_{10}$ 0
4.90	5.6073	27.8450	0.2014	0.4167	0.0667	+2.7076$_{10}$ 0
4.91	5.6264	27.9594	0.2012	0.4165	0.0662	+2.7154$_{10}$ 0
4.92	5.6455	28.0741	0.2011	0.4164	0.0657	+2.7232$_{10}$ 0
4.93	5.6647	28.1890	0.2010	0.4163	0.0652	+2.7309$_{10}$ 0
4.94	5.6839	28.3042	0.2008	0.4161	0.0647	+2.7387$_{10}$ 0
4.95	5.7032	28.4196	0.2007	0.4160	0.0642	+2.7465$_{10}$ 0
4.96	5.7224	28.5352	0.2005	0.4158	0.0637	+2.7542$_{10}$ 0
4.97	5.7418	28.6510	0.2004	0.4157	0.0632	+2.7620$_{10}$ 0
4.98	5.7611	28.7671	0.2003	0.4155	0.0627	+2.7697$_{10}$ 0
4.99	5.7806	28.8834	0.2001	0.4154	0.0622	+2.7775$_{10}$ 0
5.00	5.8000	29.0000	0.2000	0.4152	0.0617	+2.7852$_{10}$ 0

TABLE 3 STEADY ADIABATIC FLOW OF AIR IN A CONSTANT AREA DUCT WITH FRICTION—FANNO FLOW

Flow and fluid parameters are calculated with the local Mach Number (M) as the independent variable for the range, M = 0 → (0·01) → 4 → (0·1) → 7 → (0·5) → 10. The summary of equations used is given in Section 5·3. The values of the following parameters are tabulated:

- T/T_* ratio of local static temperature to that at reference sonic condition
- p/p_* ratio of local static pressure to that at reference sonic condition
- $q/q_* = \rho_*/\rho$, ratio of local flow velocity to that at reference sonic condition
- p_o/p_{o*} ratio of local total head pressure to that at reference sonic condition
- $4C_f(L_*/D)$ friction-length parameter for distance from sonic condition
- $(s - s_*)/R$ parameter for entropy change due to friction

(Note: (5·35) gives the simple logarithmic relation between the entropy and total head pressure parameters)

M	T/T_*	p/p_*	$q/q_* = \rho_*/\rho$	p_o/p_{o*}	$4C_fL_*/D$	$(s-s_*)/R$
0	1·2000	∞	0	∞	+∞	−∞
0.01	1.2000	109.5434	0.0110	57.8738	+7.1344$_{10}$ 3	-4.0583$_{10}$ 0
0.02	1.1999	54.7701	0.0219	28.9421	+1.7784$_{10}$ 3	-3.3653$_{10}$ 0
0.03	1.1998	36.5116	0.0329	19.3005	+7.8708$_{10}$ 2	-2.9601$_{10}$ 0
0.04	1.1996	27.3817	0.0438	14.4815	+4.4035$_{10}$ 2	-2.6729$_{10}$ 0
0.05	1.1994	21.9034	0.0548	11.5914	+2.8002$_{10}$ 2	-2.4503$_{10}$ 0
0.06	1.1991	18.2508	0.0657	9.6659	+1.9303$_{10}$ 2	-2.2686$_{10}$ 0
0.07	1.1988	15.6416	0.0766	8.2915	+1.4066$_{10}$ 2	-2.1152$_{10}$ 0
0.08	1.1985	13.6843	0.0876	7.2616	+1.0672$_{10}$ 2	-1.9826$_{10}$ 0
0.09	1.1981	12.1618	0.0985	6.4613	+8.3496$_{10}$ 1	-1.8658$_{10}$ 0
0.10	1.1976	10.9435	0.1094	5.8218	+6.6922$_{10}$ 1	-1.7616$_{10}$ 0
0.11	1.1971	9.9466	0.1204	5.2992	+5.4683$_{10}$ 1	-1.6676$_{10}$ 0
0.12	1.1966	9.1156	0.1313	4.8643	+4.5408$_{10}$ 1	-1.5819$_{10}$ 0
0.13	1.1960	8.4123	0.1422	4.4969	+3.8207$_{10}$ 1	-1.5034$_{10}$ 0
0.14	1.1953	7.8093	0.1531	4.1824	+3.2511$_{10}$ 1	-1.4309$_{10}$ 0
0.15	1.1946	7.2865	0.1639	3.9103	+2.7932$_{10}$ 1	-1.3636$_{10}$ 0
0.16	1.1939	6.8291	0.1748	3.6727	+2.4198$_{10}$ 1	-1.3009$_{10}$ 0
0.17	1.1931	6.4253	0.1857	3.4635	+2.1115$_{10}$ 1	-1.2423$_{10}$ 0
0.18	1.1923	6.0662	0.1965	3.2779	+1.8543$_{10}$ 1	-1.1872$_{10}$ 0
0.19	1.1914	5.7448	0.2074	3.1123	+1.6375$_{10}$ 1	-1.1353$_{10}$ 0
0.20	1.1905	5.4554	0.2182	2.9635	+1.4533$_{10}$ 1	-1.0864$_{10}$ 0
0.21	1.1895	5.1936	0.2290	2.8293	+1.2956$_{10}$ 1	-1.0400$_{10}$ 0
0.22	1.1885	4.9554	0.2398	2.7076	+1.1596$_{10}$ 1	-9.9606$_{10}$ -1
0.23	1.1874	4.7378	0.2506	2.5968	+1.0416$_{10}$ 1	-9.5428$_{10}$ -1
0.24	1.1863	4.5383	0.2614	2.4956	+9.3865$_{10}$ 0	-9.1451$_{10}$ -1
0.25	1.1852	4.3546	0.2722	2.4027	+8.4834$_{10}$ 0	-8.7660$_{10}$ -1
0.26	1.1840	4.1851	0.2829	2.3173	+7.6876$_{10}$ 0	-8.4040$_{10}$ -1
0.27	1.1828	4.0279	0.2936	2.2385	+6.9832$_{10}$ 0	-8.0579$_{10}$ -1
0.28	1.1815	3.8820	0.3043	2.1656	+6.3572$_{10}$ 0	-7.7268$_{10}$ -1
0.29	1.1801	3.7460	0.3150	2.0979	+5.7989$_{10}$ 0	-7.4095$_{10}$ -1
0.30	1.1788	3.6191	0.3257	2.0351	+5.2993$_{10}$ 0	-7.1053$_{10}$ -1
0.31	1.1774	3.5002	0.3364	1.9765	+4.8507$_{10}$ 0	-6.8133$_{10}$ -1
0.32	1.1759	3.3837	0.3470	1.9219	+4.4467$_{10}$ 0	-6.5329$_{10}$ -1
0.33	1.1744	3.2840	0.3576	1.8707	+4.0821$_{10}$ 0	-6.2634$_{10}$ -1
0.34	1.1729	3.1853	0.3682	1.8229	+3.7520$_{10}$ 0	-6.0042$_{10}$ -1
0.35	1.1713	3.0922	0.3788	1.7780	+3.4525$_{10}$ 0	-5.7547$_{10}$ -1
0.36	1.1697	3.0042	0.3893	1.7358	+3.1801$_{10}$ 0	-5.5146$_{10}$ -1
0.37	1.1680	2.9209	0.3999	1.6961	+2.9320$_{10}$ 0	-5.2832$_{10}$ -1
0.38	1.1663	2.8420	0.4104	1.6587	+2.7054$_{10}$ 0	-5.0603$_{10}$ -1
0.39	1.1646	2.7671	0.4209	1.6234	+2.4983$_{10}$ 0	-4.8454$_{10}$ -1

M	T/T_*	p/p_*	$q/q_* = \rho_*/\rho$	p_o/p_{o*}	$4C_fL_*/D$	$(s-s_*)/R$
0.40	1.1628	2.6958	0.4313	1.5901	+2.3085$_{10}$ 0	-4.6382$_{10}$ -1
0.41	1.1610	2.6280	0.4418	1.5587	+2.1344$_{10}$ 0	-4.4384$_{10}$ -1
0.42	1.1591	2.5634	0.4522	1.5289	+1.9744$_{10}$ 0	-4.2455$_{10}$ -1
0.43	1.1572	2.5017	0.4626	1.5007	+1.8272$_{10}$ 0	-4.0594$_{10}$ -1
0.44	1.1553	2.4428	0.4729	1.4740	+1.6915$_{10}$ 0	-3.8798$_{10}$ -1
0.45	1.1533	2.3865	0.4833	1.4487	+1.5664$_{10}$ 0	-3.7065$_{10}$ -1
0.46	1.1513	2.3326	0.4936	1.4246	+1.4509$_{10}$ 0	-3.5391$_{10}$ -1
0.47	1.1492	2.2809	0.5038	1.4018	+1.3441$_{10}$ 0	-3.3775$_{10}$ -1
0.48	1.1471	2.2313	0.5141	1.3801	+1.2453$_{10}$ 0	-3.2215$_{10}$ -1
0.49	1.1450	2.1838	0.5243	1.3595	+1.1539$_{10}$ 0	-3.0709$_{10}$ -1
0.50	1.1429	2.1381	0.5345	1.3398	+1.0691$_{10}$ 0	-2.9255$_{10}$ -1
0.51	1.1407	2.0942	0.5447	1.3212	+9.9041$_{10}$ -1	-2.7852$_{10}$ -1
0.52	1.1384	2.0519	0.5548	1.3034	+9.1742$_{10}$ -1	-2.6497$_{10}$ -1
0.53	1.1362	2.0112	0.5649	1.2865	+8.4962$_{10}$ -1	-2.5189$_{10}$ -1
0.54	1.1339	1.9719	0.5750	1.2703	+7.8663$_{10}$ -1	-2.3927$_{10}$ -1
0.55	1.1315	1.9341	0.5851	1.2549	+7.2805$_{10}$ -1	-2.2709$_{10}$ -1
0.56	1.1292	1.8975	0.5951	1.2403	+6.7357$_{10}$ -1	-2.1535$_{10}$ -1
0.57	1.1268	1.8623	0.6051	1.2263	+6.2287$_{10}$ -1	-2.0402$_{10}$ -1
0.58	1.1244	1.8282	0.6150	1.2130	+5.7568$_{10}$ -1	-1.9310$_{10}$ -1
0.59	1.1219	1.7952	0.6249	1.2003	+5.3174$_{10}$ -1	-1.8258$_{10}$ -1
0.60	1.1194	1.7634	0.6348	1.1882	+4.9082$_{10}$ -1	-1.7244$_{10}$ -1
0.61	1.1169	1.7325	0.6447	1.1767	+4.5271$_{10}$ -1	-1.6267$_{10}$ -1
0.62	1.1143	1.7026	0.6545	1.1656	+4.1720$_{10}$ -1	-1.5328$_{10}$ -1
0.63	1.1117	1.6736	0.6643	1.1552	+3.8412$_{10}$ -1	-1.4423$_{10}$ -1
0.64	1.1091	1.6456	0.6740	1.1451	+3.5330$_{10}$ -1	-1.3553$_{10}$ -1
0.65	1.1065	1.6183	0.6837	1.1356	+3.2459$_{10}$ -1	-1.2718$_{10}$ -1
0.66	1.1038	1.5919	0.6934	1.1265	+2.9785$_{10}$ -1	-1.1915$_{10}$ -1
0.67	1.1011	1.5662	0.7031	1.1179	+2.7295$_{10}$ -1	-1.1144$_{10}$ -1
0.68	1.0984	1.5413	0.7127	1.1097	+2.4978$_{10}$ -1	-1.0405$_{10}$ -1
0.69	1.0957	1.5170	0.7223	1.1018	+2.2820$_{10}$ -1	-9.6965$_{10}$ -2
0.70	1.0929	1.4935	0.7318	1.0944	+2.0814$_{10}$ -1	-9.0181$_{10}$ -2
0.71	1.0901	1.4705	0.7413	1.0873	+1.8948$_{10}$ -1	-8.3692$_{10}$ -2
0.72	1.0873	1.4482	0.7508	1.0806	+1.7215$_{10}$ -1	-7.7490$_{10}$ -2
0.73	1.0844	1.4265	0.7602	1.0742	+1.5605$_{10}$ -1	-7.1569$_{10}$ -2
0.74	1.0815	1.4054	0.7696	1.0681	+1.4112$_{10}$ -1	-6.5923$_{10}$ -2
0.75	1.0787	1.3848	0.7789	1.0624	+1.2728$_{10}$ -1	-6.0547$_{10}$ -2
0.76	1.0757	1.3647	0.7883	1.0570	+1.1447$_{10}$ -1	-5.5434$_{10}$ -2
0.77	1.0728	1.3451	0.7975	1.0519	+1.0262$_{10}$ -1	-5.0580$_{10}$ -2
0.78	1.0698	1.3261	0.8068	1.0471	+9.1672$_{10}$ -2	-4.5979$_{10}$ -2
0.79	1.0668	1.3074	0.8160	1.0425	+8.1580$_{10}$ -2	-4.1627$_{10}$ -2

M	T/T_*	p/p_*	$q/q_* = \rho_*/\rho$	p_o/p_{o*}	$4C_fL_*/D$	$(s-s_*)/R$
0.80	1.0638	1.2893	0.8251	1.0332	+7.2290$_{10}$ -2	-3.7517$_{10}$ -2
0.81	1.0608	1.2715	0.8343	1.0342	+6.3755$_{10}$ -2	-3.3646$_{10}$ -2
0.82	1.0578	1.2542	0.843[illegible]	1.0305	+5.5932$_{10}$ -2	-3.0009$_{10}$ -2
0.83	1.0547	1.2373	0.8524	1.0270	+4.8778$_{10}$ -2	-2.6602$_{10}$ -2
0.84	1.0516	1.2208	0.8614	1.0237	+4.2256$_{10}$ -2	-2.3419$_{10}$ -2
0.85	1.0485	1.2047	0.8704	1.0207	+3.6330$_{10}$ -2	-2.0458$_{10}$ -2
0.86	1.0454	1.1889	0.8793	1.0179	+3.0965$_{10}$ -2	-1.7713$_{10}$ -2
0.87	1.0422	1.1734	0.8882	1.0153	+2.6130$_{10}$ -2	-1.5181$_{10}$ -2
0.88	1.0391	1.1583	0.8970	1.0129	+2.1795$_{10}$ -2	-1.2858$_{10}$ -2
0.89	1.0359	1.1436	0.9058	1.0108	+1.7931$_{10}$ -2	-1.0740$_{10}$ -2
0.90	1.0327	1.1291	0.9146	1.0089	+1.4512$_{10}$ -2	-8.8238$_{10}$ -3
0.91	1.0295	1.1150	0.9233	1.0071	+1.1514$_{10}$ -2	-7.1054$_{10}$ -3
0.92	1.0263	1.1011	0.9320	1.0056	+8.9133$_{10}$ -3	-5.5815$_{10}$ -3
0.93	1.0230	1.0876	0.9407	1.0043	+6.6873$_{10}$ -3	-4.2486$_{10}$ -3
0.94	1.0198	1.0743	0.9493	1.0031	+4.8154$_{10}$ -3	-3.1035$_{10}$ -3
0.95	1.0165	1.0613	0.9578	1.0021	+3.2782$_{10}$ -3	-2.1429$_{10}$ -3
0.96	1.0132	1.0485	0.9663	1.0014	+2.0571$_{10}$ -3	-1.3636$_{10}$ -3
0.97	1.0099	1.0360	0.9748	1.0008	+1.1348$_{10}$ -3	-7.6271$_{10}$ -4
0.98	1.0066	1.0238	0.9833	1.0003	+4.9470$_{10}$ -4	-3.3708$_{10}$ -4
0.99	1.0033	1.0118	0.9916	1.0001	+1.2133$_{10}$ -4	-8.3799$_{10}$ -5
1	1	1	1	1	1	0
1.01	0.9967	0.9884	1.0083	1.0001	+1.1683$_{10}$ -4	-8.2873$_{10}$ -5
1.02	0.9933	0.9771	1.0166	1.0003	+4.5869$_{10}$ -4	-3.2967$_{10}$ -4
1.03	0.9900	0.9660	1.0248	1.0007	+1.0132$_{10}$ -3	-7.3770$_{10}$ -4
1.04	0.9866	0.9551	1.0330	1.0013	+1.7685$_{10}$ -3	-1.3043$_{10}$ -3
1.05	0.9832	0.9443	1.0411	1.0020	+2.7136$_{10}$ -3	-2.0270$_{10}$ -3
1.06	0.9798	0.9338	1.0492	1.0029	+3.8379$_{10}$ -3	-2.9032$_{10}$ -3
1.07	0.9764	0.9235	1.0573	1.0039	+5.1313$_{10}$ -3	-3.9304$_{10}$ -3
1.08	0.9730	0.9133	1.0653	1.0051	+6.5846$_{10}$ -3	-5.1061$_{10}$ -3
1.09	0.9696	0.9034	1.0733	1.0064	+8.1887$_{10}$ -3	-6.4282$_{10}$ -3
1.10	0.9662	0.8936	1.0812	1.0079	+9.9350$_{10}$ -3	-7.8941$_{10}$ -3
1.11	0.9628	0.8840	1.0891	1.0095	+1.1816$_{10}$ -2	-9.5016$_{10}$ -3
1.12	0.9593	0.8745	1.0970	1.0113	+1.3823$_{10}$ -2	-1.1249$_{10}$ -2
1.13	0.9559	0.8652	1.1048	1.0132	+1.5949$_{10}$ -2	-1.3133$_{10}$ -2
1.14	0.9524	0.8561	1.1126	1.0153	+1.8188$_{10}$ -2	-1.5152$_{10}$ -2
1.15	0.9490	0.8471	1.1203	1.0175	+2.0533$_{10}$ -2	-1.7304$_{10}$ -2
1.16	0.9455	0.8383	1.1280	1.0198	+2.2977$_{10}$ -2	-1.9587$_{10}$ -2
1.17	0.9421	0.8296	1.1356	1.0222	+2.5516$_{10}$ -2	-2.1998$_{10}$ -2
1.18	0.9386	0.8210	1.1432	1.0248	+2.8142$_{10}$ -2	-2.4537$_{10}$ -2
1.19	0.9351	0.8126	1.1508	1.0276	+3.0851$_{10}$ -2	-2.7200$_{10}$ -2
1.20	0.9317	0.8044	1.1583	1.0304	+3.3638$_{10}$ -2	-2.9986$_{10}$ -2
1.21	0.9282	0.7962	1.1658	1.0334	+3.6498$_{10}$ -2	-3.2893$_{10}$ -2
1.22	0.9247	0.7882	1.1732	1.0366	+3.9426$_{10}$ -2	-3.5919$_{10}$ -2
1.23	0.9212	0.7803	1.1806	1.0398	+4.2418$_{10}$ -2	-3.9062$_{10}$ -2
1.24	0.9178	0.7726	1.1879	1.0432	+4.5471$_{10}$ -2	-4.2321$_{10}$ -2
1.25	0.9143	0.7649	1.1952	1.0468	+4.8579$_{10}$ -2	-4.5693$_{10}$ -2
1.26	0.9108	0.7574	1.2025	1.0504	+5.1739$_{10}$ -2	-4.9177$_{10}$ -2
1.27	0.9073	0.7500	1.2097	1.0542	+5.4947$_{10}$ -2	-5.2772$_{10}$ -2
1.28	0.9038	0.7427	1.2169	1.0581	+5.8201$_{10}$ -2	-5.6474$_{10}$ -2
1.29	0.9003	0.7356	1.2240	1.0621	+6.1497$_{10}$ -2	-6.0284$_{10}$ -2
1.30	0.8969	0.7285	1.2311	1.0663	+6.4832$_{10}$ -2	-6.4199$_{10}$ -2
1.31	0.8934	0.7215	1.2382	1.0706	+6.8203$_{10}$ -2	-6.8217$_{10}$ -2
1.32	0.8899	0.7147	1.2452	1.0750	+7.1607$_{10}$ -2	-7.2338$_{10}$ -2
1.33	0.8864	0.7079	1.2522	1.0796	+7.5041$_{10}$ -2	-7.6558$_{10}$ -2
1.34	0.8329	0.7012	1.2591	1.0842	+7.8504$_{10}$ -2	-8.0878$_{10}$ -2
1.35	0.8794	0.6947	1.2660	1.0890	+8.1991$_{10}$ -2	-8.5295$_{10}$ -2
1.36	0/8760	0.6882	1.2729	1.0940	+8.5503$_{10}$ -2	-8.9808$_{10}$ -2
1.37	0.8725	0.6818	1.2797	1.0990	+8.9035$_{10}$ -2	-9.4415$_{10}$ -2
1.38	0.8690	0.6755	1.2864	1.1042	+9.2586$_{10}$ -2	-9.9115$_{10}$ -2
1.39	0.8655	0.6693	1.2932	1.1095	+9.6155$_{10}$ -2	-1.0391$_{10}$ -1
1.40	0.8621	0.6632	1.2999	1.1149	+9.9738$_{10}$ -2	-1.0879$_{10}$ -1
1.41	0.8586	0.6572	1.3065	1.1205	+1.0334$_{10}$ -1	-1.1376$_{10}$ -1
1.42	0.8551	0.6512	1.3131	1.1262	+1.0694$_{10}$ -1	-1.1882$_{10}$ -1
1.43	0.8517	0.6454	1.3197	1.1320	+1.1056$_{10}$ -1	-1.2396$_{10}$ -1
1.44	0.8482	0.6396	1.3262	1.1379	+1.1419$_{10}$ -1	-1.2919$_{10}$ -1
1.45	0.8448	0.6339	1.3327	1.1440	+1.1782$_{10}$ -1	-1.3450$_{10}$ -1
1.46	0.8413	0.6282	1.3392	1.1501	+1.2146$_{10}$ -1	-1.3989$_{10}$ -1
1.47	0.8379	0.6227	1.3456	1.1565	+1.2511$_{10}$ -1	-1.4537$_{10}$ -1
1.48	0.8344	0.6172	1.3520	1.1629	+1.2875$_{10}$ -1	-1.5092$_{10}$ -1
1.49	0.8310	0.6118	1.3583	1.1695	+1.3240$_{10}$ -1	-1.5655$_{10}$ -1
1.50	0.8276	0.6065	1.3646	1.1762	+1.3605$_{10}$ -1	-1.6226$_{10}$ -1
1.51	0.8242	0.6012	1.3708	1.1830	+1.3970$_{10}$ -1	-1.6805$_{10}$ -1
1.52	0.8207	0.5960	1.3770	1.1899	+1.4335$_{10}$ -1	-1.7391$_{10}$ -1
1.53	0.8173	0.5909	1.3832	1.1970	+1.4699$_{10}$ -1	-1.7984$_{10}$ -1
1.54	0.8139	0.5858	1.3894	1.2042	+1.5063$_{10}$ -1	-1.8584$_{10}$ -1
1.55	0.8105	0.5808	1.3955	1.2116	+1.5427$_{10}$ -1	-1.9192$_{10}$ -1
1.56	0.8071	0.5759	1.4015	1.2190	+1.5790$_{10}$ -1	-1.9807$_{10}$ -1
1.57	0.8038	0.5710	1.4075	1.2266	+1.6152$_{10}$ -1	-2.0428$_{10}$ -1
1.58	0.8004	0.5662	1.4135	1.2344	+1.6514$_{10}$ -1	-2.1057$_{10}$ -1
1.59	0.7970	0.5615	1.4195	1.2422	+1.6875$_{10}$ -1	-2.1692$_{10}$ -1
1.60	0.7937	0.5568	1.4254	1.2502	+1.7236$_{10}$ -1	-2.2333$_{10}$ -1
1.61	0.7903	0.5522	1.4313	1.2584	+1.7595$_{10}$ -1	-2.2981$_{10}$ -1
1.62	0.7869	0.5476	1.4371	1.2666	+1.7954$_{10}$ -1	-2.3636$_{10}$ -1
1.63	0.7836	0.5431	1.4429	1.2750	+1.8311$_{10}$ -1	-2.4296$_{10}$ -1
1.64	0.7803	0.5386	1.4487	1.2836	+1.8667$_{10}$ -1	-2.4963$_{10}$ -1
1.65	0.7770	0.5342	1.4544	1.2922	+1.9023$_{10}$ -1	-2.5636$_{10}$ -1
1.66	0.7736	0.5299	1.4601	1.3010	+1.9377$_{10}$ -1	-2.6315$_{10}$ -1
1.67	0.7703	0.5256	1.4657	1.3100	+1.9729$_{10}$ -1	-2.7000$_{10}$ -1
1.68	0.7670	0.5213	1.4713	1.3190	+2.0081$_{10}$ -1	-2.7690$_{10}$ -1
1.69	0.7637	0.5171	1.4769	1.3283	+2.0431$_{10}$ -1	-2.8386$_{10}$ -1
1.70	0.7605	0.5130	1.4825	1.3376	+2.0780$_{10}$ -1	-2.9088$_{10}$ -1
1.71	0.7572	0.5089	1.4880	1.3471	+2.1128$_{10}$ -1	-2.9795$_{10}$ -1
1.72	0.7539	0.5048	1.4935	1.3567	+2.1474$_{10}$ -1	-3.0508$_{10}$ -1
1.73	0.7507	0.5008	1.4989	1.3665	+2.1819$_{10}$ -1	-3.1226$_{10}$ -1
1.74	0.7474	0.4969	1.5043	1.3764	+2.2162$_{10}$ -1	-3.1949$_{10}$ -1
1.75	0.7442	0.4929	1.5097	1.3865	+2.2504$_{10}$ -1	-3.2678$_{10}$ -1
1.76	0.7410	0.4891	1.5150	1.3967	+2.2844$_{10}$ -1	-3.3411$_{10}$ -1
1.77	0.7377	0.4853	1.5203	1.4070	+2.3182$_{10}$ -1	-3.4149$_{10}$ -1
1.78	0.7345	0.4815	1.5256	1.4175	+2.3519$_{10}$ -1	-3.4893$_{10}$ -1
1.79	0.7313	0.4778	1.5308	1.4282	+2.3855$_{10}$ -1	-3.5641$_{10}$ -1
1.80	0.7282	0.4741	1.5360	1.4390	+2.4189$_{10}$ -1	-3.6394$_{10}$ -1
1.81	0.7250	0.4704	1.5411	1.4499	+2.4521$_{10}$ -1	-3.7151$_{10}$ -1
1.82	0.7218	0.4668	1.5463	1.4610	+2.4851$_{10}$ -1	-3.7913$_{10}$ -1
1.83	0.7187	0.4632	1.5514	1.4723	+2.5180$_{10}$ -1	-3.8680$_{10}$ -1
1.84	0.7155	0.4597	1.5564	1.4836	+2.5507$_{10}$ -1	-3.9450$_{10}$ -1
1.85	0.7124	0.4562	1.5614	1.4952	+2.5832$_{10}$ -1	-4.0226$_{10}$ -1
1.86	0.7093	0.4528	1.5664	1.5069	+2.6156$_{10}$ -1	-4.1005$_{10}$ -1
1.87	0.7061	0.4494	1.5714	1.5187	+2.6478$_{10}$ -1	-4.1789$_{10}$ -1
1.88	0.7030	0.4460	1.5763	1.5308	+2.6798$_{10}$ -1	-4.2576$_{10}$ -1
1.89	0.6999	0.4427	1.5812	1.5429	+2.7116$_{10}$ -1	-4.3368$_{10}$ -1

M	T/T_*	p/p_*	$q/q_* = \rho_*/\rho$	p_0/p_{0*}	$4C_fL_*/D$	$(s-s_*)/R$
1.90	0.6969	0.4394	1.5861	1.5553	+2.7433$_{10}$ -1	-4.4164$_{10}$ -1
1.91	0.6938	0.4361	1.5909	1.5677	+2.7748$_{10}$ -1	-4.4964$_{10}$ -1
1.92	0.6907	0.4329	1.5957	1.5804	+2.8061$_{10}$ -1	-4.5767$_{10}$ -1
1.93	0.6877	0.4297	1.6005	1.5932	+2.8372$_{10}$ -1	-4.6574$_{10}$ -1
1.94	0.6847	0.4265	1.605	1.6062	+2.8681$_{10}$ -1	-4.7385$_{10}$ -1
1.95	0.6816	0.4234	1.6099	1.6193	+2.8989$_{10}$ -1	-4.8200$_{10}$ -1
1.96	0.6786	0.4203	1.6146	1.6326	+2.9295$_{10}$ -1	-4.9018$_{10}$ -1
1.97	0.6756	0.4172	1.6192	1.6461	+2.9599$_{10}$ -1	-4.9840$_{10}$ -1
1.98	0.6726	0.4142	1.6239	1.6597	+2.9901$_{10}$ -1	-5.0665$_{10}$ -1
1.99	0.6696	0.4112	1.6284	1.6735	+3.0201$_{10}$ -1	-5.1493$_{10}$ -1
2.00	0.6667	0.4082	1.6330	1.6875	+3.0500$_{10}$ -1	-5.2325$_{10}$ -1
2.01	0.6637	0.4053	1.6375	1.7016	+3.0796$_{10}$ -1	-5.3160$_{10}$ -1
2.02	0.6608	0.4024	1.6420	1.7160	+3.1091$_{10}$ -1	-5.3998$_{10}$ -1
2.03	0.6578	0.3995	1.6465	1.7305	+3.1384$_{10}$ -1	-5.4839$_{10}$ -1
2.04	0.6549	0.3967	1.6509	1.7451	+3.1676$_{10}$ -1	-5.5683$_{10}$ -1
2.05	0.6520	0.3939	1.6553	1.7600	+3.1965$_{10}$ -1	-5.6531$_{10}$ -1
2.06	0.6491	0.3911	1.6597	1.7750	+3.2253$_{10}$ -1	-5.7381$_{10}$ -1
2.07	0.6462	0.3883	1.6640	1.7902	+3.2538$_{10}$ -1	-5.8234$_{10}$ -1
2.08	0.6433	0.3856	1.6683	1.8056	+3.2822$_{10}$ -1	-5.9090$_{10}$ -1
2.09	0.6405	0.3829	1.6726	1.8212	+3.3105$_{10}$ -1	-5.9949$_{10}$ -1
2.10	0.6376	0.3802	1.6769	1.8369	+3.3385$_{10}$ -1	-6.0810$_{10}$ -1
2.11	0.6348	0.3776	1.6811	1.8529	+3.3664$_{10}$ -1	-6.1674$_{10}$ -1
2.12	0.6320	0.3750	1.6853	1.8690	+3.3940$_{10}$ -1	-6.2541$_{10}$ -1
2.13	0.6291	0.3724	1.6895	1.8853	+3.4215$_{10}$ -1	-6.3411$_{10}$ -1
2.14	0.6263	0.3698	1.6936	1.9018	+3.4489$_{10}$ -1	-6.4282$_{10}$ -1
2.15	0.6235	0.3673	1.6977	1.9185	+3.4760$_{10}$ -1	-6.5157$_{10}$ -1
2.16	0.6208	0.3648	1.7018	1.9354	+3.5030$_{10}$ -1	-6.6033$_{10}$ -1
2.17	0.6180	0.3623	1.7059	1.9525	+3.5298$_{10}$ -1	-6.6912$_{10}$ -1
2.18	0.6152	0.3598	1.7099	1.9698	+3.5564$_{10}$ -1	-6.7794$_{10}$ -1
2.19	0.6125	0.3574	1.7139	1.9873	+3.5828$_{10}$ -1	-6.8677$_{10}$ -1
2.20	0.6098	0.3549	1.7179	2.0050	+3.6091$_{10}$ -1	-6.9563$_{10}$ -1
2.21	0.6070	0.3525	1.7219	2.0229	+3.6352$_{10}$ -1	-7.0451$_{10}$ -1
2.22	0.6043	0.3502	1.7258	2.0409	+3.6611$_{10}$ -1	-7.1341$_{10}$ -1
2.23	0.6016	0.3478	1.7297	2.0592	+3.6869$_{10}$ -1	-7.2233$_{10}$ -1
2.24	0.5989	0.3455	1.7336	2.0777	+3.7124$_{10}$ -1	-7.3128$_{10}$ -1
2.25	0.5963	0.3432	1.7374	2.0964	+3.7378$_{10}$ -1	-7.4024$_{10}$ -1
2.26	0.5936	0.3409	1.7412	2.1153	+3.7631$_{10}$ -1	-7.4922$_{10}$ -1
2.27	0.5910	0.3387	1.7450	2.1345	+3.7881$_{10}$ -1	-7.5822$_{10}$ -1
2.28	0.5883	0.3364	1.7483	2.1538	+3.8130$_{10}$ -1	-7.6724$_{10}$ -1
2.29	0.5857	0.3342	1.7526	2.1734	+3.8377$_{10}$ -1	-7.7628$_{10}$ -1
2.30	0.5831	0.3320	1.7563	2.1931	+3.8623$_{10}$ -1	-7.8533$_{10}$ -1
2.31	0.5805	0.3298	1.7600	2.2131	+3.8867$_{10}$ -1	-7.9440$_{10}$ -1
2.32	0.5779	0.3277	1.7637	2.2333	+3.9109$_{10}$ -1	-8.0349$_{10}$ -1
2.33	0.5753	0.3255	1.7673	2.2538	+3.9350$_{10}$ -1	-8.1260$_{10}$ -1
2.34	0.5728	0.3234	1.7709	2.2744	+3.9589$_{10}$ -1	-8.2172$_{10}$ -1
2.35	0.5702	0.3213	1.7745	2.2953	+3.9826$_{10}$ -1	-8.3085$_{10}$ -1
2.36	0.5677	0.3193	1.7781	2.3164	+4.0062$_{10}$ -1	-8.4001$_{10}$ -1
2.37	0.5651	0.3172	1.7817	2.3377	+4.0296$_{10}$ -1	-8.4917$_{10}$ -1
2.38	0.5626	0.3152	1.7852	2.3593	+4.0529$_{10}$ -1	-8.5835$_{10}$ -1
2.39	0.5601	0.3131	1.7887	2.3811	+4.0760$_{10}$ -1	-8.6755$_{10}$ -1
2.40	0.5576	0.3111	1.7922	2.4031	+4.0989$_{10}$ -1	-8.7676$_{10}$ -1
2.41	0.5551	0.3092	1.7956	2.4254	+4.1217$_{10}$ -1	-8.8598$_{10}$ -1
2.42	0.5527	0.3072	1.7991	2.4479	+4.1443$_{10}$ -1	-8.9522$_{10}$ -1
2.43	0.5502	0.3053	1.8025	2.4706	+4.1668$_{10}$ -1	-9.0447$_{10}$ -1
2.44	0.5478	0.3033	1.8059	2.4936	+4.1891$_{10}$ -1	-9.1373$_{10}$ -1

M	T/T_*	p/p_*	$q/q_* = \rho_*/\rho$	p_0/p_{0*}	$4C_fL_*/D$	$(s-s_*)/R$
2.45	0.5453	0.3014	1.8092	2.5168	+4.2112$_{10}$ -1	-9.2300$_{10}$ -1
2.46	0.5429	0.2995	1.8126	2.5403	+4.2332$_{10}$ -1	-9.3229$_{10}$ -1
2.47	0.5405	0.2976	1.8159	2.5640	+4.2551$_{10}$ -1	-9.4158$_{10}$ -1
2.48	0.5381	0.2958	1.8192	2.5830	+4.2768$_{10}$ -1	-9.5089$_{10}$ -1
2.49	0.5357	0.2939	1.8225	2.6122	+4.2984$_{10}$ -1	-9.6021$_{10}$ -1
2.50	0.5333	0.2921	1.8257	2.6367	+4.3198$_{10}$ -1	-9.6954$_{10}$ -1
2.51	0.5310	0.2903	1.8290	2.6615	+4.3410$_{10}$ -1	-9.7837$_{10}$ -1
2.52	0.5286	0.28[illegible]5	1.8322	2.6865	+4.3621$_{10}$ -1	-9.8[illegible]$_{10}$ -1
2.53	0.5263	0.2867	1.8354	2.7117	+4.3831$_{10}$ -1	-9.9758$_{10}$ -1
2.54	0.5239	0.2850	1.8386	2.7372	+4.4039$_{10}$ -1	-1.0069$_{10}$ 0
2.55	0.5216	0.2832	1.8417	2.7630	+4.4246$_{10}$ -1	-1.0163$_{10}$ 0
2.56	0.5193	0.2815	1.8448	2.7891	+4.4451$_{10}$ -1	-1.0257$_{10}$ 0
2.57	0.5170	0.2798	1.8479	2.8154	+4.4655$_{10}$ -1	-1.0351$_{10}$ 0
2.58	0.5147	0.2781	1.8510	2.8420	+4.4858$_{10}$ -1	-1.0445$_{10}$ 0
2.59	0.5125	0.2764	1.8541	2.8688	+4.5059$_{10}$ -1	-1.0539$_{10}$ 0
2.60	0.5102	0.2747	1.8571	2.8960	+4.5259$_{10}$ -1	-1.0633$_{10}$ 0
2.61	0.5080	0.2731	1.8602	2.9234	+4.5457$_{10}$ -1	-1.0727$_{10}$ 0
2.62	0.5057	0.2714	1.8632	2.9511	+4.5654$_{10}$ -1	-1.0822$_{10}$ 0
2.63	0.5035	0.2698	1.8662	2.9791	+4.5850$_{10}$ -1	-1.0916$_{10}$ 0
2.64	0.5013	0.2682	1.8691	3.0073	+4.6044$_{10}$ -1	-1.1011$_{10}$ 0
2.65	0.4991	0.2665	1.8721	3.0359	+4.6237$_{10}$ -1	-1.1105$_{10}$ 0
2.66	0.4969	0.2650	1.8750	3.0647	+4.6429$_{10}$ -1	-1.1200$_{10}$ 0
2.67	0.4947	0.2634	1.8779	3.0938	+4.6619$_{10}$ -1	-1.1294$_{10}$ 0
2.68	0.4925	0.2619	1.8808	3.1233	+4.6808$_{10}$ -1	-1.1389$_{10}$ 0
2.69	0.4904	0.2603	1.8837	3.1530	+4.6996$_{10}$ -1	-1.1484$_{10}$ 0
2.70	0.4832	0.2583	1.8865	3.1830	+4.7182$_{10}$ -1	-1.1578$_{10}$ 0
2.71	0.4861	0.2573	1.8894	3.2133	+4.7367$_{10}$ -1	-1.1673$_{10}$ 0
2.72	0.4839	0.2558	1.8922	3.2440	+4.7551$_{10}$ -1	-1.1768$_{10}$ 0
2.73	0.4818	0.2543	1.8950	3.2749	+4.7733$_{10}$ -1	-1.1863$_{10}$ 0
2.74	0.4797	0.2528	1.8978	3.3061	+4.7915$_{10}$ -1	-1.1958$_{10}$ 0
2.75	0.4776	0.2513	1.9005	3.3377	+4.8095$_{10}$ -1	-1.2053$_{10}$ 0
2.76	0.4755	0.2498	1.9033	3.3695	+4.8273$_{10}$ -1	-1.2148$_{10}$ 0
2.77	0.4735	0.2484	1.9060	3.4017	+4.8451$_{10}$ -1	-1.2243$_{10}$ 0
2.78	0.4714	0.2470	1.9087	3.4342	+4.8627$_{10}$ -1	-1.2338$_{10}$ 0
2.79	0.4693	0.2455	1.9114	3.4670	+4.8803$_{10}$ -1	-1.2433$_{10}$ 0
2.80	0.4673	0.2441	1.9140	3.5001	+4.8976$_{10}$ -1	-1.2528$_{10}$ 0
2.81	0.4653	0.2427	1.9167	3.5336	+4.9149$_{10}$ -1	-1.2623$_{10}$ 0
2.82	0.4632	0.2414	1.9193	3.5674	+4.9321$_{10}$ -1	-1.2718$_{10}$ 0
2.83	0.4612	0.2400	1.9219	3.6015	+4.9491$_{10}$ -1	-1.2813$_{10}$ 0
2.84	0.4592	0.2386	1.9246	3.6359	+4.9660$_{10}$ -1	-1.2909$_{10}$ 0
2.85	0.4572	0.2373	1.9271	3.6707	+4.9828$_{10}$ -1	-1.3004$_{10}$ 0
2.86	0.4552	0.2359	1.9297	3.7058	+4.9995$_{10}$ -1	-1.3099$_{10}$ 0
2.87	0.4533	0.2346	1.9323	3.7413	+5.0161$_{10}$ -1	-1.3194$_{10}$ 0
2.88	0.4513	0.2333	1.9348	3.7771	+5.0326$_{10}$ -1	-1.3290$_{10}$ 0
2.89	0.4494	0.2320	1.9373	3.8133	+5.0489$_{10}$ -1	-1.3385$_{10}$ 0
2.90	0.4474	0.2307	1.9398	3.8498	+5.0652$_{10}$ -1	-1.3480$_{10}$ 0
2.91	0.4455	0.2294	1.9423	3.8866	+5.0813$_{10}$ -1	-1.3575$_{10}$ 0
2.92	0.4436	0.2281	1.9448	3.9238	+5.0973$_{10}$ -1	-1.3671$_{10}$ 0
2.93	0.4417	0.2268	1.9472	3.9614	+5.1132$_{10}$ -1	-1.3766$_{10}$ 0
2.94	0.4398	0.2256	1.9497	3.9993	+5.1290$_{10}$ -1	-1.3861$_{10}$ 0
2.95	0.4379	0.2243	1.9521	4.0376	+5.1447$_{10}$ -1	-1.3957$_{10}$ 0
2.96	0.4360	0.2231	1.9545	4.0763	+5.1603$_{10}$ -1	-1.4052$_{10}$ 0
2.97	0.4341	0.2218	1.9569	4.1153	+5.1758$_{10}$ -1	-1.4147$_{10}$ 0
2.98	0.4323	0.2206	1.9593	4.1547	+5.1912$_{10}$ -1	-1.4242$_{10}$ 0
2.99	0.4304	0.2194	1.9616	4.1944	+5.2064$_{10}$ -1	-1.4338$_{10}$ 0

M	T/T_*	p/p_*	$q/q_* = \rho_*/\rho$	p_o/p_{o*}	$4C_fL_*/D$	$(s-s_*)/R$
3.00	0.4286	0.2182	1.9640	4.2346	+5.2216$_{10}$ -1	-1.4433$_{10}$ 0
3.01	0.4267	0.2170	1.9663	4.2751	+5.2367$_{10}$ -1	-1.4528$_{10}$ 0
3.02	0.4249	0.2158	1.9686	4.3160	+5.2516$_{10}$ -1	-1.4623$_{10}$ 0
3.03	0.4231	0.2147	1.9709	4.3573	+5.2665$_{10}$ -1	-1.4718$_{10}$ 0
3.04	0.4213	0.2135	1.9732	4.3989	+5.2813$_{10}$ -1	-1.4814$_{10}$ 0
3.05	0.4195	0.2124	1.9755	4.4410	+5.2959$_{10}$ -1	-1.4909$_{10}$ 0
3.06	0.4177	0.2112	1.9777	4.4835	+5.3105$_{10}$ -1	-1.5004$_{10}$ 0
3.07	0.4159	0.2101	1.9800	4.5263	+5.3249$_{10}$ -1	-1.5099$_{10}$ 0
3.08	0.4142	0.2090	1.9822	4.5696	+5.3393$_{10}$ -1	-1.5194$_{10}$ 0
3.09	0.4124	0.2078	1.9844	4.6132	+5.3536$_{10}$ -1	-1.5289$_{10}$ 0
3.10	0.4107	0.2067	1.9866	4.6573	+5.3678$_{10}$ -1	-1.5384$_{10}$ 0
3.11	0.4089	0.2056	1.9838	4.7018	+5.3818$_{10}$ -1	-1.5479$_{10}$ 0
3.12	0.4072	0.2045	1.9910	4.7467	+5.3958$_{10}$ -1	-1.5574$_{10}$ 0
3.13	0.4055	0.2034	1.9931	4.7920	+5.4097$_{10}$ -1	-1.5669$_{10}$ 0
3.14	0.4038	0.2024	1.9953	4.8377	+5.4235$_{10}$ -1	-1.5764$_{10}$ 0
3.15	0.4021	0.2013	1.9974	4.8838	+5.4372$_{10}$ -1	-1.5859$_{10}$ 0
3.16	0.4004	0.2002	1.9995	4.9304	+5.4509$_{10}$ -1	-1.5954$_{10}$ 0
3.17	0.3987	0.1992	2.0016	4.9774	+5.4644$_{10}$ -1	-1.6049$_{10}$ 0
3.18	0.3970	0.1981	2.0037	5.0248	+5.4778$_{10}$ -1	-1.6144$_{10}$ 0
3.19	0.3954	0.1971	2.0058	5.0727	+5.4912$_{10}$ -1	-1.6239$_{10}$ 0
3.20	0.3937	0.1961	2.0079	5.1210	+5.5044$_{10}$ -1	-1.6333$_{10}$ 0
3.21	0.3921	0.1951	2.0099	5.1697	+5.5176$_{10}$ -1	-1.6428$_{10}$ 0
3.22	0.3904	0.1940	2.0119	5.2189	+5.5307$_{10}$ -1	-1.6523$_{10}$ 0
3.23	0.3888	0.1930	2.0140	5.2685	+5.5437$_{10}$ -1	-1.6617$_{10}$ 0
3.24	0.3872	0.1920	2.0160	5.3186	+5.5566$_{10}$ -1	-1.6712$_{10}$ 0
3.25	0.3855	0.1911	2.0180	5.3691	+5.5694$_{10}$ -1	-1.6807$_{10}$ 0
3.26	0.3839	0.1901	2.0200	5.4201	+5.5822$_{10}$ -1	-1.6901$_{10}$ 0
3.27	0.3823	0.1891	2.0220	5.4715	+5.5948$_{10}$ -1	-1.6996$_{10}$ 0
3.28	0.3807	0.1881	2.0239	5.5234	+5.6074$_{10}$ -1	-1.7090$_{10}$ 0
3.29	0.3792	0.1872	2.0259	5.5758	+5.6199$_{10}$ -1	-1.7184$_{10}$ 0
3.30	0.3776	0.1862	2.0278	5.6286	+5.6323$_{10}$ -1	-1.7279$_{10}$ 0
3.31	0.3760	0.1853	2.0297	5.6820	+5.6446$_{10}$ -1	-1.7373$_{10}$ 0
3.32	0.3745	0.1843	2.0317	5.7358	+5.6569$_{10}$ -1	-1.7467$_{10}$ 0
3.33	0.3729	0.1834	2.0336	5.7900	+5.6691$_{10}$ -1	-1.7561$_{10}$ 0
3.34	0.3714	0.1825	2.0355	5.8418	+5.6812$_{10}$ -1	-1.7656$_{10}$ 0
3.35	0.3699	0.1815	2.0373	5.9000	+5.6932$_{10}$ -1	-1.7750$_{10}$ 0
3.36	0.3683	0.1806	2.0392	5.9558	+5.7051$_{10}$ -1	-1.7844$_{10}$ 0
3.37	0.3668	0.1797	2.0411	6.0120	+5.7170$_{10}$ -1	-1.7938$_{10}$ 0
3.38	0.3653	0.1788	2.0429	6.0687	+5.7287$_{10}$ -1	-1.8031$_{10}$ 0
3.39	0.3638	0.1779	2.0447	6.1260	+5.7404$_{10}$ -1	-1.8125$_{10}$ 0
3.40	0.3623	0.1770	2.0466	6.1837	+5.7521$_{10}$ -1	-1.8219$_{10}$ 0
3.41	0.3608	0.1762	2.0484	6.2419	+5.7636$_{10}$ -1	-1.8313$_{10}$ 0
3.42	0.3594	0.1753	2.0502	6.3007	+5.7751$_{10}$ -1	-1.8407$_{10}$ 0
3.43	0.3579	0.1744	2.0520	6.3600	+5.7865$_{10}$ -1	-1.8500$_{10}$ 0
3.44	0.3564	0.1736	2.0537	6.4198	+5.7978$_{10}$ -1	-1.8594$_{10}$ 0
3.45	0.3550	0.1727	2.0555	6.4801	+5.8091$_{10}$ -1	-1.8687$_{10}$ 0
3.46	0.3535	0.1718	2.0573	6.5409	+5.8203$_{10}$ -1	-1.8781$_{10}$ 0
3.47	0.3521	0.1710	2.0590	6.6023	+5.8314$_{10}$ -1	-1.8874$_{10}$ 0
3.48	0.3507	0.1702	2.0607	6.6642	+5.8424$_{10}$ -1	-1.8967$_{10}$ 0
3.49	0.3492	0.1693	2.0625	6.7266	+5.8534$_{10}$ -1	-1.9061$_{10}$ 0
3.50	0.3478	0.1685	2.0642	6.7896	+5.8643$_{10}$ -1	-1.9154$_{10}$ 0
3.51	0.3464	0.1677	2.0659	6.8532	+5.8751$_{10}$ -1	-1.9247$_{10}$ 0
3.52	0.3450	0.1669	2.0676	6.9172	+5.8859$_{10}$ -1	-1.9340$_{10}$ 0
3.53	0.3436	0.1661	2.0693	6.9819	+5.8966$_{10}$ -1	-1.9433$_{10}$ 0
3.54	0.3422	0.1653	2.0709	7.0471	+5.9072$_{10}$ -1	-1.9526$_{10}$ 0

M	T/T_*	p/p_*	$q/q_* = \rho_*/\rho$	p_o/p_{o*}	$4C_fL_*/D$	$(s-s_*)/R$
3.55	0.3409	0.1645	2.0726	7.1128	+5.9178$_{10}$ -1	-1.9619$_{10}$ 0
3.56	0.3395	0.1637	2.0743	7.1791	+5.9282$_{10}$ -1	-1.9712$_{10}$ 0
3.57	0.3381	0.1629	2.0759	7.2460	+5.9387$_{10}$ -1	-1.9805$_{10}$ 0
3.58	0.3368	0.1621	2.0775	7.3135	+5.9490$_{10}$ -1	-1.9897$_{10}$ 0
3.59	0.3354	0.1613	2.0792	7.3815	+5.9593$_{10}$ -1	-1.9990$_{10}$ 0
3.60	0.3341	0.1606	2.0808	7.4501	+5.9695$_{10}$ -1	-2.0082$_{10}$ 0
3.61	0.3327	0.1598	2.0824	7.5193	+5.9797$_{10}$ -1	-2.0175$_{10}$ 0
3.62	0.3314	0.1590	2.0840	7.5891	+5.9898$_{10}$ -1	-2.0267$_{10}$ 0
3.63	0.3301	0.1583	2.0856	7.6595	+5.9998$_{10}$ -1	-2.0359$_{10}$ 0
3.64	0.3288	0.1575	2.0871	7.7305	+6.0098$_{10}$ -1	-2.0452$_{10}$ 0
3.65	0.3275	0.1568	2.0887	7.8020	+6.0197$_{10}$ -1	-2.0544$_{10}$ 0
3.66	0.3262	0.1560	2.0903	7.8742	+6.0296$_{10}$ -1	-2.0636$_{10}$ 0
3.67	0.3249	0.1553	2.0918	7.9470	+6.0394$_{10}$ -1	-2.0728$_{10}$ 0
3.68	0.3236	0.1546	2.0933	8.0204	+6.0491$_{10}$ -1	-2.0820$_{10}$ 0
3.69	0.3223	0.1539	2.0949	8.0944	+6.0588$_{10}$ -1	-2.0912$_{10}$ 0
3.70	0.3210	0.1531	2.0964	8.1691	+6.0684$_{10}$ -1	-2.1004$_{10}$ 0
3.71	0.3198	0.1524	2.0979	8.2443	+6.0779$_{10}$ -1	-2.1095$_{10}$ 0
3.72	0.3185	0.1517	2.0994	8.3202	+6.0874$_{10}$ -1	-2.1187$_{10}$ 0
3.73	0.3172	0.1510	2.1009	8.3968	+6.0968$_{10}$ -1	-2.1278$_{10}$ 0
3.74	0.3160	0.1503	2.1024	8.4739	+6.1062$_{10}$ -1	-2.1370$_{10}$ 0
3.75	0.3148	0.1496	2.1039	8.5517	+6.1155$_{10}$ -1	-2.1461$_{10}$ 0
3.76	0.3135	0.1489	2.1053	8.6302	+6.1247$_{10}$ -1	-2.1553$_{10}$ 0
3.77	0.3123	0.1482	2.1068	8.7093	+6.1339$_{10}$ -1	-2.1644$_{10}$ 0
3.78	0.3111	0.1475	2.1082	8.7891	+6.1431$_{10}$ -1	-2.1735$_{10}$ 0
3.79	0.3099	0.1469	2.1097	8.8695	+6.1522$_{10}$ -1	-2.1826$_{10}$ 0
3.80	0.3086	0.1462	2.1111	8.9506	+6.1612$_{10}$ -1	-2.1917$_{10}$ 0
3.81	0.3074	0.1455	2.1125	9.0323	+6.1702$_{10}$ -1	-2.2008$_{10}$ 0
3.82	0.3062	0.1449	2.1140	9.1148	+6.1791$_{10}$ -1	-2.2099$_{10}$ 0
3.83	0.3051	0.1442	2.1154	9.1979	+6.1879$_{10}$ -1	-2.2190$_{10}$ 0
3.84	0.3039	0.1436	2.1168	9.2817	+6.1968$_{10}$ -1	-2.2280$_{10}$ 0
3.85	0.3027	0.1429	2.1182	9.3661	+6.2055$_{10}$ -1	-2.2371$_{10}$ 0
3.86	0.3015	0.1423	2.1195	9.4513	+6.2142$_{10}$ -1	-2.2462$_{10}$ 0
3.87	0.3003	0.1416	2.1209	9.5372	+6.2229$_{10}$ -1	-2.2552$_{10}$ 0
3.88	0.2992	0.1410	2.1223	9.6237	+6.2315$_{10}$ -1	-2.2642$_{10}$ 0
3.89	0.2980	0.1403	2.1236	9.7110	+6.2400$_{10}$ -1	-2.2733$_{10}$ 0
3.90	0.2969	0.1397	2.1250	9.7990	+6.2485$_{10}$ -1	-2.2823$_{10}$ 0
3.91	0.2957	0.1391	2.1263	9.8877	+6.2569$_{10}$ -1	-2.2913$_{10}$ 0
3.92	0.2946	0.1385	2.1277	9.9771	+6.2653$_{10}$ -1	-2.3003$_{10}$ 0
3.93	0.2935	0.1378	2.1290	10.0672	+6.2737$_{10}$ -1	-2.3093$_{10}$ 0
3.94	0.2923	0.1372	2.1303	10.1581	+6.2819$_{10}$ -1	-2.3183$_{10}$ 0
3.95	0.2912	0.1366	2.1316	10.2496	+6.2902$_{10}$ -1	-2.3272$_{10}$ 0
3.96	0.2901	0.1360	2.1329	10.3420	+6.2984$_{10}$ -1	-2.3362$_{10}$ 0
3.97	0.2890	0.1354	2.1342	10.4350	+6.3065$_{10}$ -1	-2.3452$_{10}$ 0
3.98	0.2879	0.1348	2.1355	10.5289	+6.3146$_{10}$ -1	-2.3541$_{10}$ 0
3.99	0.2868	0.1342	2.1368	10.6234	+6.3227$_{10}$ -1	-2.3631$_{10}$ 0
4.00	0.2857	0.1336	2.1381	10.7187	+6.3306$_{10}$ -1	-2.3720$_{10}$ 0
4.10	0.2751	0.1279	2.1505	11.7147	+6.4080$_{10}$ -1	-2.4608$_{10}$ 0
4.20	0.2650	0.1226	2.1622	12.7916	+6.4810$_{10}$ -1	-2.5488$_{10}$ 0
4.30	0.2554	0.1175	2.1732	13.9549	+6.5499$_{10}$ -1	-2.6358$_{10}$ 0
4.40	0.2453	0.1128	2.1837	15.2099	+6.6149$_{10}$ -1	-2.7219$_{10}$ 0
4.50	0.2376	0.1083	2.1936	16.5026	+6.6763$_{10}$ -1	-2.8071$_{10}$ 0
4.60	0.2294	0.1041	2.2030	18.0178	+6.7345$_{10}$ -1	-2.8914$_{10}$ 0
4.70	0.2215	0.1001	2.2119	19.5828	+6.7895$_{10}$ -1	-2.9747$_{10}$ 0
4.80	0.2140	0.0964	2.2204	21.2637	+6.8417$_{10}$ -1	-3.0570$_{10}$ 0
4.90	0.2068	0.0928	2.2284	23.0671	+6.8911$_{10}$ -1	-3.1384$_{10}$ 0

M	T/T*	p/p*	q/q* = ρ*/ρ	p₀/p₀*	4C_fL*/D	(s − s*)/R
5.00	0.2000	0.0894	2.2361	25.0000	$+6.9380_{10}$ -1	-3.2189_{10} 0
5.10	0.1935	0.0862	2.2433	27.0696	$+6.9826_{10}$ -1	-3.2984_{10} 0
5.20	0.1873	0.0832	2.2503	29.2833	$+7.0249_{10}$ -1	-3.3770_{10} 0
5.30	0.1813	0.0803	2.2569	31.6491	$+7.0652_{10}$ -1	-3.4547_{10} 0
5.40	0.1756	0.0776	2.2631	34.1748	$+7.1035_{10}$ -1	-3.5315_{10} 0
5.50	0.1702	0.0750	2.2691	36.8690	$+7.1400_{10}$ -1	-3.6074_{10} 0
5.60	0.1650	0.0725	2.2748	39.7402	$+7.1748_{10}$ -1	-3.6824_{10} 0
5.70	0.1600	0.0702	2.2803	42.7974	$+7.2080_{10}$ -1	-3.7565_{10} 0
5.80	0.1553	0.0679	2.2855	46.0500	$+7.2397_{10}$ -1	-3.8297_{10} 0
5.90	0.1507	0.0658	2.2905	49.5075	$+7.2699_{10}$ -1	-3.9021_{10} 0
6.00	0.1463	0.0638	2.2953	53.1798	$+7.2988_{10}$ -1	-3.9737_{10} 0
6.10	0.1421	0.0618	2.2998	57.0772	$+7.3264_{10}$ -1	-4.0444_{10} 0
6.20	0.1381	0.0599	2.3042	61.2102	$+7.3528_{10}$ -1	-4.1143_{10} 0
6.30	0.1343	0.0582	2.3084	65.5899	$+7.3780_{10}$ -1	-4.1834_{10} 0
6.40	0.1305	0.0565	2.3124	70.2[illegible]74	$+7.4022_{10}$ -1	-4.2517_{10} 0
6.50	0.1270	0.0548	2.3163	75.1343	$+7.4254_{10}$ -1	-4.3193_{10} 0
6.60	0.1236	0.0533	2.3200	80.3227	$+7.4477_{10}$ -1	-4.3861_{10} 0
6.70	0.1203	0.0518	2.3235	85.8049	$+7.4690_{10}$ -1	-4.4521_{10} 0
6.80	0.1171	0.0503	2.3269	91.5935	$+7.4895_{10}$ -1	-4.5174_{10} 0
6.90	0.1140	0.0489	2.3302	97.7017	$+7.5091_{10}$ -1	-4.5819_{10} 0

M	T/T*	p/p*	q/q* = ρ*/ρ	p₀/p₀*	4C_fL*/D	(s − s*)/R
7.00	0.1111	0.0476	2.3333	104.1429	$+7.5280_{10}$ -1	-4.6458_{10} 0
7.50	0.0980	0.0417	2.3474	141.8415	$+7.6121_{10}$ -1	-4.9547_{10} 0
8.00	0.0870	0.0369	2.3591	190.1094	$+7.6819_{10}$ -1	-5.2476_{10} 0
8.50	0.0777	0.0328	2.3689	251.0862	$+7.7404_{10}$ -1	-5.5258_{10} 0
9.00	0.0698	0.0293	2.3772	327.1893	$+7.7899_{10}$ -1	-5.7905_{10} 0
9.50	0.0630	0.0264	2.3843	421.1314	$+7.8320_{10}$ -1	-6.0429_{10} 0
10.00	0.0571	0.0239	2.3905	535.9375	$+7.8683_{10}$ -1	-6.2840_{10} 0

TABLE 4 STEADY FRICTIONLESS FLOW OF AIR IN A CONSTANT AREA DUCT WITH HEAT ENERGY TRANSFER–RAYLEIGH FLOW

Flow and fluid parameters are calculated with the local Mach number (M) as the independent variable for the range, M = 0 → (0·01) → 4 → (0·1) → 7 → (0·5) → 10. The summary of equations used is given in Section 6·3. The values of the following parameters are tabulated:

T/T_* ratio of local static temperature to that at reference sonic condition
p/p_* ratio of local static pressure to that at reference sonic condition
$q/q_* = \rho_*/\rho$, ratio of local flow velocity to that at reference sonic condition
p_o/p_{o*} ratio of local total head pressure to that at reference sonic condition
T_o/T_{o*} ratio of local total head temperature to that at reference sonic condition
$(s - s_*)/R$ parameter for entropy change due to heat energy transfer

M	T/T_*	p/p_*	$q/q_* = \rho_*/\rho$	p_o/p_{o*}	T_o/T_{o*}	$(s - s_*)/R$
0	0	2·4000	0	1·2679	0	−∞
0.01	0.0006	2.3997	0.0002	1.2678	0.0005	-2.6984$_{10}$ 1
0.02	0.0023	2.3987	0.0010	1.2675	0.0019	-2.2135$_{10}$ 1
0.03	0.0052	2.3970	0.0022	1.2671	0.0043	-1.9301$_{10}$ 1
0.04	0.0092	2.3946	0.0038	1.2665	0.0076	-1.7293$_{10}$ 1
0.05	0.0143	2.3916	0.0060	1.2657	0.0119	-1.5738$_{10}$ 1
0.06	0.0205	2.3880	0.0086	1.2647	0.0171	-1.4471$_{10}$ 1
0.07	0.0278	2.3836	0.0117	1.2636	0.0232	-1.3403$_{10}$ 1
0.08	0.0362	2.3787	0.0152	1.2623	0.0302	-1.2481$_{10}$ 1
0.09	0.0456	2.3731	0.0192	1.2608	0.0381	-1.1670$_{10}$ 1
0.10	0.0560	2.3669	0.0237	1.2591	0.0468	-1.0949$_{10}$ 1
0.11	0.0674	2.3600	0.0286	1.2573	0.0563	-1.0299$_{10}$ 1
0.12	0.0797	2.3526	0.0339	1.2554	0.0666	-9.708 $_{10}$ 0
0.13	0.0929	2.3445	0.0396	1.2533	0.0777	-9.1690$_{10}$ 0
0.14	0.1069	2.3359	0.0458	1.2510	0.0895	-8.6724$_{10}$ 0
0.15	0.1218	2.3267	0.0524	1.2486	0.1020	-8.2131$_{10}$ 0
0.16	0.1374	2.3170	0.0593	1.2461	0.1151	-7.7865$_{10}$ 0
0.17	0.1538	2.3067	0.0667	1.2434	0.1289	-7.38[illegible]9$_{10}$ 0
0.18	0.1708	2.2959	0.0744	1.2406	0.1432	-7.0169$_{10}$ 0
0.19	0.1884	2.2845	0.0825	1.2377	0.1581	-6.6681$_{10}$ 0
0.20	0.2066	2.2727	0.0909	1.2346	0.1736	-6.3402$_{10}$ 0
0.21	0.2253	2.2604	0.0997	1.2314	0.1894	-6.0312$_{10}$ 0
0.22	0.2445	2.2477	0.108[illegible]	1.2281	0.2057	-5.7395$_{10}$ 0
0.23	0.2641	2.2345	0.1182	1.2247	0.2224	-5.4636$_{10}$ 0
0.24	0.2841	2.2209	0.1279	1.2213	0.2395	-5.2023$_{10}$ 0
0.25	0.3044	2.2069	0.1379	1.2177	0.2568	-4.9545$_{10}$ 0
0.26	0.3250	2.1925	0.1482	1.2140	0.2745	-4.7193$_{10}$ 0
0.27	0.3457	2.1777	0.1583	1.2102	0.2923	-4.4956$_{10}$ 0
0.28	0.3667	2.1626	0.1696	1.2064	0.3104	-4.2828$_{10}$ 0
0.29	0.3877	2.1472	0.1806	1.2025	0.3285	-4.0802$_{10}$ 0
0.30	0.4089	2.1314	0.1918	1.1985	0.3469	-3.8870$_{10}$ 0
0.31	0.4300	2.1154	0.2033	1.1945	0.3653	-3.7028$_{10}$ 0
0.32	0.4512	2.0991	0.2149	1.1904	0.3837	-3.5271$_{10}$ 0
0.33	0.4723	2.0825	0.2268	1.1863	0.4021	-3.3592$_{10}$ 0
0.34	0.4933	2.0657	0.2388	1.1822	0.4206	-3.1989$_{10}$ 0
0.35	0.5141	2.0487	0.2510	1.1779	0.4389	-3.0457$_{10}$ 0
0.36	0.5348	2.0314	0.2633	1.1737	0.4572	-2.8992$_{10}$ 0
0.37	0.5553	2.0140	0.2757	1.1695	0.4754	-2.7590$_{10}$ 0
0.38	0.5755	1.9964	0.2883	1.1652	0.4935	-2.6250$_{10}$ 0
0.39	0.5955	1.9787	0.3010	1.1609	0.5113	-2.4967$_{10}$ 0
0.40	0.6151	1.9608	0.3137	1.1566	0.5290	-2.3740$_{10}$ 0
0.41	0.6345	1.9428	0.3266	1.1523	0.5465	-2.2565$_{10}$ 0
0.42	0.6535	1.9247	0.3395	1.1480	0.5638	-2.1439$_{10}$ 0
0.43	0.6721	1.9065	0.3525	1.1437	0.5808	-2.0362$_{10}$ 0
0.44	0.6903	1.8382	0.3656	1.1394	0.5975	-1.9331$_{10}$ 0

M	T/T_*	p/p_*	$q/q_* = \rho_*/\rho$	p_o/p_{o*}	T_o/T_{o*}	$(s - s_*)/R$
0.45	0.7080	1.8699	0.3787	1.1351	0.6139	-1.8343$_{10}$ 0
0.46	0.7254	1.8515	0.3918	1.1308	0.6301	-1.7397$_{10}$ 0
0.47	0.7423	1.8331	0.4049	1.1266	0.6459	-1.6491$_{10}$ 0
0.48	0.7587	1.8147	0.4181	1.1224	0.6614	-1.5624$_{10}$ 0
0.49	0.7747	1.7962	0.4313	1.1182	0.6765	-1.4793$_{10}$ 0
0.50	0.7901	1.7778	0.4444	1.1141	0.6914	-1.3998$_{10}$ 0
0.51	0.8051	1.7594	0.4576	1.1099	0.7058	-1.3237$_{10}$ 0
0.52	0.8196	1.7409	0.4708	1.1059	0.7199	-1.2509$_{10}$ 0
0.53	0.8335	1.722[illegible]	0.4839	1.1019	0.7336	-1.1812$_{10}$ 0
0.54	0.8469	1.7043	0.4970	1.0979	0.7470	-1.1145$_{10}$ 0
0.55	0.8599	1.6860	0.5100	1.0940	0.7599	-1.0508$_{10}$ 0
0.56	0.8723	1.6678	0.5230	1.0901	0.7725	-9.8977$_{10}$ -1
0.57	0.8842	1.6496	0.5360	1.0863	0.7847	-9.3148$_{10}$ -1
0.58	0.8955	1.6316	0.5489	1.0826	0.7965	-8.7577$_{10}$ -1
0.59	0.9064	1.6136	0.5617	1.0789	0.8079	-8.2255$_{10}$ -1
0.60	0.9167	1.5957	0.5745	1.0753	0.8189	-7.7174$_{10}$ -1
0.61	0.9265	1.5780	0.5872	1.0717	0.8296	-7.2323$_{10}$ -1
0.62	0.9358	1.5[illegible]03	0.5998	1.0682	0.8398	-6.7696$_{10}$ -1
0.63	0.9447	1.5428	0.6123	1.0648	0.8497	-6.3284$_{10}$ -1
0.64	0.9530	1.5253	0.6248	1.0615	0.8592	-5.9078$_{10}$ -1
0.65	0.9608	1.5080	0.6371	1.0582	0.8683	-5.5073$_{10}$ -1
0.66	0.9682	1.4908	0.6494	1.0550	0.8771	-5.1260$_{10}$ -1
0.67	0.9750	1.4738	0.6616	1.0519	0.8355	-4.7634$_{10}$ -1
0.68	0.9814	1.4569	0.6737	1.0489	0.8935	-4.4187$_{10}$ -1
0.69	0.9874	1.4[illegible]01	0.685[illegible]	1.0460	0.9012	-4.0913$_{10}$ -1
0.70	0.9929	1.4235	0.[illegible]975	1.0431	0.9085	-3.7807$_{10}$ -1
0.71	0.9980	1.4070	0.7093	1.0403	0.9155	-3.4861$_{10}$ -1
0.72	1.002[illegible]	1.3907	0.7209	1.0376	0.9221	-3.2072$_{10}$ -1
0.73	1.0068	1.3745	0.7325	1.0350	0.9284	-2.9433$_{10}$ -1
0.74	1.0106	1.3585	0.7439	1.0325	0.9344	-2.6940$_{10}$ -1
0.75	1.0140	1.3427	0.7552	1.0301	0.9401	-2.4587$_{10}$ -1
0.76	1.0171	1.3270	0.7665	1.0278	0.9455	-2.2370$_{10}$ -1
0.77	1.0197	1.3114	0.7775	1.0255	0.9505	-2.0283$_{10}$ -1
0.78	1.0220	1.2961	0.7835	1.0234	0.9553	-1.8324$_{10}$ -1
0.79	1.0239	1.2809	0.7994	1.0213	0.9598	-1.6486$_{10}$ -1
0.80	1.0255	1.2658	0.8101	1.0193	0.9639	-1.4767$_{10}$ -1
0.81	1.0267	1.2510	0.8207	1.0175	0.9679	-1.3162$_{10}$ -1
0.82	1.0276	1.2362	0.8313	1.0157	0.9715	-1.1668$_{10}$ -1
0.83	1.0282	1.2217	0.8416	1.0140	0.9749	-1.0280$_{10}$ -1
0.84	1.0285	1.2073	0.8519	1.0124	0.9781	-8.9953$_{10}$ -2
0.85	1.0285	1.1931	0.8620	1.0109	0.9810	-7.8104$_{10}$ -2
0.86	1.0283	1.1791	0.8721	1.0095	0.9836	-6.7220$_{10}$ -2
0.87	1.0277	1.1652	0.8820	1.0082	0.9861	-5.7268$_{10}$ -2
0.88	1.0269	1.1515	0.8918	1.0070	0.9883	-4.8217$_{10}$ -2
0.89	1.0258	1.1380	0.9014	1.0059	0.9903	-4.0037$_{10}$ -2

M	T/T_*	p/p_*	$q/q_* = \rho_*/\rho$	p_0/p_{0*}	T_0/T_{0*}	$(s - s_*)/R$
0.90	1.0245	1.1246	0.9110	1.0049	0.9921	-3.2700$_{10}$ -2
0.91	1.0230	1.1115	0.9204	1.0039	0.9937	-2.6178$_{10}$ -2
0.92	1.0212	1.0984	0.9297	1.0031	0.9951	-2.0441$_{10}$ -2
0.93	1.0192	1.0856	0.9389	1.0024	0.9963	-1.5472$_{10}$ -2
0.94	1.0170	1.0728	0.9480	1.0017	0.9973	-1.1237$_{10}$ -2
0.95	1.0146	1.0603	0.9569	1.0012	0.9981	-7.7144$_{10}$ -3
0.96	1.0121	1.0479	0.9658	1.0008	0.9988	-4.8812$_{10}$ -3
0.97	1.0093	1.0357	0.9745	1.0004	0.9993	-2.7147$_{10}$ -3
0.98	1.0064	1.0236	0.9831	1.0002	0.9997	-1.1930$_{10}$ -3
0.99	1.0033	1.0117	0.9916	1.0000	0.9999	-2.9493$_{10}$ -4
1	1	1	1	1	1	0
1.01	0.9966	0.9884	1.0083	1.0000	0.9999	-2.8845$_{10}$ -4
1.02	0.9930	0.9770	1.0164	1.0002	0.9997	-1.1412$_{10}$ -3
1.03	0.9894	0.9657	1.0245	1.0004	0.9994	-2.5397$_{10}$ -3
1.04	0.9855	0.9546	1.0325	1.0008	0.9989	-4.4660$_{10}$ -3
1.05	0.9816	0.9436	1.0403	1.0012	0.9984	-6.9029$_{10}$ -3
1.06	0.9776	0.9327	1.0480	1.0017	0.9977	-9.8337$_{10}$ -3
1.07	0.9734	0.9221	1.0557	1.0024	0.9969	-1.3242$_{10}$ -2
1.08	0.9691	0.9115	1.0632	1.0031	0.9960	-1.7112$_{10}$ -2
1.09	0.9648	0.9011	1.0706	1.0039	0.9950	-2.1430$_{10}$ -2
1.10	0.9603	0.8909	1.0780	1.0049	0.9939	-2.6179$_{10}$ -2
1.11	0.9558	0.8808	1.0852	1.0059	0.9927	-3.1346$_{10}$ -2
1.12	0.9512	0.8708	1.0923	1.0070	0.9915	-3.6917$_{10}$ -2
1.13	0.9465	0.8609	1.0993	1.0082	0.9901	-4.2879$_{10}$ -2
1.14	0.9417	0.8512	1.1063	1.0095	0.9887	-4.9219$_{10}$ -2
1.15	0.9369	0.8417	1.1131	1.0109	0.9872	-5.5925$_{10}$ -2
1.16	0.9320	0.8322	1.1198	1.0124	0.9856	-6.2984$_{10}$ -2
1.17	0.9270	0.8229	1.1265	1.0140	0.9840	-7.0385$_{10}$ -2
1.18	0.9220	0.8137	1.1330	1.0157	0.9823	-7.8116$_{10}$ -2
1.19	0.9169	0.8047	1.1395	1.0175	0.9805	-8.6166$_{10}$ -2
1.20	0.9118	0.7958	1.1459	1.0194	0.9787	-9.4525$_{10}$ -2
1.21	0.9067	0.7870	1.1522	1.0214	0.9768	-1.0318$_{10}$ -1
1.22	0.9015	0.7783	1.1584	1.0235	0.9749	-1.1213$_{10}$ -1
1.23	0.8963	0.7697	1.1645	1.0257	0.9729	-1.2135$_{10}$ -1
1.24	0.8911	0.7613	1.1705	1.0279	0.9709	-1.3085$_{10}$ -1
1.25	0.8858	0.7529	1.1765	1.0303	0.9689	-1.4060$_{10}$ -1
1.26	0.8805	0.7447	1.1823	1.0328	0.9668	-1.5061$_{10}$ -1
1.27	0.8752	0.7366	1.1881	1.0354	0.9646	-1.6086$_{10}$ -1
1.28	0.8699	0.7287	1.1938	1.0380	0.9624	-1.7135$_{10}$ -1
1.29	0.8645	0.7208	1.1994	1.0408	0.9602	-1.8206$_{10}$ -1
1.30	0.8592	0.7130	1.2050	1.0437	0.9580	-1.9299$_{10}$ -1
1.31	0.8538	0.7054	1.2105	1.0466	0.9557	-2.0413$_{10}$ -1
1.32	0.8484	0.6978	1.2159	1.0497	0.9534	-2.1548$_{10}$ -1
1.33	0.8430	0.6904	1.2212	1.0528	0.9511	-2.2702$_{10}$ -1
1.34	0.8377	0.6830	1.2264	1.0561	0.9487	-2.3876$_{10}$ -1
1.35	0.8323	0.6758	1.2316	1.0594	0.9464	-2.5068$_{10}$ -1
1.36	0.8269	0.6686	1.2367	1.0629	0.9440	-2.6277$_{10}$ -1
1.37	0.8215	0.6616	1.2417	1.0664	0.9416	-2.7504$_{10}$ -1
1.38	0.8161	0.6546	1.2467	1.0701	0.9391	-2.8747$_{10}$ -1
1.39	0.8108	0.6478	1.2516	1.0738	0.9367	-3.0006$_{10}$ -1
1.40	0.8054	0.6410	1.2564	1.0777	0.9343	-3.1281$_{10}$ -1
1.41	0.8000	0.6344	1.2612	1.0816	0.9318	-3.2570$_{10}$ -1
1.42	0.7947	0.6278	1.2659	1.0856	0.9293	-3.3874$_{10}$ -1
1.43	0.7894	0.6213	1.2705	1.0898	0.9268	-3.5191$_{10}$ -1
1.44	0.7840	0.6149	1.2751	1.0940	0.9243	-3.6522$_{10}$ -1

M	T/T_*	p/p_*	$q/q_* = \rho_*/\rho$	p_0/p_{0*}	T_0/T_{0*}	$(s - s_*)/R$
1.45	0.7787	0.6086	1.2796	1.0983	0.9218	-3.7865$_{10}$ -1
1.46	0.7735	0.6024	1.2840	1.1028	0.9193	-3.9221$_{10}$ -1
1.47	0.7682	0.5962	1.2884	1.1073	0.9168	-4.0589$_{10}$ -1
1.48	0.7629	0.5902	1.2927	1.1120	0.9143	-4.1968$_{10}$ -1
1.49	0.7577	0.5842	1.2970	1.1167	0.9118	-4.3358$_{10}$ -1
1.50	0.7525	0.5783	1.3012	1.1215	0.9093	-4.4758$_{10}$ -1
1.51	0.7473	0.5725	1.3054	1.1265	0.9068	-4.6169$_{10}$ -1
1.52	0.7422	0.5668	1.3095	1.1315	0.9042	-4.7589$_{10}$ -1
1.53	0.7370	0.5611	1.3135	1.1367	0.9017	-4.9019$_{10}$ -1
1.54	0.7319	0.5555	1.3175	1.1419	0.8992	-5.0458$_{10}$ -1
1.55	0.7268	0.5500	1.3214	1.1473	0.8967	-5.1905$_{10}$ -1
1.56	0.7217	0.5446	1.3253	1.1527	0.8942	-5.3361$_{10}$ -1
1.57	0.7167	0.5392	1.3291	1.1583	0.8917	-5.4824$_{10}$ -1
1.58	0.7117	0.5339	1.3329	1.1640	0.8892	-5.6295$_{10}$ -1
1.59	0.7067	0.5287	1.3366	1.1697	0.8867	-5.7774$_{10}$ -1
1.60	0.7017	0.5236	1.3403	1.1756	0.8842	-5.9259$_{10}$ -1
1.61	0.6968	0.5185	1.3439	1.1816	0.8817	-6.0752$_{10}$ -1
1.62	0.6919	0.5135	1.3475	1.1877	0.8792	-6.2250$_{10}$ -1
1.63	0.6870	0.5085	1.3511	1.1939	0.8768	-6.3755$_{10}$ -1
1.64	0.6822	0.5036	1.3546	1.2002	0.8743	-6.5265$_{10}$ -1
1.65	0.6774	0.4988	1.3580	1.2066	0.8718	-6.6781$_{10}$ -1
1.66	0.6726	0.4940	1.3614	1.2131	0.8694	-6.8303$_{10}$ -1
1.67	0.6678	0.4894	1.3647	1.2197	0.8670	-6.9829$_{10}$ -1
1.68	0.6631	0.4847	1.3681	1.2264	0.8645	-7.1360$_{10}$ -1
1.69	0.6584	0.4801	1.3713	1.2333	0.8621	-7.2896$_{10}$ -1
1.70	0.6538	0.4756	1.3746	1.2402	0.8597	-7.4436$_{10}$ -1
1.71	0.6491	0.4712	1.3777	1.2473	0.8573	-7.5981$_{10}$ -1
1.72	0.6445	0.4668	1.3809	1.2545	0.8549	-7.7529$_{10}$ -1
1.73	0.6400	0.4624	1.3840	1.2618	0.8526	-7.9081$_{10}$ -1
1.74	0.6355	0.4581	1.3870	1.2692	0.8502	-8.0636$_{10}$ -1
1.75	0.6310	0.4539	1.3901	1.2767	0.8478	-8.2195$_{10}$ -1
1.76	0.6265	0.4497	1.3931	1.2843	0.8455	-8.3757$_{10}$ -1
1.77	0.6221	0.4456	1.3960	1.2920	0.8432	-8.5322$_{10}$ -1
1.78	0.6176	0.4415	1.3989	1.2999	0.8409	-8.6889$_{10}$ -1
1.79	0.6133	0.4375	1.4018	1.3078	0.8386	-8.8459$_{10}$ -1
1.80	0.6089	0.4335	1.4046	1.3159	0.8363	-9.0031$_{10}$ -1
1.81	0.6046	0.4296	1.4074	1.3241	0.8340	-9.1606$_{10}$ -1
1.82	0.6004	0.4257	1.4102	1.3324	0.8317	-9.3183$_{10}$ -1
1.83	0.5961	0.4219	1.4129	1.3409	0.8295	-9.4761$_{10}$ -1
1.84	0.5919	0.4181	1.4156	1.3494	0.8273	-9.6342$_{10}$ -1
1.85	0.5877	0.4144	1.4183	1.3581	0.8250	-9.7924$_{10}$ -1
1.86	0.5836	0.4107	1.4209	1.3669	0.8228	-9.9507$_{10}$ -1
1.87	0.5795	0.4071	1.4235	1.3758	0.8206	-1.0109$_{10}$ 0
1.88	0.5754	0.4035	1.4261	1.3849	0.8185	-1.0268$_{10}$ 0
1.89	0.5714	0.3999	1.4286	1.3940	0.8163	-1.0426$_{10}$ 0
1.90	0.5673	0.3964	1.4311	1.4033	0.8141	-1.0585$_{10}$ 0
1.91	0.5634	0.3930	1.4336	1.4127	0.8120	-1.0744$_{10}$ 0
1.92	0.5594	0.3895	1.4360	1.4222	0.8099	-1.0903$_{10}$ 0
1.93	0.5555	0.3862	1.4384	1.4319	0.8078	-1.1062$_{10}$ 0
1.94	0.5516	0.3828	1.4408	1.4417	0.8057	-1.1221$_{10}$ 0
1.95	0.5477	0.3795	1.4432	1.4516	0.8036	-1.1380$_{10}$ 0
1.96	0.5439	0.3763	1.4455	1.4616	0.8015	-1.1539$_{10}$ 0
1.97	0.5401	0.3731	1.4478	1.4718	0.7995	-1.1698$_{10}$ 0
1.98	0.5364	0.3699	1.4501	1.4821	0.7974	-1.1858$_{10}$ 0
1.99	0.5326	0.3667	1.4523	1.4925	0.7954	-1.2017$_{10}$ 0

M	T/T_*	p/p_*	$q/q_* = \rho_*/\rho$	p_o/p_{o*}	T_o/T_{o*}	$(s - s_*)/R$
2.00	0.5289	0.3636	1.4545	1.5031	0.7934	-1.2176D 0
2.01	0.5253	0.3606	1.4567	1.5138	0.7914	-1.2335D 0
2.02	0.5216	0.3575	1.4589	1.5246	0.7894	-1.2494D 0
2.03	0.5180	0.3545	1.4610	1.5356	0.7874	-1.2653D 0
2.04	0.5144	0.3516	1.4632	1.5467	0.7855	-1.2812D 0
2.05	0.5109	0.3487	1.4652	1.5579	0.7835	-1.2971D 0
2.06	0.5074	0.3458	1.4673	1.5693	0.7816	-1.3130D 0
2.07	0.5039	0.3429	1.4693	1.5808	0.7797	-1.3288D 0
2.08	0.5004	0.3401	1.4714	1.5924	0.7778	-1.3447D 0
2.09	0.4970	0.3373	1.4734	1.6042	0.7759	-1.3606D 0
2.10	0.4936	0.3345	1.4753	1.6162	0.7741	-1.3764D 0
2.11	0.4902	0.3318	1.4773	1.6282	0.7722	-1.3922D 0
2.12	0.4868	0.3291	1.4792	1.6404	0.7704	-1.4081D 0
2.13	0.4835	0.3265	1.4811	1.6528	0.7685	-1.4239D 0
2.14	0.4802	0.3238	1.4830	1.6653	0.7667	-1.4397D 0
2.15	0.4770	0.3212	1.4848	1.6780	0.7649	-1.4555D 0
2.16	0.4737	0.3186	1.4867	1.6908	0.7631	-1.4713D 0
2.17	0.4705	0.3161	1.4885	1.7037	0.7614	-1.4870D 0
2.18	0.4673	0.3136	1.4903	1.7168	0.7596	-1.5028D 0
2.19	0.4642	0.3111	1.4921	1.7300	0.7579	-1.5185D 0
2.20	0.4611	0.3086	1.4938	1.7434	0.7561	-1.5342D 0
2.21	0.4580	0.3062	1.4956	1.7570	0.7544	-1.5499D 0
2.22	0.4549	0.3038	1.4973	1.7707	0.7527	-1.5656D 0
2.23	0.4518	0.3014	1.4990	1.7846	0.7510	-1.5813D 0
2.24	0.4488	0.2991	1.5007	1.7986	0.7493	-1.5970D 0
2.25	0.4458	0.2968	1.5023	1.8128	0.7477	-1.6126D 0
2.26	0.4428	0.2945	1.5040	1.8271	0.7460	-1.6282D 0
2.27	0.4399	0.2922	1.5056	1.8416	0.7444	-1.6438D 0
2.28	0.4370	0.2899	1.5072	1.8562	0.7428	-1.6594D 0
2.29	0.4341	0.2877	1.5088	1.8710	0.7411	-1.6750D 0
2.30	0.4312	0.2855	1.5103	1.8860	0.7395	-1.6905D 0
2.31	0.4284	0.2833	1.5119	1.9012	0.7380	-1.7060D 0
2.32	0.4256	0.2812	1.5134	1.9165	0.7364	-1.7215D 0
2.33	0.4228	0.2791	1.5150	1.9319	0.7348	-1.7370D 0
2.34	0.4200	0.2769	1.5165	1.9476	0.7333	-1.7525D 0
2.35	0.4172	0.2749	1.5180	1.9634	0.7317	-1.7679D 0
2.36	0.4145	0.2728	1.5194	1.9794	0.7302	-1.7833D 0
2.37	0.4118	0.2708	1.5209	1.9955	0.7287	-1.7987D 0
2.38	0.4091	0.2688	1.5223	2.0119	0.7272	-1.8141D 0
2.39	0.4065	0.2668	1.5237	2.0284	0.7257	-1.8294D 0
2.40	0.4038	0.2648	1.5252	2.0451	0.7242	-1.8448D 0
2.41	0.4012	0.2628	1.5265	2.0619	0.7227	-1.8601D 0
2.42	0.3986	0.2609	1.5279	2.0789	0.7213	-1.8754D 0
2.43	0.3961	0.2590	1.5293	2.0962	0.7198	-1.8906D 0
2.44	0.3935	0.2571	1.5306	2.1136	0.7184	-1.9059D 0
2.45	0.3910	0.2552	1.5320	2.1311	0.7170	-1.9211D 0
2.46	0.3885	0.2534	1.5333	2.1489	0.7156	-1.9363D 0
2.47	0.3860	0.2515	1.5346	2.1669	0.7142	-1.9514D 0
2.48	0.3836	0.2497	1.5359	2.1850	0.7128	-1.9666D 0
2.49	0.3811	0.2479	1.5372	2.2033	0.7114	-1.9817D 0
2.50	0.3787	0.2462	1.5385	2.2218	0.7101	-1.9968D 0
2.51	0.3763	0.2444	1.5397	2.2405	0.7087	-2.0118D 0
2.52	0.3739	0.2427	1.5410	2.2594	0.7074	-2.0269D 0
2.53	0.3716	0.2409	1.5422	2.2785	0.7060	-2.0419D 0
2.54	0.3692	0.2392	1.5434	2.2978	0.7047	-2.0569D 0

M	T/T_*	p/p_*	$q/q_* = \rho_*/\rho$	p_o/p_{o*}	T_o/T_{o*}	$(s - s_*)/R$
2.55	0.3669	0.2375	1.5446	2.3173	0.7034	-2.0718D 0
2.56	0.3646	0.2359	1.5458	2.3370	0.7021	-2.0868D 0
2.57	0.3623	0.2342	1.5470	2.3569	0.7008	-2.1017D 0
2.58	0.3601	0.2326	1.5482	2.3770	0.6995	-2.1166D 0
2.59	0.3578	0.2310	1.5493	2.3972	0.6983	-2.1314D 0
2.60	0.3556	0.2294	1.5505	2.4177	0.6970	-2.1463D 0
2.61	0.3534	0.2278	1.5516	2.4384	0.6957	-2.1611D 0
2.62	0.3512	0.2262	1.5527	2.4593	0.6945	-2.1758D 0
2.63	0.3491	0.2246	1.5538	2.4805	0.6933	-2.1906D 0
2.64	0.3469	0.2231	1.5549	2.5018	0.6921	-2.2053D 0
2.65	0.3448	0.2216	1.5560	2.5233	0.6908	-2.2200D 0
2.66	0.3427	0.2201	1.5571	2.5451	0.6896	-2.2347D 0
2.67	0.3406	0.2186	1.5582	2.5671	0.6885	-2.2493D 0
2.68	0.3385	0.2171	1.5592	2.5892	0.6873	-2.2640D 0
2.69	0.3364	0.2156	1.5603	2.6117	0.6861	-2.2786D 0
2.70	0.3344	0.2142	1.5613	2.6343	0.6849	-2.2931D 0
2.71	0.3324	0.2127	1.5623	2.6571	0.6838	-2.3077D 0
2.72	0.3304	0.2113	1.5634	2.6802	0.6826	-2.3222D 0
2.73	0.3284	0.2099	1.5644	2.7035	0.6815	-2.3367D 0
2.74	0.3264	0.2085	1.5654	2.7270	0.6804	-2.3511D 0
2.75	0.3244	0.2071	1.5663	2.7508	0.6793	-2.3655D 0
2.76	0.3225	0.2058	1.5673	2.7748	0.6781	-2.3799D 0
2.77	0.3205	0.2044	1.5683	2.7990	0.6770	-2.3943D 0
2.78	0.3186	0.2030	1.5693	2.8235	0.6760	-2.4087D 0
2.79	0.3167	0.2017	1.5702	2.8482	0.6749	-2.4230D 0
2.80	0.3149	0.2004	1.5711	2.8731	0.6738	-2.4373D 0
2.81	0.3130	0.1991	1.5721	2.8982	0.6727	-2.4515D 0
2.82	0.3111	0.1978	1.5730	2.9237	0.6717	-2.4658D 0
2.83	0.3093	0.1965	1.5739	2.9493	0.6706	-2.4800D 0
2.84	0.3075	0.1953	1.5748	2.9752	0.6696	-2.4942D 0
2.85	0.3057	0.1940	1.5757	3.0014	0.6685	-2.5083D 0
2.86	0.3039	0.1927	1.5766	3.0278	0.6675	-2.5225D 0
2.87	0.3021	0.1915	1.5775	3.0544	0.6665	-2.5366D 0
2.88	0.3004	0.1903	1.5784	3.0813	0.6655	-2.5506D 0
2.89	0.2986	0.1891	1.5792	3.1084	0.6645	-2.5647D 0
2.90	0.2969	0.1879	1.5801	3.1359	0.6635	-2.5787D 0
2.91	0.2951	0.1867	1.5809	3.1635	0.6625	-2.5927D 0
2.92	0.2934	0.1855	1.5818	3.1914	0.6615	-2.6066D 0
2.93	0.2918	0.1843	1.5826	3.2196	0.6606	-2.6206D 0
2.94	0.2901	0.1832	1.5834	3.2481	0.6596	-2.6345D 0
2.95	0.2884	0.1820	1.5843	3.2768	0.6586	-2.6483D 0
2.96	0.2868	0.1809	1.5851	3.3058	0.6577	-2.6622D 0
2.97	0.2851	0.1798	1.5859	3.3350	0.6568	-2.6760D 0
2.98	0.2835	0.1787	1.5867	3.3646	0.6558	-2.6898D 0
2.99	0.2819	0.1776	1.5875	3.3944	0.6549	-2.7036D 0
3.00	0.2803	0.1765	1.5882	3.4245	0.6540	-2.7173D 0
3.01	0.2787	0.1754	1.5890	3.4548	0.6531	-2.7310D 0
3.02	0.2771	0.1743	1.5898	3.4854	0.6522	-2.7447D 0
3.03	0.2756	0.1732	1.5905	3.5164	0.6513	-2.7584D 0
3.04	0.2740	0.1722	1.5913	3.5476	0.6504	-2.7720D 0
3.05	0.2725	0.1711	1.5920	3.5790	0.6495	-2.7856D 0
3.06	0.2709	0.1701	1.5928	3.6108	0.6486	-2.7992D 0
3.07	0.2694	0.1691	1.5935	3.6429	0.6477	-2.8127D 0
3.08	0.2679	0.1681	1.5942	3.6752	0.6469	-2.8262D 0
3.09	0.2664	0.1670	1.5950	3.7079	0.6460	-2.8397D 0

M	T/T_*	p/p_*	$q/q_* = \rho_*/\rho$	p_0/p_{0*}	T_0/T_{0*}	$(s - s_*)/R$
3.10	0.2650	0.1660	1.5957	3.7408	0.6452	-2.8532D 0
3.11	0.2635	0.1651	1.5964	3.7741	0.6443	-2.8666D 0
3.12	0.2620	0.1641	1.5971	3.8076	0.6435	-2.8800D 0
3.13	0.2606	0.1631	1.5978	3.8415	0.6426	-2.8934D 0
3.14	0.2592	0.1621	1.5985	3.8756	0.6418	-2.9068D 0
3.15	0.2577	0.1612	1.5992	3.9101	0.6410	-2.9201D 0
3.16	0.2563	0.1602	1.5998	3.9449	0.6402	-2.9334D 0
3.17	0.2549	0.1593	1.6005	3.9800	0.6394	-2.9467D 0
3.18	0.2535	0.1583	1.6012	4.0154	0.6386	-2.9599D 0
3.19	0.2522	0.1574	1.6018	4.0511	0.6378	-2.9732D 0
3.20	0.2508	0.1565	1.6025	4.0871	0.6370	-2.9863D 0
3.21	0.2494	0.1556	1.6032	4.1235	0.6362	-2.9995D 0
3.22	0.2481	0.1547	1.6038	4.1602	0.6354	-3.0127D 0
3.23	0.2467	0.1538	1.6044	4.1972	0.6347	-3.0258D 0
3.24	0.2454	0.1529	1.6051	4.2345	0.6339	-3.0389D 0
3.25	0.2441	0.1520	1.6057	4.2721	0.6331	-3.0519D 0
3.26	0.2428	0.1511	1.6063	4.3101	0.6324	-3.0649D 0
3.27	0.2415	0.1503	1.6069	4.3485	0.6316	-3.0780D 0
3.28	0.2402	0.1494	1.6076	4.3871	0.6309	-3.0909D 0
3.29	0.2389	0.1486	1.6082	4.4261	0.6301	-3.1039D 0
3.30	0.2377	0.1477	1.6088	4.4655	0.6294	-3.1168D 0
3.31	0.2364	0.1469	1.6094	4.5052	0.6287	-3.1297D 0
3.32	0.2352	0.1461	1.6100	4.5452	0.6280	-3.1426D 0
3.33	0.2339	0.1452	1.6105	4.5856	0.6272	-3.1554D 0
3.34	0.2327	0.1444	1.6111	4.6263	0.6265	-3.1683D 0
3.35	0.2315	0.1436	1.6117	4.6674	0.6258	-3.1810D 0
3.36	0.2303	0.1428	1.6123	4.7089	0.6251	-3.1938D 0
3.37	0.2290	0.1420	1.6128	4.7507	0.6244	-3.2066D 0
3.38	0.2279	0.1412	1.6134	4.7929	0.6237	-3.2193D 0
3.39	0.2267	0.1404	1.6140	4.8354	0.6230	-3.2320D 0
3.40	0.2255	0.1397	1.6145	4.8783	0.6224	-3.2446D 0
3.41	0.2243	0.1389	1.6151	4.9216	0.6217	-3.2573D 0
3.42	0.2232	0.1381	1.6156	4.9652	0.6210	-3.2699D 0
3.43	0.2220	0.1374	1.6162	5.0092	0.6203	-3.2825D 0
3.44	0.2209	0.1366	1.6167	5.0536	0.6197	-3.2950D 0
3.45	0.2197	0.1359	1.6172	5.0984	0.6190	-3.3076D 0
3.46	0.2186	0.1351	1.6178	5.1435	0.6184	-3.3201D 0
3.47	0.2175	0.1344	1.6183	5.1891	0.6177	-3.3326D 0
3.48	0.2164	0.1337	1.6188	5.2350	0.6171	-3.3450D 0
3.49	0.2153	0.1329	1.6193	5.2813	0.6164	-3.3575D 0
3.50	0.2142	0.1322	1.6198	5.3280	0.6158	-3.3699D 0
3.51	0.2131	0.1315	1.6203	5.3751	0.6152	-3.3823D 0
3.52	0.2120	0.1308	1.6208	5.4226	0.6145	-3.3946D 0
3.53	0.2110	0.1301	1.6213	5.4705	0.6139	-3.4069D 0
3.54	0.2099	0.1294	1.6218	5.5188	0.6133	-3.4193D 0
3.55	0.2088	0.1287	1.6223	5.5676	0.6127	-3.4315D 0
3.56	0.2078	0.1280	1.6228	5.6167	0.6121	-3.4438D 0
3.57	0.2068	0.1274	1.6233	5.6662	0.6115	-3.4560D 0
3.58	0.2057	0.1267	1.6238	5.7162	0.6109	-3.4682D 0
3.59	0.2047	0.1260	1.6243	5.7665	0.6103	-3.4804D 0

M	T/T_*	p/p_*	$q/q_* = \rho_*/\rho$	p_0/p_{0*}	T_0/T_{0*}	$(s - s_*)/R$
3.60	0.2037	0.1254	1.6247	5.8173	0.6097	-3.4926D 0
3.61	0.2027	0.1247	1.6252	5.8685	0.6091	-3.5047D 0
3.62	0.2017	0.1241	1.6257	5.9201	0.6085	-3.5168D 0
3.63	0.2007	0.1234	1.6261	5.9722	0.6080	-3.5289D 0
3.64	0.1997	0.1228	1.6266	6.0247	0.6074	-3.5410D 0
3.65	0.1987	0.1221	1.6271	6.0776	0.6068	-3.5530D 0
3.66	0.1977	0.1215	1.6275	6.1310	0.6062	-3.5650D 0
3.67	0.1968	0.1209	1.6280	6.1848	0.6057	-3.5770D 0
3.68	0.1958	0.1202	1.6284	6.2390	0.6051	-3.5890D 0
3.69	0.1949	0.1196	1.6288	6.2937	0.6046	-3.6009D 0
3.70	0.1939	0.1190	1.6293	6.3488	0.6040	-3.6128D 0
3.71	0.1930	0.1184	1.6297	6.4044	0.6035	-3.6247D 0
3.72	0.1920	0.1178	1.6301	6.4605	0.6029	-3.6366D 0
3.73	0.1911	0.1172	1.6306	6.5170	0.6024	-3.6485D 0
3.74	0.1902	0.1166	1.6310	6.5739	0.6018	-3.6603D 0
3.75	0.1893	0.1160	1.6314	6.6314	0.6013	-3.6721D 0
3.76	0.1884	0.1154	1.6318	6.6893	0.6008	-3.6838D 0
3.77	0.1875	0.1148	1.6323	6.7476	0.6003	-3.6956D 0
3.78	0.1866	0.1143	1.6327	6.8065	0.5997	-3.7073D 0
3.79	0.1857	0.1137	1.6331	6.8658	0.5992	-3.7190D 0
3.80	0.1848	0.1131	1.6335	6.9256	0.5987	-3.7307D 0
3.81	0.1839	0.1126	1.6339	6.9858	0.5982	-3.7424D 0
3.82	0.1830	0.1120	1.6343	7.0466	0.5977	-3.7540D 0
3.83	0.1822	0.1114	1.6347	7.1078	0.5972	-3.7656D 0
3.84	0.1813	0.1109	1.6351	7.1696	0.5967	-3.7772D 0
3.85	0.1805	0.1103	1.6355	7.2318	0.5962	-3.7888D 0
3.86	0.1796	0.1098	1.6359	7.2945	0.5957	-3.8003D 0
3.87	0.1788	0.1093	1.6362	7.3578	0.5952	-3.8118D 0
3.88	0.1779	0.1087	1.6366	7.4215	0.5947	-3.8233D 0
3.89	0.1771	0.1082	1.6370	7.4858	0.5942	-3.8348D 0
3.90	0.1763	0.1077	1.6374	7.5505	0.5937	-3.8463D 0
3.91	0.1754	0.1071	1.6378	7.6158	0.5933	-3.8577D 0
3.92	0.1746	0.1066	1.6381	7.6816	0.5928	-3.8691D 0
3.93	0.1738	0.1061	1.6385	7.7479	0.5923	-3.8805D 0
3.94	0.1730	0.1056	1.6389	7.8147	0.5918	-3.8918D 0
3.95	0.1722	0.1051	1.6392	7.8820	0.5914	-3.9032D 0
3.96	0.1714	0.1046	1.6396	7.9499	0.5909	-3.9145D 0
3.97	0.1706	0.1041	1.6400	8.0184	0.5904	-3.9258D 0
3.98	0.1699	0.1036	1.6403	8.0873	0.5900	-3.9371D 0
3.99	0.1691	0.1031	1.6407	8.1568	0.5895	-3.9483D 0
4.00	0.16831	0.10256	1.6410	8.227	0.5891	-3.9595D 0
4.10	0.16086	0.09782	1.6444	8.958	0.5847	-4.0706D 0
4.20	0.15388	0.09340	1.6476	9.747	0.5807	-4.1796D 0
4.30	0.14734	0.08927	1.6505	10.599	0.5768	-4.2865D 0
4.40	0.14119	0.08540	1.6533	11.516	0.5732	-4.3914D 0
4.50	0.13540	0.08177	1.6559	12.502	0.5698	-4.4944D 0
4.60	0.12996	0.07837	1.6583	13.563	0.5666	-4.5955D 0
4.70	0.12483	0.07517	1.6606	14.702	0.5636	-4.6948D 0
4.80	0.12000	0.07217	1.6627	15.923	0.5608	-4.7923D 0
4.90	0.11543	0.06934	1.6648	17.232	0.5581	-4.8881D 0

M	T/T_*	p/p_*	$q/q_* = \rho_*/\rho$	p_0/p_{0*}	T_0/T_{0*}	$(s - s_*)/R$
5.00	0.11111	0.06667	1.6667	18.634	0.5556	-4.9822$_{10}$ 0
5.10	0.10703	0.06415	1.6685	20.133	0.5532	-5.0748$_{10}$ 0
5.20	0.10316	0.06177	1.6702	21.734	0.5509	-5.1658$_{10}$ 0
5.30	0.09950	0.05951	1.6718	23.444	0.5487	-5.2552$_{10}$ 0
5.40	0.09602	0.05738	1.6733	25.268	0.5467	-5.3432$_{10}$ 0
5.50	0.09272	0.05536	1.6747	27.211	0.5447	-5.4298$_{10}$ 0
5.60	0.08958	0.05345	1.6761	29.281	0.5429	-5.5150$_{10}$ 0
5.70	0.08660	0.05163	1.6774	31.482	0.5411	-5.5988$_{10}$ 0
5.80	0.08376	0.04990	1.6786	33.822	0.5394	-5.6814$_{10}$ 0
5.90	0.08106	0.04826	1.6798	36.308	0.5378	-5.7627$_{10}$ 0
6.00	0.07849	0.04669	1.6809	38.946	0.5363	-5.8427$_{10}$ 0
6.10	0.07603	0.04520	1.6820	41.744	0.5349	-5.9215$_{10}$ 0
6.20	0.07369	0.04378	1.6830	44.708	0.5335	-5.9992$_{10}$ 0
6.30	0.07145	0.04243	1.6840	47.848	0.5322	-6.0758$_{10}$ 0
6.40	0.06931	0.04114	1.6849	51.170	0.5309	-6.1512$_{10}$ 0
6.50	0.06726	0.03990	1.6858	54.683	0.5297	-6.2256$_{10}$ 0
6.60	0.06531	0.03872	1.6866	58.395	0.5285	-6.2990$_{10}$ 0
6.70	0.06343	0.03759	1.6874	62.315	0.5274	-6.3713$_{10}$ 0
6.80	0.06164	0.03651	1.6882	66.452	0.5264	-6.4426$_{10}$ 0
6.90	0.05991	0.03547	1.6889	70.815	0.5254	-6.5130$_{10}$ 0

M	T/T_*	p/p_*	$q/q_* = \rho_*/\rho$	p_0/p_{0*}	T_0/T_{0*}	$(s - s_*)/R$
7.00	0.05826	0.03448	1.6897	75.414	0.5244	-6.5824$_{10}$ 0
7.50	0.05094	0.03009	1.6928	102.287	0.5200	-6.9162$_{10}$ 0
8.00	0.04491	0.02649	1.6954	136.624	0.5165	-7.2298$_{10}$ 0
8.50	0.03988	0.02349	1.6975	179.924	0.5135	-7.5254$_{10}$ 0
9.00	0.03565	0.02098	1.6993	233.884	0.5110	-7.8048$_{10}$ 0
9.50	0.03205	0.01885	1.7008	300.407	0.5088	-8.0698$_{10}$ 0
10.00	0.02897	0.01702	1.7021	381.615	0.5070	-8.3217$_{10}$ 0

TABLE 5 STEADY ADIABATIC FLOW OF AIR THROUGH A PLANE OBLIQUE SHOCK WAVE

Flow and fluid parameters are calculated with the incident Mach number (M_1) as an independent variable for the range, $M_1 = 1{\cdot}02 \rightarrow (0{\cdot}02) \rightarrow 2{\cdot}2 \rightarrow (0{\cdot}1) \rightarrow 2.4$
The other independent variable, (δ) the angular (degrees) change in flow direction across the shock, is assigned values for each value of M_1 in the range,

$$\delta = 0, 1 \rightarrow (+2) \rightarrow \delta_*, \delta_m \rightarrow (-2) \rightarrow 1, 0$$

where δ_* and δ_m are, respectively, the values of δ at the sonic condition ($M_2 = 1$) and at the tangency (maximum deflection) condition.
Tabulated data above δ_m in Table 5 correspond to the *weak shock* solution ($\delta = 0$ gives the isentropic Mach wave solution of Table 1); while tabulated data below δ_m in Table 5 correspond to the *strong shock* solution ($\delta = 0$ gives the plane normal shock wave solution of Table 2).
The summary of equations used is given in Section 7·3. The values of the following parameters are tabulated:

- θ angle (degrees) which the oblique shock wave makes with the incident flow direction
- T_2/T_1 static temperature ratio across the shock
- p_2/p_1 static pressure ratio across the shock
- p_{o2}/p_{o1} total head pressure ratio across the shock
- M_2 Mach number downstream of shock
- q_2/q_* ratio of local velocity downstream of shock to that at reference sonic condition
- u_2/q_* X–direction component of q_2/q_*
- v_2/q_* Y–direction component of q_2/q_*
- $(s_2 - s_1)/R$ parameter for entropy change across shock

(Note: (1) (7·26) gives the simple logarithmic relation between the entropy and total head pressure parameters
(2) (q_1/q_*) is the ratio of local velocity upstream of shock to that at reference condition and is given for each value of M_1

M_1	δ	θ	T_2/T_1	p_2/p_1	p_{o2}/p_{o1}	M_2	q_2/q_*	u_2/q_*	v_2/q_*	$(s_2 - s_1)/R$
1.02 (1.017)	0.0	78.635	1.0000	1.0000	1.0000	1.0200	1.017	1.017	0.000	0.0000
	δ_* = 0.1331	82.024	1.0067	1.0238	1.0000	1.0000	1.000	1.000	0.002	+1.3220$_{10}$ -6
	δ_m = 0.1447	83.487	1.0089	1.0315	1.0000	0.9935	0.995	0.994	0.003	+3.0484$_{10}$ -6
	0.0	90.000	1.0132	1.0471	1.0000	0.9805	0.984	0.984	0.000	+9.9682$_{10}$ -6
1.04 (1.033)	0.0	74.058	1.0000	1.0000	1.0000	1.0400	1.033	1.033	0.000	0.0000
	δ_* = 0.3713	78.897	1.0136	1.0484	1.0000	1.0000	1.000	0.999	0.006	+1.0776$_{10}$ -5
	δ_m = 0.4026	80.930	1.0178	1.0638	1.0000	0.9874	0.989	0.983	0.007	+2.4187$_{10}$ -5
	0.0	90.000	1.0263	1.0952	0.9999	0.9620	0.968	0.968	0.000	+7.6721$_{10}$ -5
1.06 (1.049)	0.0	70.630	1.0000	1.0000	1.0000	1.0600	1.049	1.049	0.000	0.0000
	δ_* = 0.6725	76.613	1.0206	1.0739	1.0000	1.0000	1.000	0.997	0.012	+3.7029$_{10}$ -5
	δ_m = 0.7277	79.061	1.0268	1.0970	0.9999	0.9816	0.985	0.982	0.012	+8.0936$_{10}$ -5
	0.0	90.000	1.0393	1.1442	0.9998	0.9444	0.953	0.953	0.000	+2.4931$_{10}$ -4
1.08 (1.065)	0.0	67.808	1.0000	1.0000	1.0000	1.0800	1.065	1.065	0.000	0.0000
	1.0	74.458	1.0267	1.0964	0.9999	1.0031	1.003	0.998	0.017	+7.9610$_{10}$ -5
	δ_* = 1.0207	74.783	1.0277	1.1004	0.9999	1.0000	1.000	0.995	0.018	+8.9292$_{10}$ -5
	δ_m = 1.1021	77.558	1.0358	1.1310	0.9998	0.9762	0.980	0.976	0.019	+1.9016$_{10}$ -4
	1.0	80.857	1.0434	1.1598	0.9997	0.9540	0.961	0.958	0.017	+3.3214$_{10}$ -4
	0.0	90.000	1.0522	1.1941	0.9994	0.9277	0.939	0.939	0.000	+5.6946$_{10}$ -4
1.10 (1.081)	0.0	65.380	1.0000	1.0000	1.0000	1.1000	1.081	1.081	0.000	0.0000
	1.0	69.803	1.0214	1.0767	1.0000	1.0393	1.032	1.026	0.018	+4.1239$_{10}$ -5
	δ_* = 1.4062	73.250	1.0350	1.1278	0.9998	1.0000	1.000	0.992	0.024	+1.7727$_{10}$ -4
	δ_m = 1.5152	76.297	1.0449	1.1658	0.9996	0.9711	0.976	0.969	0.026	+3.6805$_{10}$ -4
	1.0	83.574	1.0605	1.2273	0.9991	0.9250	0.936	0.934	0.016	+8.7607$_{10}$ -4
	0.0	90.000	1.0649	1.2450	0.9989	0.9118	0.925	0.925	0.000	+1.0726$_{10}$ -3
1.12 (1.097)	0.0	63.234	1.0000	1.0000	1.0000	1.1200	1.097	1.097	0.000	0.0000
	1.0	66.737	1.0191	1.0685	1.0000	1.0663	1.054	1.046	0.018	+2.9705$_{10}$ -5
	δ_* = 1.8223	71.935	1.0424	1.1561	0.9997	1.0000	1.000	0.989	0.031	+3.1112$_{10}$ -4
	δ_m = 1.9592	75.209	1.0540	1.2014	0.9994	0.9664	0.972	0.962	0.033	+6.3007$_{10}$ -4
	1.0	84.852	1.0748	1.2850	0.9984	0.9051	0.919	0.918	0.016	+1.6068$_{10}$ -3
	0.0	90.000	1.0776	1.2968	0.9982	0.8966	0.912	0.912	0.000	+1.7885$_{10}$ -3
1.14 (1.113)	0.0	61.306	1.0000	1.0000	1.0000	1.1400	1.113	1.113	0.000	0.0000
	1.0	64.260	1.0178	1.0636	1.0000	1.0907	1.074	1.065	0.019	+2.3886$_{10}$ -5
	2.0	68.709	1.0407	1.1496	0.9997	1.0262	1.022	1.007	0.035	+2.7653$_{10}$ -4
	δ_* = 2.2636	70.785	1.0499	1.1854	0.9995	1.0000	1.000	0.985	0.039	+5.0136$_{10}$ -4
	δ_m = 2.4283	74.257	1.0632	1.2379	0.9990	0.9619	0.968	0.956	0.041	+9.9097$_{10}$ -4
	2.0	80.179	1.0797	1.3054	0.9981	0.9135	0.926	0.920	0.032	+1.9287$_{10}$ -3
	1.0	85.671	1.0882	1.3409	0.9974	0.8852	0.904	0.903	0.016	+2.5703$_{10}$ -3
	0.0	90.000	1.0903	1.3495	0.9973	0.8820	0.899	0.899	0.000	+2.7426$_{10}$ -3
1.16 (1.128)	0.0	59.550	1.0000	1.0000	1.0000	1.1600	1.128	1.128	0.000	0.0000
	1.0	62.133	1.0168	1.0602	1.0000	1.1138	1.092	1.082	0.019	+2.0385$_{10}$ -5
	2.0	65.557	1.0367	1.1344	0.9998	1.0586	1.048	1.031	0.036	+2.0458$_{10}$ -4

M_1	δ	θ	T_2/T_1	p_2/p_1	p_{o2}/p_{o1}	M_2	q_2/q_*	u_2/q_*	v_2/q_*	$(s_2-s_1)/R$
	δ_* = 2.7260	69.777	1.0576	1.2156	0.9992	1.0000	1.000	0.981	0.047	+7.5884$_{10}$ -4
	δ_m = 2.9180	73.412	1.0724	1.2753	0.9985	0.9577	0.964	0.949	0.048	+1.4648$_{10}$ -3
	2.0	81.929	1.0956	1.3723	0.9968	0.8898	0.906	0.901	0.031	+3.2262$_{10}$ -3
	1.0	86.254	1.1013	1.3965	0.9962	0.8728	0.891	0.890	0.016	+3.7909$_{10}$ -3
	0.0	90.000	1.1029	1.4032	0.9961	0.8682	0.887	0.887	0.000	+3.9560$_{10}$ -3
1.18	0.0	57.936	1.0000	1.0000	1.0000	1.1800	1.143	1.143	0.000	0.0000
(1.143)	1.0	60.248	1.0162	1.0578	1.0000	1.1361	1.110	1.098	0.019	+1.8071$_{10}$ -5
	2.0	63.115	1.0344	1.1256	0.9998	1.0861	1.070	1.051	0.037	+1.6900$_{10}$ -4
	δ_* = 3.2060	68.879	1.0654	1.2469	0.9989	1.0000	1.000	0.977	0.055	+1.0947$_{10}$ -3
	δ_m = 3.4243	72.657	1.0817	1.3134	0.9979	0.9539	0.961	0.941	0.056	+2.0647$_{10}$ -3
	2.0	83.048	1.1100	1.4340	0.9952	0.8712	0.889	0.885	0.031	+4.7648$_{10}$ -3
	1.0	86.694	1.1142	1.4524	0.9947	0.8586	0.878	0.877	0.015	+5.2872$_{10}$ -3
	0.0	90.000	1.1154	1.4578	0.9946	0.8549	0.875	0.875	0.000	+5.4462$_{10}$ -3
1.20	0.0	56.443	1.0000	1.0000	1.0000	1.2000	1.158	1.158	0.000	0.0000
(1.158)	1.0	58.548	1.0157	1.0559	1.0000	1.1578	1.126	1.114	0.019	+1.6451$_{10}$ -5
	2.0	61.050	1.0329	1.1197	0.9999	1.1113	1.090	1.068	0.037	+1.4746$_{10}$ -4
	δ_* = 3.7008	68.076	1.0733	1.2791	0.9985	1.0000	1.000	0.972	0.063	+1.5201$_{10}$ -3
	δ_m = 3.9442	71.977	1.0910	1.3525	0.9972	0.9502	0.958	0.934	0.064	+2.8033$_{10}$ -3
	2.0	83.861	1.1237	1.4941	0.9934	0.8551	0.875	0.871	0.030	+6.5811$_{10}$ -3
	1.0	87.040	1.1270	1.5089	0.9930	0.8452	0.866	0.865	0.015	+7.0741$_{10}$ -3
	0.0	90.000	1.1280	1.5133	0.9928	0.8422	0.863	0.863	0.000	+7.2277$_{10}$ -3
1.22	0.0	55.052	1.0000	1.0000	1.0000	1.2200	1.173	1.173	0.000	0.0000
(1.173)	1.0	56.994	1.0153	1.0545	1.0000	.1793	1.143	1.130	0.020	+1.5272$_{10}$ -5
	2.0	59.235	1.0318	1.1155	0.9999	1.1352	1.109	1.085	0.038	+1.3310$_{10}$ -4
	4.0	66.029	1.0743	1.2832	0.9984	1.0196	1.016	0.982	0.069	+1.5795$_{10}$ -3
	δ_* = 4.2078	67.355	1.0814	1.3124	0.9980	1.0000	1.000	0.967	0.071	+2.0465$_{10}$ -3
	δ_m = 4.4748	71.361	1.1004	1.3924	0.9963	0.9469	0.955	0.927	0.073	+3.6924$_{10}$ -3
	4.0	76.844	1.1205	1.4798	0.9939	0.8893	0.905	0.888	0.062	+6.1214$_{10}$ -3
	2.0	84.489	1.1370	1.5538	0.9913	0.8406	0.862	0.858	0.030	+8.6933$_{10}$ -3
	1.0	87.320	1.1397	1.5660	0.9909	0.8325	0.855	0.854	0.015	+9.1636$_{10}$ -3
	0.0	90.000	1.1405	1.5698	0.9907	0.8300	0.852	0.852	0.000	+9.3123$_{10}$ -3
1.24	0.0	53.751	1.0000	1.0000	1.0000	1.2400	1.188	1.188	0.000	0.0000
(1.188)	1.0	55.559	1.0150	1.0534	1.0000	1.2004	1.159	1.145	0.020	+1.4392$_{10}$ -5
	2.0	57.605	1.0309	1.1123	0.9999	1.1583	1.127	1.101	0.038	+1.2296$_{10}$ -4
	4.0	63.150	1.0690	1.2613	0.9987	1.0564	1.046	1.007	0.070	+1.2744$_{10}$ -3
	δ_* = 4.7247	66.707	1.0896	1.3467	0.9973	1.0000	1.000	0.961	0.079	+2.6853$_{10}$ -3
	δ_m = 5.0139	70.801	1.1098	1.4332	0.9953	0.9438	0.953	0.920	0.081	+4.7429$_{10}$ -3
	4.0	78.711	1.1380	1.5585	0.9912	0.8629	0.882	0.867	0.061	+8.8718$_{10}$ -3
	2.0	84.994	1.1501	1.6135	0.9889	0.8272	0.850	0.847	0.030	+1.1114$_{10}$ -2
	1.0	87.551	1.1524	1.6239	0.9885	0.8204	0.844	0.843	0.015	+1.1565$_{10}$ -2
	0.0	90.000	1.1531	1.6272	0.9884	0.8183	0.842	0.842	0.000	+1.1710$_{10}$ -2
1.26	0.0	52.528	1.0000	1.0000	1.0000	1.2600	1.202	1.202	0.000	0.0000
(1.202)	1.0	54.227	1.0148	1.0526	1.0000	1.2214	1.174	1.159	0.020	+1.3724$_{10}$ -5
	2.0	56.117	1.0303	1.1099	0.9999	1.1807	1.144	1.116	0.039	+1.1555$_{10}$ -4
	4.0	60.934	1.0658	1.2484	0.9989	1.0868	1.071	1.027	0.072	+1.1128$_{10}$ -3
	δ_* = 5.2498	66.122	1.0979	1.3820	0.9966	1.0000	1.000	0.955	0.088	+3.4478$_{10}$ -3
	δ_m = 5.5593	70.290	1.1193	1.4749	0.9941	0.9409	0.950	0.913	0.089	+5.9651$_{10}$ -3
	4.0	79.958	1.1535	1.6292	0.9883	0.8431	0.864	0.851	0.060	+1.1799$_{10}$ -2
	2.0	85.409	1.1631	1.6737	0.9862	0.8147	0.839	0.836	0.029	+1.3851$_{10}$ -2
	1.0	87.745	1.1650	1.6827	0.9858	0.8089	0.833	0.833	0.015	+1.4287$_{10}$ -2
	0.0	90.000	1.1657	1.6855	0.9857	0.8071	0.832	0.832	0.000	+1.4427$_{10}$ -2
1.28	0.0	51.375	1.0000	1.0000	1.0000	1.2800	1.217	1.217	0.000	0.0000
(1.217)	1.0	52.981	1.0146	1.0519	1.0000	1.2422	1.190	1.174	0.020	+1.3213$_{10}$ -5
	2.0	54.747	1.0298	1.1080	0.9999	1.2027	1.160	1.131	0.039	+1.1001$_{10}$ -4
	4.0	59.065	1.0636	1.2397	0.9990	1.1142	1.092	1.044	0.073	+1.0108$_{10}$ -3
	6.0	67.382	1.1164	1.4621	0.9944	0.9727	0.977	0.929	0.098	+5.5741$_{10}$ -3
	δ_* = 5.7811	65.593	1.1064	1.4184	0.9957	1.0000	1.000	0.949	0.096	+4.3454$_{10}$ -3
	δ_m = 6.1092	69.823	1.1289	1.5174	0.9927	0.9383	0.948	0.905	0.097	+7.3687$_{10}$ -3
	6.0	72.276	1.1401	1.5676	0.9908	0.9071	0.921	0.885	0.093	+9.2277$_{10}$ -3
	4.0	80.898	1.1681	1.6970	0.9851	0.8265	0.849	0.838	0.059	+1.4992$_{10}$ -2
	2.0	85.759	1.1760	1.7343	0.9832	0.8029	0.828	0.825	0.029	+1.6913$_{10}$ -2
	1.0	87.910	1.1777	1.7423	0.9828	0.7979	0.823	0.823	0.014	+1.7333$_{10}$ -2
	0.0	90.000	1.1783	1.7448	0.9827	0.7963	0.822	0.822	0.000	+1.7470$_{10}$ -2
1.30	0.0	50.285	1.0000	1.0000	1.0000	1.3000	1.231	1.231	0.000	0.0000
(1.231)	1.0	51.812	1.0144	1.0514	1.0000	1.2629	1.205	1.188	0.021	+1.2822$_{10}$ -5
	2.0	53.474	1.0294	1.1065	0.9999	1.2244	1.176	1.145	0.040	+1.0582$_{10}$ -4
	4.0	57.423	1.0621	1.2334	0.9991	1.1398	1.112	1.060	0.074	+9.4085$_{10}$ -4
	6.0	63.458	1.1048	1.4113	0.9958	1.0274	1.023	0.964	0.101	+4.1610$_{10}$ -3
	δ_* = 6.3173	65.115	1.1150	1.4559	0.9946	1.0000	1.000	0.943	0.104	+5.3890$_{10}$ -3
	δ_m = 6.6621	69.395	1.1386	1.5608	0.9911	0.9359	0.946	0.898	0.105	+8.9623$_{10}$ -3
	6.0	75.372	1.1643	1.6793	0.9860	0.8636	0.883	0.854	0.090	+1.4121$_{10}$ -2
	4.0	81.649	1.1822	1.7634	0.9817	0.8118	0.836	0.825	0.058	+1.8482$_{10}$ -2
	2.0	86.058	1.1889	1.7957	0.9799	0.7918	0.818	0.815	0.028	+2.0302$_{10}$ -2
	1.0	88.053	1.1904	1.8027	0.9795	0.7874	0.814	0.813	0.014	+2.0710$_{10}$ -2
	0.0	90.000	1.1909	1.8050	0.9794	0.7860	0.812	0.812	0.000	+2.0842$_{10}$ -2
1.32	0.0	49.251	1.0000	1.0000	1.0000	1.3200	1.245	1.245	0.000	0.0000
(1.245)	1.0	50.709	1.0143	1.0510	1.0000	1.2835	1.219	1.202	0.021	+1.2523$_{10}$ -5
	2.0	52.284	1.0291	1.1054	0.9999	1.2458	1.192	1.159	0.040	+1.0263$_{10}$ -4
	4.0	55.946	1.0609	1.2287	0.9991	1.1642	1.131	1.075	0.075	+8.9074$_{10}$ -4
	6.0	61.028	1.0996	1.3892	0.9964	1.0638	1.052	0.986	0.104	+3.6153$_{10}$ -3
	δ_* = 6.8568	64.682	1.1237	1.4944	0.9934	1.0000	1.000	0.936	0.113	+6.5894$_{10}$ -3
	δ_m = 7.2165	69.002	1.1483	1.6051	0.9893	0.9337	0.944	0.891	0.113	+1.0754$_{10}$ -2
	6.0	77.006	1.1822	1.7634	0.9817	0.8387	0.860	0.835	0.080	+1.8479$_{10}$ -2
	4.0	82.269	1.1959	1.8294	0.9780	0.7986	0.824	0.814	0.057	+2.2285$_{10}$ -2
	2.0	86.316	1.2018	1.8577	0.9763	0.7812	0.808	0.806	0.028	+2.4022$_{10}$ -2
	1.0	88.177	1.2031	1.8641	0.9759	0.7773	0.804	0.804	0.014	+2.4418$_{10}$ -2
	0.0	90.000	1.2035	1.8661	0.9758	0.7760	0.803	0.803	0.000	+2.4547$_{10}$ -2
1.34	0.0	48.268	1.0000	1.0000	1.0000	1.3400	1.259	1.259	0.000	0.0000
(1.259)	1.0	49.667	1.0142	1.0506	1.0000	1.3040	1.234	1.215	0.021	+1.2299$_{10}$ -5
	2.0	51.167	1.0288	1.1045	0.9999	1.2669	1.208	1.173	0.041	+1.0023$_{10}$ -4
	4.0	54.599	1.0600	1.2252	0.9991	1.1878	1.149	1.089	0.076	+8.5394$_{10}$ -4
	6.0	59.089	1.0964	1.3754	0.9967	1.0946	1.077	1.004	0.105	+3.2963$_{10}$ -3
	δ_* = 7.3985	64.290	1.1326	1.5340	0.9921	1.0000	1.000	0.930	0.121	+7.9572$_{10}$ -3
	δ_m = 7.7713	68.640	1.1581	1.6503	0.9873	0.9316	0.942	0.884	0.121	+1.2752$_{10}$ -2
	6.0	78.176	1.1982	1.8402	0.9773	0.8195	0.843	0.820	0.086	+2.2944$_{10}$ -2
	4.0	82.794	1.2095	1.8952	0.9739	0.7864	0.813	0.804	0.056	+2.6407$_{10}$ -2
	2.0	86.541	1.2147	1.9206	0.9723	0.7711	0.799	0.796	0.028	+2.8075$_{10}$ -2
	1.0	88.287	1.2159	1.9263	0.9719	0.7676	0.795	0.795	0.014	+2.8460$_{10}$ -2
	0.0	90.000	1.2162	1.9282	0.9718	0.7664	0.794	0.794	0.000	+2.8586$_{10}$ -2
1.36	0.0	47.332	1.0000	1.0000	1.0000	1.3600	1.273	1.273	0.000	0.0000
(1.273)	1.0	48.678	1.0142	1.0504	1.0000	1.3244	1.248	1.229	0.021	+1.2136$_{10}$ -5
	2.0	50.114	1.0287	1.1039	0.9999	1.2879	1.223	1.186	0.041	+9.8435$_{10}$ -5
	4.0	53.355	1.0593	1.2225	0.9992	1.2108	1.166	1.103	0.077	+8.2658$_{10}$ -4
	6.0	57.430	1.0941	1.3659	0.9969	1.1226	1.099	1.020	0.107	+3.0857$_{10}$ -3
	8.0	64.286	1.1439	1.5850	0.9901	0.9940	0.995	0.919	0.129	+9.9194$_{10}$ -3
	δ_* = 7.941	63.935	1.1416	1.5746	0.9905	1.0000	1.000	0.923	0.129	+9.5023$_{10}$ -3
	δ_m = 8.3253	68.307	1.1680	1.6964	0.9851	0.9298	0.940	0.876	0.128	+1.4962$_{10}$ -2

M_1	δ	θ	T_2/T_1	p_2/p_1	p_{o2}/p_{o1}	M_2	q_2/q_*	u_2/q_*	v_2/q_*	$(s_2-s_1)/R$
	8.0	72.305	1.1881	1.7918	0.9801	0.8747	0.892	0.844	0.119	+2.0081$_{10}$ -2
	6.0	79.094	1.2133	1.9140	0.9727	0.8033	0.828	0.807	0.085	+2.7635$_{10}$ -2
	4.0	83.246	1.2230	1.9614	0.9696	0.7751	0.802	0.794	0.055	+3.0853$_{10}$ -2
	2.0	86.740	1.2276	1.9842	0.9681	0.7614	0.790	0.787	0.027	+3.2461$_{10}$ -2
	1.0	88.383	1.2287	1.9895	0.9677	0.7582	0.787	0.786	0.014	+3.2836$_{10}$ -2
	0.0	90.000	1.2290	1.9912	0.9676	0.7572	0.786	0.786	0.000	+3.2959$_{10}$ -2
1.38 (1.286)	0.0	46.439	1.0000	1.0000	1.0000	1.3800	1.286	1.286	0.000	0.0000
	1.0	47.738	1.0141	1.0502	1.0000	1.3447	1.262	1.242	0.022	+1.2023$_{10}$ -5
	2.0	49.118	1.0285	1.1034	0.9999	1.3088	1.237	1.199	0.042	+9.7138$_{10}$ -5
	4.0	52.199	1.0588	1.2205	0.9992	1.2332	1.183	1.116	0.078	+8.0625$_{10}$ -4
	6.0	55.961	1.0925	1.3590	0.9971	1.1488	1.119	1.034	0.109	+2.9384$_{10}$ -3
	8.0	61.427	1.1355	1.5469	0.9916	1.0395	1.033	0.944	0.133	+8.4337$_{10}$ -3
	δ_* = 8.4838	63.615	1.1507	1.6163	0.9888	1.0000	1.000	0.916	0.137	+1.1235$_{10}$ -2
	δ_m = 8.8775	67.999	1.1779	1.7433	0.9828	0.9281	0.939	0.869	0.136	+1.7391$_{10}$ -2
	8.0	74.488	1.2097	1.8962	0.9739	0.8411	0.862	0.821	0.115	+2.6470$_{10}$ -2
	6.0	79.848	1.2280	1.9861	0.9679	0.7891	0.815	0.795	0.084	+3.2595$_{10}$ -2
	4.0	83.640	1.2364	2.0279	0.9650	0.7645	0.792	0.784	0.055	+3.5626$_{10}$ -2
	2.0	86.917	1.2405	2.0487	0.9635	0.7521	0.781	0.779	0.027	+3.7180$_{10}$ -2
	1.0	88.470	1.2415	2.0535	0.9632	0.7492	0.778	0.778	0.014	+3.7545$_{10}$ -2
	0.0	90.000	1.2418	2.0551	0.9630	0.7483	0.777	0.777	0.000	+3.7665$_{10}$ -2
1.40 (1.300)	0.0	45.585	1.0000	1.0000	1.0000	1.4000	1.300	1.300	0.000	0.0000
	1.0	46.842	1.0141	1.0501	1.0000	1.3650	1.276	1.255	0.022	+1.1953$_{10}$ -5
	2.0	48.173	1.0284	1.1030	0.9999	1.3295	1.252	1.212	0.042	+9.6251$_{10}$ -5
	4.0	51.117	1.0584	1.2190	0.9992	1.2553	1.199	1.129	0.079	+7.9132$_{10}$ -4
	6.0	54.633	1.0913	1.3539	0.9972	1.1737	1.138	1.048	0.110	+2.8325$_{10}$ -3
	8.0	59.367	1.1309	1.5263	0.9923	1.0744	1.061	0.963	0.135	+7.6830$_{10}$ -3
	δ_* = 9.0255	63.325	1.1600	1.6592	0.9869	1.0000	1.000	0.909	0.144	+1.3163$_{10}$ -2
	δ_m = 9.4272	67.716	1.1880	1.7912	0.9802	0.9266	0.938	0.862	0.143	+2.0044$_{10}$ -2
	8.0	75.893	1.2276	1.9842	0.9681	0.8184	0.842	0.804	0.113	+3.2457$_{10}$ -2
	6.0	80.485	1.2423	2.0575	0.9629	0.7763	0.803	0.785	0.082	+3.7845$_{10}$ -2
	4.0	83.988	1.2497	2.0949	0.9601	0.7545	0.783	0.775	0.054	+4.0725$_{10}$ -2
	2.0	87.075	1.2535	2.1140	0.9586	0.7432	0.773	0.771	0.027	+4.2231$_{10}$ -2
	1.0	88.547	1.2544	2.1185	0.9583	0.7406	0.770	0.770	0.013	+4.2588$_{10}$ -2
	0.0	90.000	1.2547	2.1200	0.9582	0.7397	0.769	0.769	0.000	+4.2705$_{10}$ -2
1.42 (1.313)	0.0	44.767	1.0000	1.0000	1.0000	1.4200	1.313	1.313	0.000	0.0000
	1.0	45.987	1.0141	1.0501	1.0000	1.3853	1.290	1.268	0.022	+1.1919$_{10}$ -5
	2.0	47.274	1.0284	1.1028	0.9999	1.3501	1.266	1.224	0.043	+9.5706$_{10}$ -5
	4.0	50.100	1.0582	1.2179	0.9992	1.2770	1.215	1.141	0.080	+7.8065$_{10}$ -4
	6.0	53.416	1.0904	1.3502	0.9972	1.1977	1.157	1.061	0.111	+2.7554$_{10}$ -3
	8.0	57.669	1.1279	1.5129	0.9928	1.1049	1.085	0.979	0.138	+7.2137$_{10}$ -3
	δ_* = 9.5654	63.064	1.1694	1.7031	0.9848	1.0000	1.000	0.902	0.152	+1.5297$_{10}$ -2
	δ_m = 9.9734	67.454	1.1981	1.8399	0.9773	0.9253	0.937	0.855	0.150	+2.2926$_{10}$ -2
	8.0	76.958	1.2440	2.0660	0.9622	0.8002	0.825	0.791	0.111	+3.8492$_{10}$ -2
	6.0	81.034	1.2564	2.1287	0.9575	0.7645	0.792	0.775	0.081	+4.3397$_{10}$ -2
	4.0	84.297	1.2631	2.1626	0.9549	0.7450	0.774	0.767	0.054	+4.6150$_{10}$ -2
	2.0	87.217	1.2666	2.1803	0.9535	0.7347	0.765	0.763	0.027	+4.7612$_{10}$ -2
	1.0	88.617	1.2674	2.1844	0.9532	0.7322	0.762	0.762	0.013	+4.7960$_{10}$ -2
	0.0	90.000	1.2676	2.1858	0.9531	0.7314	0.762	0.762	0.000	+4.8074$_{10}$ -2
1.44 (1.326)	0.0	43.983	1.0000	1.0000	1.0000	1.4400	1.326	1.326	0.000	0.0000
	1.0	45.170	1.0141	1.0501	1.0000	1.4055	1.304	1.281	0.022	+1.1916$_{10}$ -5
	2.0	46.417	1.0284	1.1027	0.9999	1.3706	1.280	1.237	0.043	+9.5452$_{10}$ -5
	4.0	49.140	1.0580	1.2171	0.9992	1.2984	1.230	1.153	0.081	+7.7344$_{10}$ -4
	6.0	52.289	1.0898	1.3474	0.9973	1.2210	1.174	1.073	0.113	+2.6996$_{10}$ -3
	8.0	56.193	1.1258	1.5036	0.9931	1.1328	1.107	0.993	0.140	+6.8960$_{10}$ -3
	10.0	62.306	1.1751	1.7300	0.9835	1.0097	1.008	0.901	0.159	+1.6684$_{10}$ -2
	δ_* = 10.1028	62.828	1.1789	1.7480	0.9825	1.0000	1.000	0.894	0.159	+1.7644$_{10}$ -2
	δ_m = 10.5157	67.211	1.2084	1.8896	0.9743	0.9241	0.936	0.848	0.157	+2.6041$_{10}$ -2
	10.0	72.021	1.2352	2.0220	0.9654	0.8525	0.873	0.810	0.143	+3.5195$_{10}$ -2
	8.0	77.821	1.2596	2.1449	0.9563	0.7847	0.811	0.779	0.109	+4.4703$_{10}$ -2
	6.0	81.514	1.2704	2.1999	0.9519	0.7537	0.782	0.766	0.080	+4.9254$_{10}$ -2
	4.0	84.574	1.2765	2.2309	0.9494	0.7359	0.766	0.759	0.053	+5.1898$_{10}$ -2
	2.0	87.346	1.2797	2.2473	0.9481	0.7265	0.757	0.755	0.026	+5.3321$_{10}$ -2
	1.0	88.680	1.2804	2.2512	0.9478	0.7242	0.755	0.754	0.013	+5.3660$_{10}$ -2
	0.0	90.000	1.2807	2.2525	0.9476	0.7235	0.754	0.754	0.000	+5.3772$_{10}$ -2
1.46 (1.339)	0.0	43.230	1.0000	1.0000	1.0000	1.4600	1.339	1.339	0.000	0.0000
	1.0	44.386	1.0141	1.0501	1.0000	1.4257	1.317	1.293	0.023	+1.1941$_{10}$ -5
	2.0	45.599	1.0284	1.1027	0.9999	1.3910	1.294	1.249	0.044	+9.5450$_{10}$ -5
	4.0	48.230	1.0579	1.2167	0.9992	1.3197	1.245	1.164	0.081	+7.6909$_{10}$ -4
	6.0	51.239	1.0893	1.3454	0.9973	1.2438	1.191	1.084	0.114	+2.6601$_{10}$ -3
	8.0	54.874	1.1243	1.4969	0.9933	1.1589	1.127	1.006	0.141	+6.6731$_{10}$ -3
	10.0	59.981	1.1683	1.6978	0.9851	1.0509	1.042	0.922	0.162	+1.5032$_{10}$ -2
	δ_* = 10.6370	62.616	1.1886	1.7941	0.9800	1.0000	1.000	0.887	0.167	+2.0271$_{10}$ -2
	δ_m = 11.0533	66.987	1.2187	1.9401	0.9710	0.9230	0.935	0.841	0.164	+2.9393$_{10}$ -2
	10.0	73.778	1.2559	2.1261	0.9577	0.8237	0.847	0.791	0.139	+4.3193$_{10}$ -2
	8.0	78.545	1.2747	2.2221	0.9501	0.7711	0.799	0.768	0.108	+5.1144$_{10}$ -2
	6.0	81.940	1.2843	2.2713	0.9461	0.7436	0.773	0.757	0.080	+5.5419$_{10}$ -2
	4.0	84.823	1.2899	2.3000	0.9437	0.7273	0.758	0.751	0.053	+5.7968$_{10}$ -2
	2.0	87.463	1.2928	2.3153	0.9424	0.7185	0.749	0.748	0.026	+5.9353$_{10}$ -2
	1.0	88.738	1.2935	2.3190	0.9421	0.7164	0.747	0.747	0.013	+5.9685$_{10}$ -2
	0.0	90.000	1.2938	2.3202	0.9420	0.7157	0.747	0.747	0.000	+5.9795$_{10}$ -2
1.48 (1.352)	0.0	42.507	1.0000	1.0000	1.0000	1.4800	1.352	1.352	0.000	0.0000
	1.0	43.635	1.0141	1.0502	1.0000	1.4458	1.330	1.305	0.023	+1.1990$_{10}$ -5
	2.0	44.815	1.0284	1.1028	0.9999	1.4114	1.307	1.261	0.044	+9.5667$_{10}$ -5
	4.0	47.366	1.0578	1.2165	0.9992	1.3407	1.260	1.176	0.082	+7.6714$_{10}$ -4
	6.0	50.254	1.0890	1.3441	0.9974	1.2660	1.207	1.095	0.115	+2.6335$_{10}$ -3
	8.0	53.675	1.1232	1.4921	0.9935	1.1839	1.146	1.018	0.143	+6.5151$_{10}$ -3
	10.0	58.190	1.1642	1.6788	0.9860	1.0845	1.069	0.938	0.165	+1.4098$_{10}$ -2
	δ_* = 11.1673	62.427	1.1984	1.8413	0.9773	1.0000	1.000	0.860	0.174	+2.3006$_{10}$ -2
	δ_m = 11.5858	66.780	1.2291	1.9916	0.9676	0.9221	0.934	0.834	0.171	+3.2986$_{10}$ -2
	10.0	75.019	1.2739	2.2180	0.9505	0.8026	0.828	0.777	0.137	+5.0796$_{10}$ -2
	8.0	79.167	1.2896	2.2985	0.9438	0.7588	0.787	0.759	0.107	+5.7840$_{10}$ -2
	6.0	82.319	1.2982	2.3432	0.9400	0.7340	0.764	0.749	0.079	+6.1892$_{10}$ -2
	4.0	85.049	1.3033	2.3698	0.9377	0.7191	0.750	0.744	0.052	+6.4356$_{10}$ -2
	2.0	87.570	1.3061	2.3842	0.9364	0.7109	0.742	0.741	0.026	+6.5707$_{10}$ -2
	1.0	88.790	1.3067	2.3877	0.9361	0.7089	0.740	0.740	0.013	+6.6032$_{10}$ -2
	0.0	90.000	1.3069	2.3888	0.9360	0.7083	0.740	0.740	0.000	+6.6139$_{10}$ -2
1.50 (1.365)	0.0	41.810	1.0000	1.0000	1.0000	1.5000	1.365	1.365	0.000	0.0000
	1.0	42.913	1.0141	1.0503	1.0000	1.4660	1.343	1.318	0.023	+1.2061$_{10}$ -5
	2.0	44.065	1.0284	1.1030	0.9999	1.4316	1.321	1.272	0.044	+9.6078$_{10}$ -5
	4.0	46.543	1.0578	1.2165	0.9992	1.3615	1.274	1.187	0.083	+7.6725$_{10}$ -4
	6.0	49.326	1.0888	1.3433	0.9974	1.2879	1.223	1.106	0.116	+2.6174$_{10}$ -3
	8.0	52.571	1.1225	1.4887	0.9936	1.2079	1.164	1.029	0.145	+6.4043$_{10}$ -3
	10.0	56.679	1.1615	1.6662	0.9866	1.1144	1.093	0.951	0.168	+1.3495$_{10}$ -2
	12.0	64.359	1.2241	1.9668	0.9693	0.9607	0.967	0.849	0.181	+3.1232$_{10}$ -2
	δ_* = 11.6933	62.257	1.2083	1.8895	0.9743	1.0000	1.000	0.873	0.181	+2.6035$_{10}$ -2
	δ_m = 12.1127	66.589	1.2396	2.0439	0.9638	0.9213	0.933	0.828	0.178	+3.6821$_{10}$ -2
	12.0	68.790	1.2537	2.1147	0.9586	0.8849	0.901	0.809	0.172	+4.2286$_{10}$ -2
	10.0	75.995	1.2908	2.3046	0.9433	0.7854	0.812	0.764	0.135	+5.8385$_{10}$ -2
	8.0	79.712	1.3042	2.3746	0.9372	0.7476	0.777	0.749	0.105	+6.4807$_{10}$ -2
	6.0	82.661	1.3121	2.4155	0.9336	0.7250	0.756	0.741	0.078	+6.8670$_{10}$ -2
	4.0	85.255	1.3168	2.4404	0.9314	0.7112	0.742	0.736	0.051	+7.1058$_{10}$ -2
	2.0	87.667	1.3194	2.4540	0.9302	0.7035	0.735	0.734	0.026	+7.2377$_{10}$ -2
	1.0	88.838	1.3200	2.4573	0.9299	0.7017	0.733	0.733	0.013	+7.2695$_{10}$ -2
	0.0	90.000	1.3202	2.4583	0.9298	0.7011	0.733	0.733	0.000	+7.2800$_{10}$ -2
1.52 (1.377)	0.0	41.140	1.0000	1.0000	1.0000	1.5200	1.377	1.377	0.000	0.0000
	1.0	42.219	1.0142	1.0504	1.0000	1.4860	1.356	1.329	0.023	+1.2151$_{10}$ -5
	2.0	43.344	1.0285	1.1032	0.9999	1.4519	1.334	1.284	0.045	+9.6663$_{10}$ -5
	4.0	45.757	1.0579	1.2167	0.9992	1.3822	1.288	1.197	0.084	+7.6913$_{10}$ -4
	6.0	48.448	1.0887	1.3429	0.9974	1.3095	1.238	1.116	0.117	+2.6098$_{10}$ -3
	8.0	51.546	1.1219	1.4864	0.9937	1.2312	1.181	1.039	0.146	+6.3293$_{10}$ -3
	10.0	55.351	1.1596	1.6575	0.9870	1.1419	1.114	0.963	0.170	+1.3086$_{10}$ -2
	12.0	61.098	1.2103	1.8992	0.9737	1.0199	1.016	0.878	0.187	+2.6662$_{10}$ -2

M_1	δ	θ	T_2/T_1	p_2/p_1	p_{o2}/p_{o1}	M_2	q_2/q_*	u_2/q_*	v_2/q_*	$(s_2 - s_1)/R$
	δ_* = 12.2145	62.106	1.2184	1.9388	0.9711	1.0000	1.000	0.865	0.187	+2.9304$_{10}$ -2
	δ_m = 12.6336	66.412	1.2502	2.0972	0.9599	0.9206	0.933	0.821	0.184	+4.0902$_{10}$ -2
	12.0	71.552	1.2819	2.2589	0.9471	0.8383	0.860	0.781	0.163	+5.4328$_{10}$ -2
	10.0	76.803	1.3069	2.3883	0.9360	0.7706	0.798	0.754	0.133	+6.6093$_{10}$ -2
	8.0	80.194	1.3187	2.4506	0.9305	0.7372	0.767	0.741	0.104	+7.2050$_{10}$ -2
	6.0	82.971	1.3259	2.4884	0.9270	0.7165	0.747	0.734	0.077	+7.5753$_{10}$ -2
	4.0	85.444	1.3304	2.5118	0.9249	0.7036	0.735	0.729	0.051	+7.8071$_{10}$ -2
	2.0	87.757	1.3328	2.5247	0.9237	0.6964	0.728	0.727	0.025	+7.9360$_{10}$ -2
	1.0	88.883	1.3334	2.5278	0.9234	0.6947	0.727	0.726	0.013	+7.9672$_{10}$ -2
	0.0	90.000	1.3336	2.5286	0.9233	0.6941	0.726	0.726	0.000	+7.9775$_{10}$ -2
1.54 (1.389)	0.0	40.493	1.0000	1.0000	1.0000	1.5400	1.389	1.389	0.000	0.0000
	1.0	41.550	1.0142	1.0506	1.0000	1.5061	1.368	1.341	0.023	+1.2260$_{10}$ -5
	2.0	42.651	1.0286	1.1035	0.9999	1.4720	1.347	1.295	0.045	+9.7405$_{10}$ -5
	4.0	45.006	1.0580	1.2170	0.9992	1.4028	1.302	1.208	0.084	+7.7259$_{10}$ -4
	6.0	47.615	1.0887	1.3429	0.9974	1.3308	1.253	1.126	0.118	+2.6096$_{10}$ -3
	8.0	50.583	1.1216	1.4849	0.9937	1.2540	1.198	1.049	0.147	+6.2824$_{10}$ -3
	10.0	54.156	1.1583	1.6514	0.9873	1.1679	1.134	0.975	0.172	+1.2804$_{10}$ -2
	12.0	59.080	1.2043	1.8696	0.9755	1.0589	1.048	0.895	0.190	+2.4768$_{10}$ -2
	δ_* = 12.7304	61.971	1.2286	1.9892	0.9677	1.0000	1.000	0.858	0.194	+3.2818$_{10}$ -2
	δ_m = 13.1482	66.248	1.2609	2.1513	0.9558	0.9200	0.932	0.814	0.190	+4.5230$_{10}$ -2
	12.0	73.086	1.3026	2.3660	0.9380	0.8119	0.836	0.765	0.163	+6.4005$_{10}$ -2
	10.0	77.493	1.3225	2.4704	0.9287	0.7576	0.786	0.744	0.131	+7.3984$_{10}$ -2
	8.0	80.626	1.3332	2.5268	0.9235	0.7275	0.758	0.733	0.103	+7.9573$_{10}$ -2
	6.0	83.253	1.3398	2.5620	0.9202	0.7084	0.740	0.727	0.076	+8.3136$_{10}$ -2
	4.0	85.617	1.3440	2.5840	0.9182	0.6963	0.728	0.723	0.051	+8.5391$_{10}$ -2
	2.0	87.840	1.3463	2.5963	0.9170	0.6896	0.722	0.720	0.025	+8.6651$_{10}$ -2
	1.0	88.924	1.3468	2.5992	0.9167	0.6879	0.720	0.720	0.013	+8.6957$_{10}$ -2
	0.0	90.000	1.3470	2.6002	0.9166	0.6874	0.720	0.720	0.000	+8.7058$_{10}$ -2
1.56 (1.402)	0.0	39.868	1.0000	1.0000	1.0000	1.5600	1.402	1.402	0.000	0.0000
	1.0	40.906	1.0143	1.0508	1.0000	1.5262	1.381	1.353	0.024	+1.2385$_{10}$ -5
	2.0	41.985	1.0286	1.1038	0.9999	1.4922	1.360	1.306	0.046	+9.8288$_{10}$ -5
	4.0	44.286	1.0581	1.2175	0.9992	1.4232	1.315	1.218	0.085	+7.7743$_{10}$ -4
	6.0	46.823	1.0888	1.3432	0.9974	1.3519	1.267	1.136	0.119	+2.6156$_{10}$ -3
	8.0	49.687	1.1214	1.4842	0.9938	1.2762	1.214	1.059	0.149	+6.2583$_{10}$ -3
	10.0	53.064	1.1574	1.6473	0.9875	1.1927	1.153	0.985	0.174	+1.2613$_{10}$ -2
	12.0	57.474	1.2006	1.8517	0.9766	1.0917	1.075	0.909	0.193	+2.3649$_{10}$ -2
	δ_* = 13.2406	61.853	1.2389	2.0407	0.9641	1.0000	1.000	0.851	0.200	+3.6580$_{10}$ -2
	δ_m = 13.6562	66.097	1.2717	2.2064	0.9514	0.9195	0.932	0.808	0.196	+4.9806$_{10}$ -2
	12.0	74.220	1.3210	2.4626	0.9294	0.7919	0.818	0.752	0.160	+7.3213$_{10}$ -2
	10.0	78.095	1.3379	2.5517	0.9212	0.7458	0.775	0.735	0.130	+8.2088$_{10}$ -2
	8.0	81.015	1.3476	2.6033	0.9163	0.7184	0.749	0.726	0.102	+8.7377$_{10}$ -2
	6.0	83.511	1.3538	2.6363	0.9132	0.7007	0.732	0.720	0.076	+9.0815$_{10}$ -2
	4.0	85.777	1.3577	2.6571	0.9112	0.6893	0.722	0.716	0.050	+9.3012$_{10}$ -2
	2.0	87.917	1.3599	2.6688	0.9101	0.6829	0.715	0.714	0.025	+9.4246$_{10}$ -2
	1.0	88.962	1.3604	2.6716	0.9098	0.6814	0.714	0.714	0.012	+9.4545$_{10}$ -2
	0.0	90.000	1.3606	2.6725	0.9097	0.6809	0.714	0.714	0.000	+9.4644$_{10}$ -2
1.58 (1.414)	0.0	39.265	1.0000	1.0000	1.0000	1.5800	1.414	1.414	0.000	0.0000
	1.0	40.285	1.0143	1.0510	1.0000	1.5462	1.393	1.364	0.024	+1.2525$_{10}$ -5
	2.0	41.343	1.0287	1.1042	0.9999	1.5123	1.372	1.317	0.046	+9.9303$_{10}$ -5
	4.0	43.595	1.0582	1.2182	0.9992	1.4435	1.329	1.228	0.086	+7.8352$_{10}$ -4
	6.0	46.067	1.0889	1.3438	0.9974	1.3727	1.282	1.145	0.120	+2.6270$_{10}$ -3
	8.0	48.836	1.1214	1.4840	0.9938	1.2980	1.230	1.068	0.150	+6.2531$_{10}$ -3
	10.0	52.055	1.1569	1.6446	0.9876	1.2165	1.171	0.994	0.175	+1.2489$_{10}$ -2
	12.0	56.103	1.1981	1.8399	0.9773	1.1211	1.098	0.920	0.196	+2.2925$_{10}$ -2
	14.0	63.375	1.2628	2.1609	0.9550	0.9678	0.973	0.826	0.206	+4.6011$_{10}$ -2
	δ_* = 13.7448	61.749	1.2494	2.0933	0.9602	1.0000	1.000	0.844	0.206	+4.0596$_{10}$ -2
	δ_m = 14.1574	65.957	1.2826	2.2624	0.9468	0.9191	0.931	0.801	0.202	+5.4631$_{10}$ -2
	14.0	68.486	1.3003	2.3541	0.9390	0.8747	0.892	0.781	0.195	+6.2902$_{10}$ -2
	12.0	75.133	1.3383	2.5541	0.9210	0.7754	0.803	0.741	0.157	+8.2328$_{10}$ -2
	10.0	78.626	1.3531	2.6325	0.9135	0.7350	0.765	0.727	0.128	+9.0424$_{10}$ -2
	8.0	81.369	1.3620	2.6802	0.9090	0.7099	0.741	0.718	0.101	+9.5460$_{10}$ -2
	6.0	83.748	1.3678	2.7113	0.9059	0.6933	0.725	0.713	0.075	+9.8788$_{10}$ -2
	4.0	85.925	1.3715	2.7311	0.9040	0.6826	0.715	0.710	0.050	+1.0093$_{10}$ -1
	2.0	87.988	1.3736	2.7422	0.9029	0.6765	0.709	0.708	0.025	+1.0214$_{10}$ -1
	1.0	88.997	1.3741	2.7449	0.9026	0.6750	0.708	0.708	0.012	+1.0243$_{10}$ -1
	0.0	90.000	1.3742	2.7458	0.9026	0.6746	0.707	0.707	0.000	+1.0253$_{10}$ -1
1.60 (1.425)	0.0	38.682	1.0000	1.0000	1.0000	1.6000	1.425	1.425	0.000	0.0000
	1.0	39.684	1.0144	1.0512	1.0000	1.5662	1.405	1.375	0.024	+1.2679$_{10}$ -5
	2.0	40.724	1.0289	1.1046	0.9999	1.5323	1.385	1.328	0.046	+1.0044$_{10}$ -4
	4.0	42.931	1.0584	1.2189	0.9992	1.4638	1.342	1.238	0.087	+7.9074$_{10}$ -4
	6.0	45.344	1.0891	1.3446	0.9974	1.3934	1.295	1.155	0.121	+2.6432$_{10}$ -3
	8.0	48.030	1.1215	1.4843	0.9938	1.3195	1.245	1.077	0.151	+6.2638$_{10}$ -3
	10.0	51.115	1.1565	1.6430	0.9877	1.2397	1.188	1.003	0.177	+1.2419$_{10}$ -2
	12.0	54.889	1.1965	1.8320	0.9778	1.1483	1.119	0.931	0.198	+2.2443$_{10}$ -2
	14.0	60.537	1.2502	2.0974	0.9599	1.0232	1.019	0.850	0.212	+4.0923$_{10}$ -2
	δ_* = 14.2428	61.658	1.2600	2.1469	0.9561	1.0000	1.000	0.837	0.212	+4.4869$_{10}$ -2
	δ_m = 14.6515	65.828	1.2936	2.3192	0.9420	0.9188	0.931	0.795	0.208	+5.9706$_{10}$ -2
	14.0	70.895	1.3281	2.5001	0.9260	0.8320	0.854	0.757	0.189	+7.6903$_{10}$ -2
	12.0	75.900	1.3550	2.6428	0.9126	0.7611	0.789	0.731	0.155	+9.1496$_{10}$ -2
	10.0	79.102	1.3682	2.7132	0.9057	0.7250	0.755	0.719	0.127	+9.9001$_{10}$ -2
	8.0	81.691	1.3764	2.7576	0.9014	0.7018	0.734	0.711	0.100	+1.0382$_{10}$ -1
	6.0	83.967	1.3819	2.7870	0.8985	0.6862	0.719	0.707	0.074	+1.0705$_{10}$ -1
	4.0	86.061	1.3854	2.8059	0.8966	0.6761	0.709	0.704	0.049	+1.0914$_{10}$ -1
	2.0	88.054	1.3873	2.8166	0.8955	0.6703	0.703	0.702	0.025	+1.1032$_{10}$ -1
	1.0	89.030	1.3878	2.8191	0.8953	0.6689	0.702	0.702	0.012	+1.1061$_{10}$ -1
	0.0	90.000	1.3880	2.8200	0.8952	0.6684	0.702	0.702	0.000	+1.1071$_{10}$ -1
1.62 (1.437)	0.0	38.118	1.0000	1.0000	1.0000	1.6200	1.437	1.437	0.000	0.0000
	1.0	39.104	1.0144	1.0514	1.0000	1.5862	1.417	1.387	0.024	+1.2847$_{10}$ -5
	2.0	40.126	1.0290	1.1051	0.9999	1.5523	1.397	1.338	0.047	+1.0169$_{10}$ -4
	4.0	42.292	1.0586	1.2197	0.9992	1.4839	1.354	1.247	0.087	+7.9900$_{10}$ -4
	6.0	44.652	1.0893	1.3456	0.9973	1.4139	1.309	1.163	0.122	+2.6637$_{10}$ -3
	8.0	47.264	1.1216	1.4851	0.9937	1.3407	1.260	1.085	0.153	+6.2882$_{10}$ -3
	10.0	50.235	1.1564	1.6424	0.9877	1.2623	1.204	1.012	0.178	+1.2391$_{10}$ -2
	12.0	53.792	1.1954	1.8267	0.9781	1.1740	1.139	0.940	0.200	+2.2126$_{10}$ -2
	14.0	58.686	1.2444	2.0681	0.9621	1.0616	1.051	0.865	0.216	+3.8651$_{10}$ -2
	δ_* = 14.7343	61.580	1.2707	2.2016	0.9518	1.0000	1.000	0.829	0.218	+4.9400$_{10}$ -2
	δ_m = 15.1385	65.709	1.3047	2.3770	0.9370	0.9186	0.931	0.789	0.213	+6.5030$_{10}$ -2
	14.0	72.327	1.3494	2.6130	0.9154	0.8063	0.831	0.742	0.185	+8.8380$_{10}$ -2
	12.0	76.563	1.3712	2.7298	0.9041	0.7485	0.778	0.722	0.153	+1.0079$_{10}$ -1
	10.0	79.531	1.3832	2.7940	0.8978	0.7157	0.747	0.711	0.125	+1.0782$_{10}$ -1
	8.0	81.988	1.3909	2.8356	0.8936	0.6941	0.726	0.705	0.099	+1.1246$_{10}$ -1
	6.0	84.169	1.3960	2.8635	0.8908	0.6794	0.712	0.701	0.074	+1.1559$_{10}$ -1
	4.0	86.189	1.3993	2.8816	0.8890	0.6698	0.703	0.698	0.049	+1.1764$_{10}$ -1
	2.0	88.116	1.4012	2.8918	0.8880	0.6643	0.698	0.696	0.024	+1.1880$_{10}$ -1
	1.0	89.060	1.4017	2.8943	0.8377	0.6630	0.696	0.696	0.012	+1.1908$_{10}$ -1
	0.0	90.000	1.4018	2.8951	0.8877	0.6625	0.696	0.696	0.000	+1.1917$_{10}$ -1
1.64 (1.449)	0.0	37.572	1.0000	1.0000	1.0000	1.6400	1.449	1.449	0.000	0.0000
	1.0	38.543	1.0145	1.0516	1.0000	1.6062	1.429	1.397	0.024	+1.3028$_{10}$ -5
	2.0	39.549	1.0291	1.1055	0.9999	1.5723	1.409	1.349	0.047	+1.0304$_{10}$ -4
	4.0	41.677	1.0589	1.2207	0.9992	1.5040	1.367	1.257	0.088	+8.0821$_{10}$ -4
	6.0	43.988	1.0896	1.3468	0.9973	1.4342	1.322	1.172	0.123	+2.6881$_{10}$ -3
	8.0	46.534	1.1219	1.4862	0.9937	1.3617	1.274	1.094	0.154	+6.3245$_{10}$ -3
	10.0	49.405	1.1564	1.6426	0.9877	1.2843	1.220	1.020	0.180	+1.2398$_{10}$ -2
	12.0	52.786	1.1947	1.8234	0.9783	1.1985	1.157	0.949	0.202	+2.1930$_{10}$ -2
	14.0	57.200	1.2409	2.0504	0.9634	1.0941	1.077	0.877	0.219	+3.7308$_{10}$ -2
	δ_* = 15.2190	61.513	1.2816	2.2573	0.9472	1.0000	1.000	0.822	0.224	+5.4193$_{10}$ -2
	δ_m = 15.6181	65.599	1.3159	2.4357	0.9318	0.9185	0.931	0.783	0.219	+7.0603$_{10}$ -2
	14.0	73.405	1.3685	2.7152	0.9055	0.7867	0.813	0.730	0.182	+9.9217$_{10}$ -2
	12.0	77.145	1.3872	2.8159	0.8956	0.7370	0.767	0.714	0.152	+1.1025$_{10}$ -1
	10.0	79.920	1.3981	2.8751	0.8897	0.7070	0.738	0.704	0.124	+1.1690$_{10}$ -1
	8.0	82.261	1.4053	2.9143	0.8857	0.6868	0.719	0.699	0.098	+1.2136$_{10}$ -1
	6.0	84.357	1.4102	2.9409	0.8830	0.6729	0.706	0.695	0.073	+1.2441$_{10}$ -1
	4.0	86.307	1.4134	2.9582	0.8812	0.6637	0.697	0.692	0.048	+1.2641$_{10}$ -1
	2.0	88.173	1.4152	2.9680	0.8802	0.6585	0.692	0.691	0.024	+1.2755$_{10}$ -1
	1.0	89.089	1.4156	2.9704	0.8800	0.6572	0.691	0.690	0.012	+1.2783$_{10}$ -1
	0.0	90.000	1.4158	2.9712	0.8799	0.6568	0.690	0.690	0.000	+1.2792$_{10}$ -1

M_1	δ	θ	T_2/T_1	p_2/p_1	p_{o2}/p_{o1}	M_2	q_2/q_*	u_2/q_*	v_2/q_*	$(s_2 - s_1)/R$
1.66 (1.460)	0.0	37.043	1.0000	1.0000	1.0000	1.6600	1.460	1.460	0.000	0.0000
	1.0	38.000	1.0146	1.0519	1.0000	1.6261	1.441	1.408	0.025	+1.3220$_{10}$ -5
	2.0	38.991	1.0293	1.1061	0.9999	1.5922	1.421	1.359	0.047	+1.0450$_{10}$ -4
	4.0	41.083	1.0591	1.2217	0.9992	1.5240	1.380	1.266	0.089	+8.1832$_{10}$ -4
	6.0	43.350	1.0900	1.3482	0.9973	1.4545	1.336	1.180	0.124	+2.7159$_{10}$ -3
	8.0	45.836	1.1222	1.4877	0.9936	1.3824	1.288	1.101	0.155	+6.3713$_{10}$ -3
	10.0	48.620	1.1566	1.6434	0.9876	1.3060	1.235	1.028	0.181	+1.2435$_{10}$ -2
	12.0	51.854	1.1943	1.8217	0.9784	1.2221	1.175	0.957	0.203	+2.1826$_{10}$ -2
	14.0	55.925	1.2386	2.0390	0.9642	1.1232	1.099	0.887	0.221	+3.6455$_{10}$ -2
	16.0	63.583	1.3114	2.4118	0.9340	0.9561	0.963	0.794	0.228	+6.8321$_{10}$ -2
	δ_* = 15.6968	61.456	1.2926	2.3142	0.9425	1.0000	1.000	0.816	0.229	+5.9248$_{10}$ -2
	δ_m = 16.0902	65.498	1.3272	2.4952	0.9264	0.9184	0.931	0.777	0.224	+7.6425$_{10}$ -2
	16.0	67.377	1.3418	2.5725	0.9192	0.8831	0.900	0.762	0.218	+8.4208$_{10}$ -2
	14.0	74.282	1.3865	2.8123	0.8960	0.7704	0.798	0.720	0.179	+1.0985$_{10}$ -1
	12.0	77.664	1.4030	2.9015	0.8870	0.7266	0.757	0.706	0.150	+1.1990$_{10}$ -1
	10.0	80.276	1.4131	2.9565	0.8814	0.6989	0.731	0.698	0.123	+1.2622$_{10}$ -1
	8.0	82.513	1.4199	2.9936	0.8776	0.6798	0.712	0.693	0.097	+1.3053$_{10}$ -1
	6.0	84.532	1.4245	3.0190	0.8750	0.6666	0.700	0.689	0.072	+1.3351$_{10}$ -1
	4.0	86.418	1.4276	3.0357	0.8733	0.6578	0.691	0.687	0.048	+1.3547$_{10}$ -1
	2.0	88.227	1.4293	3.0451	0.8723	0.6528	0.686	0.685	0.024	+1.3659$_{10}$ -1
	1.0	89.116	1.4297	3.0474	0.8721	0.6516	0.685	0.685	0.012	+1.3686$_{10}$ -1
	0.0	90.000	1.4299	3.0482	0.8720	0.6512	0.685	0.685	0.000	+1.3695$_{10}$ -1
1.68 (1.471)	0.0	36.530	1.0000	1.0000	1.0000	1.6800	1.471	1.471	0.000	0.0000
	1.0	37.474	1.0146	1.0522	1.0000	1.6461	1.452	1.419	0.025	+1.3424$_{10}$ -5
	2.0	38.451	1.0294	1.1066	0.9999	1.6121	1.433	1.369	0.048	+1.0604$_{10}$ -4
	4.0	40.510	1.0594	1.2228	0.9992	1.5440	1.392	1.275	0.089	+8.2926$_{10}$ -4
	6.0	42.736	1.0903	1.3498	0.9973	1.4746	1.348	1.189	0.125	+2.7470$_{10}$ -3
	8.0	45.168	1.1226	1.4894	0.9936	1.4029	1.302	1.109	0.156	+6.4275$_{10}$ -3
	10.0	47.876	1.1569	1.6447	0.9876	1.3272	1.250	1.035	0.182	+1.2496$_{10}$ -2
	12.0	50.985	1.1942	1.8212	0.9784	1.2450	1.192	0.965	0.205	+2.1795$_{10}$ -2
	14.0	54.794	1.2371	2.0317	0.9647	1.1502	1.120	0.896	0.223	+3.5911$_{10}$ -2
	16.0	60.602	1.2962	2.3327	0.9409	1.0173	1.014	0.818	0.234	+6.0935$_{10}$ -2
	δ_* = 16.1675	61.409	1.3037	2.3720	0.9375	1.0000	1.000	0.809	0.234	+6.4566$_{10}$ -2
	δ_m = 16.5549	65.404	1.3387	2.5557	0.9208	0.9184	0.931	0.771	0.229	+8.2495$_{10}$ -2
	16.0	69.983	1.3732	2.7403	0.9031	0.8345	0.856	0.737	0.211	+1.0193$_{10}$ -1
	14.0	75.025	1.4039	2.9063	0.8865	0.7563	0.785	0.711	0.177	+1.2044$_{10}$ -1
	12.0	78.131	1.4186	2.9868	0.8783	0.7169	0.748	0.699	0.149	+1.2974$_{10}$ -1
	10.0	80.604	1.4281	3.0384	0.8730	0.6911	0.723	0.692	0.122	+1.3579$_{10}$ -1
	8.0	82.748	1.4345	3.0737	0.8694	0.6731	0.706	0.687	0.097	+1.3997$_{10}$ -1
	6.0	84.695	1.4389	3.0980	0.8669	0.6605	0.694	0.683	0.072	+1.4287$_{10}$ -1
	4.0	86.522	1.4418	3.1140	0.8652	0.6522	0.686	0.681	0.048	+1.4479$_{10}$ -1
	2.0	88.278	1.4435	3.1232	0.8643	0.6474	0.681	0.680	0.024	+1.4589$_{10}$ -1
	1.0	89.141	1.4439	3.1254	0.8640	0.6462	0.680	0.680	0.012	+1.4616$_{10}$ -1
	0.0	90.000	1.4440	3.1261	0.8639	0.6458	0.680	0.680	0.000	+1.4625$_{10}$ -1
1.70 (1.482)	0.0	36.032	1.0000	1.0000	1.0000	1.7000	1.482	1.482	0.000	0.0000
	1.0	36.964	1.0147	1.0525	1.0000	1.6660	1.463	1.429	0.025	+1.3639$_{10}$ -5
	2.0	37.927	1.0296	1.1072	0.9999	1.6320	1.444	1.379	0.048	+1.0768$_{10}$ -4
	4.0	39.957	1.0597	1.2239	0.9992	1.5638	1.404	1.284	0.090	+8.4098$_{10}$ -4
	6.0	42.145	1.0907	1.3514	0.9972	1.4946	1.361	1.196	0.126	+2.7811$_{10}$ -3
	8.0	44.528	1.1231	1.4914	0.9935	1.4232	1.315	1.116	0.157	+6.4922$_{10}$ -3
	10.0	47.167	1.1573	1.6466	0.9875	1.3482	1.265	1.042	0.184	+1.2580$_{10}$ -2
	12.0	50.168	1.1943	1.8217	0.9784	1.2673	1.208	0.972	0.207	+2.1825$_{10}$ -2
	14.0	53.771	1.2362	2.0273	0.9650	1.1757	1.140	0.904	0.225	+3.5581$_{10}$ -2
	16.0	58.794	1.2898	2.2999	0.9437	1.0569	1.047	0.831	0.238	+5.7964$_{10}$ -2
	δ_* = 16.6311	61.371	1.3150	2.4309	0.9323	1.0000	1.000	0.802	0.240	+7.0149$_{10}$ -2
	δ_m = 17.0119	65.319	1.3502	2.6171	0.9150	0.9185	0.931	0.765	0.234	+8.8812$_{10}$ -2
	16.0	71.426	1.3959	2.8629	0.8909	0.8077	0.832	0.723	0.207	+1.1552$_{10}$ -1
	14.0	75.670	1.4208	2.9985	0.8771	0.7439	0.773	0.702	0.175	+1.3110$_{10}$ -1
	12.0	78.555	1.4342	3.0722	0.8695	0.7079	0.739	0.692	0.147	+1.3980$_{10}$ -1
	10.0	80.906	1.4431	3.1208	0.8645	0.6838	0.716	0.686	0.121	+1.4560$_{10}$ -1
	8.0	82.966	1.4492	3.1544	0.8610	0.6667	0.700	0.681	0.096	+1.4966$_{10}$ -1
	6.0	84.848	1.4534	3.1778	0.8586	0.6547	0.688	0.678	0.071	+1.5250$_{10}$ -1
	4.0	86.620	1.4562	3.1933	0.8569	0.6467	0.681	0.676	0.047	+1.5438$_{10}$ -1
	2.0	88.325	1.4578	3.2021	0.8560	0.6421	0.676	0.675	0.024	+1.5546$_{10}$ -1
	1.0	89.164	1.4582	3.2043	0.8558	0.6409	0.675	0.675	0.012	+1.5572$_{10}$ -1
	0.0	90.000	1.4583	3.2050	0.8557	0.6405	0.675	0.675	0.000	+1.5581$_{10}$ -1
1.72 (1.493)	0.0	35.549	1.0000	1.0000	1.0000	1.7200	1.493	1.493	0.000	0.0000
	1.0	36.470	1.0148	1.0528	1.0000	1.6859	1.475	1.440	0.025	+1.3865$_{10}$ -5
	2.0	37.420	1.0297	1.1078	0.9999	1.6519	1.455	1.388	0.048	+1.0941$_{10}$ -4
	4.0	39.421	1.0600	1.2251	0.9991	1.5837	1.416	1.292	0.090	+8.5345$_{10}$ -4
	6.0	41.575	1.0911	1.3532	0.9972	1.5145	1.374	1.204	0.127	+2.8180$_{10}$ -3
	8.0	43.913	1.1236	1.4936	0.9935	1.4433	1.328	1.123	0.158	+6.5648$_{10}$ -3
	10.0	46.490	1.1578	1.6488	0.9874	1.3689	1.279	1.049	0.185	+1.2683$_{10}$ -2
	12.0	49.399	1.1946	1.8230	0.9783	1.2891	1.223	0.979	0.208	+2.1904$_{10}$ -2
	14.0	52.831	1.2358	2.0250	0.9652	1.2000	1.158	0.911	0.227	+3.5411$_{10}$ -2
	16.0	57.363	1.2862	2.2809	0.9453	1.0898	1.073	0.842	0.241	+5.6268$_{10}$ -2
	δ_* = 17.0874	61.341	1.3264	2.4909	0.9268	1.0000	1.000	0.795	0.244	+7.5990$_{10}$ -2
	δ_m = 17.4614	65.240	1.3619	2.6794	0.9090	0.9186	0.931	0.759	0.239	+9.5374$_{10}$ -2
	16.0	72.501	1.4161	2.9728	0.8798	0.7874	0.814	0.711	0.204	+1.2810$_{10}$ -1
	14.0	76.241	1.4374	3.0895	0.8677	0.7326	0.763	0.695	0.173	+1.4186$_{10}$ -1
	12.0	78.942	1.4498	3.1578	0.8606	0.6995	0.731	0.686	0.146	+1.5007$_{10}$ -1
	10.0	81.186	1.4581	3.2038	0.8559	0.6768	0.710	0.680	0.120	+1.5590$_{10}$ -1
	8.0	83.169	1.4639	3.2360	0.8525	0.6605	0.694	0.676	0.095	+1.5961$_{10}$ -1
	6.0	84.991	1.4680	3.2585	0.8501	0.6491	0.683	0.673	0.071	+1.6238$_{10}$ -1
	4.0	86.711	1.4707	3.2734	0.8485	0.6413	0.675	0.671	0.047	+1.6423$_{10}$ -1
	2.0	88.370	1.4722	3.2820	0.8476	0.6369	0.671	0.670	0.023	+1.6529$_{10}$ -1
	1.0	89.187	1.4726	3.2841	0.8474	0.6358	0.670	0.670	0.012	+1.6555$_{10}$ -1
	0.0	90.000	1.4727	3.2848	0.8474	0.6355	0.670	0.670	0.000	+1.6563$_{10}$ -1
1.74 (1.504)	0.0	35.080	1.0000	1.0000	1.0000	1.7400	1.504	1.504	0.000	0.0000
	1.0	35.989	1.0149	1.0531	1.0000	1.7058	1.486	1.450	0.025	+1.4100$_{10}$ -5
	2.0	36.929	1.0299	1.1084	0.9999	1.6717	1.467	1.398	0.049	+1.1122$_{10}$ -4
	4.0	38.903	1.0603	1.2264	0.9991	1.6034	1.427	1.301	0.091	+8.6663$_{10}$ -4
	6.0	41.024	1.0916	1.3551	0.9971	1.5343	1.386	1.212	0.127	+2.8575$_{10}$ -3
	8.0	43.322	1.1241	1.4960	0.9934	1.4633	1.341	1.130	0.159	+6.6444$_{10}$ -3
	10.0	45.843	1.1583	1.6514	0.9873	1.3893	1.293	1.055	0.186	+1.2804$_{10}$ -2
	12.0	48.669	1.1950	1.8250	0.9782	1.3105	1.239	0.985	0.209	+2.2025$_{10}$ -2
	14.0	51.960	1.2356	2.0243	0.9653	1.2234	1.176	0.918	0.229	+3.5363$_{10}$ -2
	16.0	56.143	1.2839	2.2692	0.9463	1.1192	1.096	0.851	0.244	+5.5232$_{10}$ -2
	δ_* = 17.5363	61.318	1.3379	2.5519	0.9212	1.0000	1.000	0.789	0.249	+8.2107$_{10}$ -2
	δ_m = 17.9033	65.168	1.3736	2.7426	0.9029	0.9188	0.931	0.753	0.243	+1.0218$_{10}$ -1
	16.0	73.374	1.4350	3.0764	0.8691	0.7709	0.798	0.702	0.201	+1.4029$_{10}$ -1
	14.0	76.751	1.4538	3.1800	0.8583	0.7223	0.753	0.688	0.171	+1.5276$_{10}$ -1
	12.0	79.298	1.4653	3.2437	0.8517	0.6916	0.724	0.680	0.144	+1.6056$_{10}$ -1
	10.0	81.446	1.4732	3.2874	0.8471	0.6701	0.703	0.674	0.119	+1.6595$_{10}$ -1
	8.0	83.360	1.4788	3.3183	0.8438	0.6546	0.688	0.670	0.094	+1.6980$_{10}$ -1
	6.0	85.126	1.4827	3.3400	0.8415	0.6436	0.678	0.668	0.070	+1.7251$_{10}$ -1
	4.0	86.797	1.4853	3.3545	0.8400	0.6362	0.670	0.666	0.047	+1.7433$_{10}$ -1
	2.0	88.412	1.4868	3.3628	0.8391	0.6319	0.666	0.665	0.023	+1.7537$_{10}$ -1
	1.0	89.208	1.4871	3.3649	0.8389	0.6309	0.665	0.665	0.012	+1.7563$_{10}$ -1
	0.0	90.000	1.4873	3.3655	0.8389	0.6305	0.665	0.665	0.000	+1.7571$_{10}$ -1
1.76 (1.515)	0.0	34.624	1.0000	1.0000	1.0000	1.7600	1.515	1.515	0.000	0.0000
	1.0	35.523	1.0150	1.0534	1.0000	1.7257	1.497	1.460	0.025	+1.4346$_{10}$ -5
	2.0	36.452	1.0300	1.1090	0.9999	1.6916	1.478	1.407	0.049	+1.1311$_{10}$ -4
	4.0	38.401	1.0606	1.2277	0.9991	1.6232	1.439	1.309	0.092	+8.8048$_{10}$ -4
	6.0	40.493	1.0921	1.3571	0.9971	1.5540	1.398	1.219	0.128	+2.8995$_{10}$ -3
	8.0	42.753	1.1247	1.4987	0.9933	1.4832	1.354	1.137	0.160	+6.7307$_{10}$ -3
	10.0	45.224	1.1590	1.6544	0.9871	1.4096	1.306	1.061	0.187	+1.2940$_{10}$ -2
	12.0	47.977	1.1956	1.8277	0.9781	1.3315	1.253	0.991	0.211	+2.2184$_{10}$ -2
	14.0	51.147	1.2358	2.0250	0.9652	1.2460	1.192	0.925	0.231	+3.5413$_{10}$ -2
	16.0	55.064	1.2825	2.2621	0.9469	1.1462	1.117	0.859	0.246	+5.4608$_{10}$ -2
	18.0	61.424	1.3508	2.6203	0.9147	0.9973	0.998	0.781	0.254	+8.9150$_{10}$ -2
	δ_* = 17.9779	61.302	1.3496	2.6139	0.9153	1.0000	1.000	0.782	0.254	+8.8480$_{10}$ -2
	δ_m = 18.3375	65.102	1.3855	2.8067	0.8965	0.9190	0.931	0.748	0.248	+1.0922$_{10}$ -1
	18.0	68.628	1.4151	2.9673	0.8803	0.8500	0.870	0.722	0.235	+1.2747$_{10}$ -1
	16.0	74.113	1.4531	3.1764	0.8587	0.7566	0.785	0.693	0.199	+1.5232$_{10}$ -1
	14.0	77.213	1.4701	3.2702	0.8489	0.7129	0.744	0.681	0.170	+1.6382$_{10}$ -1
	12.0	79.627	1.4809	3.3300	0.8426	0.6842	0.717	0.674	0.143	+1.7126$_{10}$ -1
	10.0	81.689	1.4884	3.3717	0.8382	0.6637	0.697	0.669	0.118	+1.7649$_{10}$ -1
	8.0	83.538	1.4937	3.4014	0.8351	0.6489	0.683	0.665	0.094	+1.8024$_{10}$ -1
	6.0	85.252	1.4975	3.4224	0.8329	0.6383	0.672	0.663	0.070	+1.8289$_{10}$ -1
	4.0	86.879	1.5000	3.4365	0.8314	0.6312	0.665	0.661	0.046	+1.8468$_{10}$ -1
	2.0	88.452	1.5014	3.4446	0.8305	0.6271	0.661	0.660	0.023	+1.8570$_{10}$ -1
	1.0	89.227	1.5018	3.4465	0.8303	0.6260	0.660	0.660	0.012	+1.8595$_{10}$ -1
	0.0	90.000	1.5019	3.4472	0.8302	0.6257	0.660	0.660	0.000	+1.8604$_{10}$ -1

M_1	δ	θ	T_2/T_1	p_2/p_1	p_{o2}/p_{o1}	M_2	q_2/q_*	u_2/q_*	v_2/q_*	$(s_2 - s_1)/R$
1.78 (1.526)	0.0	34.180	1.0000	1.0000	1.0000	1.7800	1.526	1.526	0.000	0.0000
	1.0	35.070	1.0151	1.0537	1.0000	1.7456	1.507	1.470	0.026	+1.4602$_{10}$ -5
	2.0	35.988	1.0302	1.1097	0.9999	1.7114	1.489	1.416	0.049	+1.1508$_{10}$ -4
	4.0	37.915	1.0610	1.2291	0.9991	1.6428	1.450	1.317	0.092	+8.9499$_{10}$ -4
	6.0	39.978	1.0926	1.3592	0.9971	1.5736	1.410	1.226	0.129	+2.9439$_{10}$ -3
	8.0	42.203	1.1253	1.5014	0.9932	1.5029	1.366	1.143	0.161	+6.8233$_{10}$ -3
	10.0	44.629	1.1597	1.6576	0.9870	1.4296	1.319	1.067	0.188	+1.3091$_{10}$ -2
	12.0	47.316	1.1962	1.8309	0.9779	1.3522	1.268	0.997	0.212	+2.2375$_{10}$ -2
	14.0	50.382	1.2361	2.0268	0.9651	1.2681	1.208	0.930	0.232	+3.5545$_{10}$ -2
	16.0	54.090	1.2818	2.2582	0.9472	1.1716	1.137	0.866	0.248	+5.4271$_{10}$ -2
	18.0	59.458	1.3423	2.5752	0.9190	1.0417	1.034	0.795	0.258	+8.4487$_{10}$ -2
	$\delta_* =$ 18.4121	61.293	1.3614	2.6770	0.9093	1.0000	1.000	0.776	0.258	+9.5115$_{10}$ -2
	$\delta_m =$ 18.7642	65.042	1.3975	2.8717	0.8900	0.9192	0.931	0.742	0.252	+1.1651$_{10}$ -1
	18.0	70.273	1.4409	3.1087	0.8658	0.8180	0.842	0.707	0.230	+1.4415$_{10}$ -1
	16.0	74.755	1.4708	3.2742	0.8485	0.7440	0.773	0.685	0.197	+1.6432$_{10}$ -1
	14.0	77.634	1.4863	3.3603	0.8394	0.7041	0.736	0.675	0.168	+1.7505$_{10}$ -1
	12.0	79.931	1.4965	3.4168	0.8335	0.6771	0.710	0.668	0.142	+1.8218$_{10}$ -1
	10.0	81.915	1.5036	3.4567	0.8292	0.6576	0.691	0.664	0.117	+1.8725$_{10}$ -1
	8.0	83.706	1.5087	3.4854	0.8262	0.6434	0.677	0.660	0.093	+1.9091$_{10}$ -1
	6.0	85.372	1.5124	3.5057	0.8241	0.6332	0.667	0.658	0.069	+1.9351$_{10}$ -1
	4.0	86.955	1.5148	3.5194	0.8226	0.6263	0.661	0.657	0.046	+1.9527$_{10}$ -1
	2.0	88.489	1.5162	3.5272	0.8218	0.6223	0.657	0.656	0.023	+1.9628$_{10}$ -1
	1.0	89.246	1.5166	3.5292	0.8216	0.6214	0.656	0.656	0.011	+1.9652$_{10}$ -1
	0.0	90.000	1.5167	3.5298	0.8215	0.6210	0.656	0.656	0.000	+1.9661$_{10}$ -1
1.80 (1.536)	0.0	33.749	1.0000	1.0000	1.0000	1.8000	1.536	1.536	0.000	0.0000
	1.0	34.630	1.0152	1.0540	1.0000	1.7655	1.518	1.479	0.026	+1.4867$_{10}$ -5
	2.0	35.538	1.0304	1.1104	0.9999	1.7311	1.500	1.425	0.050	+1.1713$_{10}$ -4
	4.0	37.443	1.0613	1.2305	0.9991	1.6625	1.461	1.325	0.093	+9.1014$_{10}$ -4
	6.0	39.480	1.0931	1.3614	0.9970	1.5932	1.421	1.233	0.130	+2.9905$_{10}$ -3
	8.0	41.673	1.1260	1.5044	0.9931	1.5225	1.379	1.150	0.162	+6.9217$_{10}$ -3
	10.0	44.057	1.1604	1.6611	0.9868	1.4494	1.332	1.073	0.189	+1.3255$_{10}$ -2
	12.0	46.686	1.1970	1.8345	0.9777	1.3725	1.281	1.002	0.213	+2.2595$_{10}$ -2
	14.0	49.661	1.2367	2.0295	0.9649	1.2896	1.224	0.936	0.233	+3.5746$_{10}$ -2
	16.0	53.198	1.2815	2.2568	0.9473	1.1958	1.155	0.872	0.250	+5.4145$_{10}$ -2
	18.0	57.995	1.3379	2.5515	0.9212	1.0766	1.063	0.805	0.262	+8.2071$_{10}$ -2
	$\delta_* =$ 18.8388	61.290	1.3733	2.7410	0.9030	1.0000	1.000	0.770	0.263	+1.0201$_{10}$ -1
	$\delta_m =$ 19.1833	64.987	1.4096	2.9376	0.8834	0.9195	0.932	0.737	0.256	+1.2403$_{10}$ -1
	18.0	71.424	1.4628	3.2297	0.8531	0.7956	0.821	0.696	0.226	+1.5884$_{10}$ -1
	16.0	75.324	1.4882	3.3707	0.8383	0.7327	0.763	0.678	0.194	+1.7636$_{10}$ -1
	14.0	78.020	1.5025	3.4505	0.8299	0.6958	0.728	0.669	0.167	+1.8645$_{10}$ -1
	12.0	80.215	1.5121	3.5041	0.8242	0.6703	0.703	0.663	0.141	+1.9331$_{10}$ -1
	10.0	82.128	1.5189	3.5424	0.8202	0.6518	0.685	0.659	0.116	+1.9823$_{10}$ -1
	8.0	83.864	1.5239	3.5702	0.8172	0.6381	0.672	0.656	0.092	+2.0181$_{10}$ -1
	6.0	85.484	1.5274	3.5899	0.8152	0.6283	0.663	0.654	0.069	+2.0437$_{10}$ -1
	4.0	87.028	1.5298	3.6032	0.8138	0.6216	0.656	0.652	0.046	+2.0609$_{10}$ -1
	2.0	88.525	1.5311	3.6108	0.8129	0.6178	0.652	0.651	0.023	+2.0709$_{10}$ -1
	1.0	89.264	1.5315	3.6127	0.8127	0.6168	0.651	0.651	0.011	+2.0733$_{10}$ -1
	0.0	90.000	1.5316	3.6133	0.8127	0.6165	0.651	0.651	0.000	+2.0741$_{10}$ -1
1.82 (1.546)	0.0	33.329	1.0000	1.0000	1.0000	1.8200	1.546	1.546	0.000	0.0000
	1.0	34.202	1.0152	1.0544	1.0000	1.7854	1.528	1.489	0.026	+1.5141$_{10}$ -5
	2.0	35.101	1.0306	1.1111	0.9999	1.7509	1.510	1.434	0.050	+1.1925$_{10}$ -4
	4.0	36.985	1.0617	1.2320	0.9991	1.6820	1.472	1.333	0.093	+9.2589$_{10}$ -4
	6.0	38.998	1.0936	1.3637	0.9970	1.6127	1.433	1.240	0.130	+3.0392$_{10}$ -3
	8.0	41.161	1.1267	1.5074	0.9930	1.5420	1.391	1.156	0.162	+7.0256$_{10}$ -3
	10.0	43.506	1.1612	1.6649	0.9867	1.4691	1.345	1.079	0.190	+1.3432$_{10}$ -2
	12.0	46.083	1.1978	1.8386	0.9774	1.3927	1.295	1.007	0.214	+2.2842$_{10}$ -2
	14.0	48.978	1.2374	2.0330	0.9646	1.3106	1.239	0.941	0.235	+3.6005$_{10}$ -2
	16.0	52.371	1.2816	2.2573	0.9473	1.2190	1.172	0.878	0.252	+5.4184$_{10}$ -2
	18.0	56.777	1.3353	2.5377	0.9225	1.1069	1.087	0.814	0.264	+8.0673$_{10}$ -2
	$\delta_* =$ 19.2581	61.291	1.3854	2.8061	0.8966	1.0000	1.000	0.763	0.267	+1.0916$_{10}$ -1
	$\delta_m =$ 19.5949	64.937	1.4218	3.0043	0.8765	0.9199	0.932	0.732	0.261	+1.3179$_{10}$ -1
	18.0	72.336	1.4830	3.3420	0.8413	0.7778	0.805	0.686	0.223	+1.7276$_{10}$ -1
	16.0	75.834	1.5053	3.4663	0.8282	0.7224	0.753	0.672	0.193	+1.8848$_{10}$ -1
	14.0	78.375	1.5187	3.5409	0.8203	0.6881	0.720	0.663	0.165	+1.9804$_{10}$ -1
	12.0	80.479	1.5278	3.5921	0.8149	0.6639	0.697	0.658	0.140	+2.0465$_{10}$ -1
	10.0	82.328	1.5344	3.6289	0.8110	0.6462	0.680	0.654	0.115	+2.0944$_{10}$ -1
	8.0	84.014	1.5391	3.6558	0.8082	0.6330	0.667	0.651	0.091	+2.1294$_{10}$ -1
	6.0	85.591	1.5425	3.6750	0.8062	0.6235	0.658	0.649	0.068	+2.1546$_{10}$ -1
	4.0	87.097	1.5448	3.6879	0.8048	0.6171	0.652	0.648	0.045	+2.1715$_{10}$ -1
	2.0	88.559	1.5462	3.6954	0.8040	0.6133	0.648	0.647	0.023	+2.1813$_{10}$ -1
	1.0	89.281	1.5465	3.6972	0.8038	0.6124	0.647	0.647	0.011	+2.1837$_{10}$ -1
	0.0	90.000	1.5466	3.6978	0.8038	0.6121	0.647	0.647	0.000	+2.1845$_{10}$ -1
1.84 (1.556)	0.0	32.921	1.0000	1.0000	1.0000	1.8400	1.556	1.556	0.000	0.0000
	1.0	33.785	1.0153	1.0547	1.0000	1.8052	1.539	1.498	0.026	+1.5424$_{10}$ -5
	2.0	34.675	1.0308	1.1118	0.9999	1.7706	1.521	1.443	0.050	+1.2144$_{10}$ -4
	4.0	36.540	1.0621	1.2335	0.9991	1.7016	1.483	1.340	0.094	+9.4224$_{10}$ -4
	6.0	38.531	1.0942	1.3661	0.9969	1.6321	1.444	1.247	0.131	+3.0901$_{10}$ -3
	8.0	40.666	1.1274	1.5106	0.9929	1.5614	1.402	1.162	0.163	+7.1348$_{10}$ -3
	10.0	42.975	1.1621	1.6688	0.9865	1.4836	1.357	1.084	0.191	+1.3619$_{10}$ -2
	12.0	45.504	1.1988	1.8430	0.9772	1.4125	1.308	1.012	0.215	+2.3112$_{10}$ -2
	14.0	48.328	1.2382	2.0372	0.9643	1.3313	1.253	0.946	0.236	+3.6316$_{10}$ -2
	16.0	51.599	1.2820	2.2592	0.9471	1.2415	1.189	0.883	0.253	+5.4356$_{10}$ -2
	18.0	55.715	1.3338	2.5298	0.9232	1.1345	1.108	0.821	0.267	+7.9878$_{10}$ -2
	$\delta_* =$ 19.6701	61.298	1.3976	2.8722	0.8900	1.0000	1.000	0.757	0.271	+1.1657$_{10}$ -1
	$\delta_m =$ 19.9990	64.892	1.4342	3.0720	0.8696	0.9203	0.932	0.727	0.264	+1.3978$_{10}$ -1
	18.0	73.099	1.5023	3.4494	0.8300	0.7628	0.791	0.678	0.220	+1.8631$_{10}$ -1
	16.0	76.296	1.5223	3.5615	0.8182	0.7130	0.744	0.665	0.191	+2.0070$_{10}$ -1
	14.0	78.705	1.5348	3.6317	0.8107	0.6808	0.713	0.658	0.164	+2.0980$_{10}$ -1
	12.0	80.726	1.5435	3.6806	0.8056	0.6578	0.691	0.653	0.139	+2.1620$_{10}$ -1
	10.0	82.516	1.5499	3.7162	0.8018	0.6408	0.675	0.649	0.114	+2.2087$_{10}$ -1
	8.0	84.154	1.5545	3.7422	0.7991	0.6281	0.662	0.647	0.091	+2.2430$_{10}$ -1
	6.0	85.692	1.5578	3.7609	0.7971	0.6189	0.653	0.645	0.068	+2.2676$_{10}$ -1
	4.0	87.162	1.5600	3.7735	0.7958	0.6126	0.647	0.643	0.045	+2.2843$_{10}$ -1
	2.0	88.591	1.5613	3.7808	0.7950	0.6090	0.644	0.643	0.022	+2.2940$_{10}$ -1
	1.0	89.297	1.5616	3.7826	0.7948	0.6081	0.643	0.643	0.011	+2.2964$_{10}$ -1
	0.0	90.000	1.5617	3.7832	0.7948	0.6078	0.643	0.643	0.000	+2.2971$_{10}$ -1
1.86 (1.566)	0.0	32.523	1.0000	1.0000	1.0000	1.8600	1.566	1.566	0.000	0.0000
	1.0	33.379	1.0154	1.0551	1.0000	1.8251	1.549	1.508	0.026	+1.5717$_{10}$ -5
	2.0	34.261	1.0310	1.1125	0.9999	1.7904	1.531	1.452	0.051	+1.2371$_{10}$ -4
	4.0	36.108	1.0625	1.2351	0.9990	1.7211	1.494	1.348	0.094	+9.5918$_{10}$ -4
	6.0	38.077	1.0948	1.3685	0.9969	1.6515	1.455	1.253	0.132	+3.1429$_{10}$ -3
	8.0	40.187	1.1281	1.5140	0.9928	1.5807	1.414	1.167	0.164	+7.2491$_{10}$ -3
	10.0	42.463	1.1630	1.6730	0.9863	1.5080	1.370	1.089	0.192	+1.3818$_{10}$ -2
	12.0	44.948	1.1997	1.8478	0.9769	1.4322	1.321	1.017	0.216	+2.3405$_{10}$ -2
	14.0	47.708	1.2392	2.0419	0.9640	1.3516	1.267	0.951	0.237	+3.6673$_{10}$ -2
	16.0	50.875	1.2826	2.2624	0.9468	1.2632	1.205	0.888	0.255	+5.4636$_{10}$ -2
	18.0	54.763	1.3330	2.5260	0.9236	1.1602	1.128	0.827	0.269	+7.9489$_{10}$ -2
	20.0	60.909	1.4055	2.9154	0.8856	1.0093	1.008	0.755	0.275	+1.2149$_{10}$ -1
	$\delta_* =$ 20.0747	61.309	1.4099	2.9393	0.8832	1.0000	1.000	0.751	0.275	+1.2423$_{10}$ -1
	$\delta_m =$ 20.3958	64.852	1.4467	3.1406	0.8624	0.9207	0.933	0.722	0.268	+1.4799$_{10}$ -1
	20.0	68.595	1.4812	3.3319	0.8424	0.8433	0.864	0.696	0.253	+1.7150$_{10}$ -1
	18.0	73.759	1.5210	3.5538	0.8190	0.7496	0.779	0.670	0.218	+1.9970$_{10}$ -1
	16.0	76.718	1.5393	3.6565	0.8081	0.7042	0.736	0.659	0.189	+2.1304$_{10}$ -1
	14.0	79.011	1.5510	3.7229	0.8011	0.6739	0.707	0.653	0.163	+2.2175$_{10}$ -1
	12.0	80.959	1.5594	3.7699	0.7962	0.6519	0.686	0.648	0.138	+2.2795$_{10}$ -1
	10.0	82.693	1.5655	3.8042	0.7925	0.6355	0.670	0.645	0.114	+2.3251$_{10}$ -1
	8.0	84.288	1.5699	3.8295	0.7899	0.6233	0.658	0.642	0.090	+2.3587$_{10}$ -1
	6.0	85.787	1.5732	3.8478	0.7880	0.6144	0.649	0.640	0.067	+2.3829$_{10}$ -1
	4.0	87.224	1.5753	3.8601	0.7867	0.6083	0.643	0.639	0.045	+2.3993$_{10}$ -1
	2.0	88.621	1.5766	3.8672	0.7859	0.6048	0.640	0.639	0.022	+2.4089$_{10}$ -1
	1.0	89.312	1.5769	3.8690	0.7857	0.6039	0.639	0.638	0.011	+2.4112$_{10}$ -1
	0.0	90.000	1.5770	3.8695	0.7857	0.6036	0.638	0.638	0.000	+2.4120$_{10}$ -1
1.88 (1.576)	0.0	32.135	1.0000	1.0000	1.0000	1.8800	1.576	1.576	0.000	0.0000
	1.0	32.984	1.0155	1.0554	1.0000	1.8449	1.559	1.517	0.026	+1.6018$_{10}$ -5
	2.0	33.858	1.0312	1.1133	0.9999	1.8101	1.541	1.460	0.051	+1.2605$_{10}$ -4
	4.0	35.688	1.0629	1.2366	0.9990	1.7406	1.505	1.355	0.095	+9.7668$_{10}$ -4
	6.0	37.637	1.0953	1.3710	0.9968	1.6708	1.466	1.259	0.132	+3.1978$_{10}$ -3
	8.0	39.722	1.1289	1.5174	0.9927	1.6000	1.425	1.173	0.165	+7.3683$_{10}$ -3
	10.0	41.969	1.1639	1.6773	0.9861	1.5273	1.382	1.094	0.193	+1.4027$_{10}$ -2
	12.0	44.413	1.2008	1.8528	0.9766	1.4517	1.334	1.022	0.217	+2.3718$_{10}$ -2
	14.0	47.117	1.2402	2.0473	0.9636	1.3716	1.281	0.955	0.238	+3.7071$_{10}$ -2
	16.0	50.192	1.2834	2.2667	0.9465	1.2844	1.220	0.893	0.256	+5.5010$_{10}$ -2
	18.0	53.895	1.3329	2.5250	0.9237	1.1845	1.147	0.832	0.270	+7.9395$_{10}$ -2
	20.0	59.210	1.3984	2.8763	0.8896	1.0503	1.041	0.766	0.279	+1.1704$_{10}$ -1
	$\delta_* =$ 20.4719	61.325	1.4224	3.0074	0.8762	1.0000	1.000	0.746	0.278	+1.3214$_{10}$ -1
	$\delta_m =$ 20.7852	64.815	1.4593	3.2101	0.8552	0.9211	0.933	0.717	0.272	+1.5044$_{10}$ -1

M_1	δ	θ	T_2/T_1	p_2/p_1	p_{o2}/p_{o1}	M_2	q_2/q_*	u_2/q_*	v_2/q_*	$(s_2 - s_1)/R$
	20.0	70.017	1.5069	3.4753	0.8273	0.8145	0.838	0.684	0.249	+1.8961$_{10}$ -1
	18.0	74.341	1.5392	3.6564	0.8081	0.7380	0.768	0.664	0.216	+2.1303$_{10}$ -1
	16.0	77.105	1.5561	3.7515	0.7381	0.6960	0.728	0.654	0.187	+2.2552$_{10}$ -1
	14.0	79.297	1.5673	3.8146	0.7915	0.6[illegible]73	0.700	0.648	0.161	+2.3388$_{10}$ -1
	12.0	81.177	1.5753	3.8598	0.7867	0.6463	0.680	0.643	0.137	+2.3990$_{10}$ -1
	10.0	82.861	1.5812	3.8931	0.7832	0.6305	0.665	0.640	0.113	+2.4435$_{10}$ -1
	8.0	84.414	1.5855	3.9177	0.7806	0.6187	0.653	0.638	0.090	+2.4765$_{10}$ -1
	6.0	85.878	1.5885	3.9355	0.7788	0.6100	0.645	0.636	0.067	+2.5003$_{10}$ -1
	4.0	87.283	1.5908	3.9475	0.7775	0.6041	0.639	0.635	0.044	+2.5165$_{10}$ -1
	2.0	88.650	1.5920	3.9545	0.7768	0.6007	0.636	0.635	0.022	+2.5259$_{10}$ -1
	1.0	89.326	1.5923	3.9562	0.7766	0.5999	0.635	0.634	0.011	+2.5282$_{10}$ -1
	0.0	90.000	1.5924	3.9568	0.7765	0.5996	0.634	0.634	0.000	+2.5290$_{10}$ -1
1.90 (1.586)	0.0	31.757	1.0000	1.0000	1.0000	1.9000	1.586	1.586	0.000	0.0000
	1.0	32.599	1.0156	1.0558	1.0000	1.8648	1.569	1.526	0.027	+1.6328$_{10}$ -5
	2.0	33.466	1.0314	1.1140	0.9999	1.8298	1.551	1.468	0.051	+1.2846$_{10}$ -4
	4.0	35.279	1.0633	1.2382	0.9990	1.7600	1.515	1.362	0.095	+9.9474$_{10}$ -4
	6.0	37.209	1.0959	1.3735	0.9968	1.6901	1.477	1.266	0.133	+3.2545$_{10}$ -3
	8.0	39.272	1.1297	1.5209	0.9925	1.6191	1.437	.178	0.166	+7.4921$_{10}$ -3
	10.0	41.490	1.1649	1.6818	0.9859	1.5464	1.393	1.099	0.194	+1.4246$_{10}$ -2
	12.0	43.898	1.2019	1.8582	0.9762	1.4709	1.346	1.026	0.218	+2.4050$_{10}$ -2
	14.0	46.550	1.2414	2.0530	0.9632	1.3913	1.294	0.959	0.239	+3.7507$_{10}$ -2
	16.0	49.54	1.2844	2.2718	0.9460	1.3052	1.235	0.897	0.257	+5.5463$_{10}$ -2
	18.0	53.095	1.3331	2.5263	0.9236	1.2077	1.164	0.837	0.272	+7.9528$_{10}$ -2
	20.0	57.901	1.3946	2.8557	0.8916	1.0835	1.068	0.775	0.282	+1.1471$_{10}$ -1
δ_*	20.8620	61.44	1.4350	3.0764	0.8691	1.0000	1.000	0.740	0.282	+1.4030$_{10}$ -1
δ_m	21.1675	64.783	1.4720	3.2805	0.8478	0.9216	0.933	0.712	0.276	+1.6510$_{10}$ -1
	20.0	71.057	1.5294	3.6012	0.8140	0.7935	0.819	0.674	0.245	+2.0583$_{10}$ -1
	18.0	74.861	1.5572	3.7578	0.7974	0.7274	0.758	0.657	0.214	+2.2635$_{10}$ -1
	16.0	77.463	1.5729	3.8466	0.7881	0.6883	0.721	0.649	0.186	+2.3813$_{10}$ -1
	14.0	79.565	1.5836	3.9068	0.7818	0.6611	0.694	0.643	0.160	+2.4619$_{10}$ -1
	12.0	81.383	1.5913	3.9504	0.7772	0.6409	0.675	0.639	0.136	+2.5204$_{10}$ -1
	10.0	83.020	1.5970	3.9828	0.7738	0.6257	0.660	0.636	0.112	+2.5640$_{10}$ -1
	8.0	84.534	1.6012	4.0068	0.7713	0.6142	0.649	0.634	0.089	+2.5963$_{10}$ -1
	6.0	85.965	1.6042	4.0241	0.7695	0.6058	0.641	0.632	0.066	+2.6198$_{10}$ -1
	4.0	87.339	1.6063	4.0359	0.7683	0.6001	0.635	0.631	0.044	+2.6357$_{10}$ -1
	2.0	88.678	1.6075	4.0428	0.7676	0.5967	0.632	0.631	0.022	+2.6450$_{10}$ -1
	1.0	89.340	1.6078	4.0444	0.7674	0.5959	0.631	0.631	0.011	+2.6473$_{10}$ -1
	0.0	90.000	1.6079	4.0450	0.7674	0.5956	0.630	0.630	0.000	+2.6480$_{10}$ -1
1.92 (1.596)	0.0	31.383	1.0000	1.0000	1.0000	1.9200	1.596	1.596	0.000	0.0000
	1.0	32.223	1.0157	1.0562	1.0000	1.8346	1.579	1.535	0.027	+1.6647$_{10}$ -5
	2.0	33.083	1.0316	1.1148	0.9999	1.8495	1.561	1.477	0.052	+1.3093$_{10}$ -4
	4.0	34.881	1.0637	1.2399	0.9990	1.7795	1.525	1.369	0.096	+1.0134$_{10}$ -3
	6.0	36.794	1.0966	1.3761	0.9967	1.7093	1.48	1.272	0.134	+3.3131$_{10}$ -3
	8.0	38.835	1.1305	1.5246	0.9924	1.6382	1.448	1.183	0.166	+7.6205$_{10}$ -3
	10.0	41.028	1.1659	1.6865	0.9856	1.5654	1.405	1.103	0.195	+1.4474$_{10}$ -2
	12.0	43.401	1.2031	1.8638	0.9759	1.4901	1.358	1.030	0.219	+2.4400$_{10}$ -2
	14.0	46.006	1.2426	2.0592	0.9627	1.4108	1.307	0.963	0.240	+3.7976$_{10}$ -2
	16.0	48.929	1.2856	2.2777	0.9456	1.3255	1.249	0.901	0.258	+5.5987$_{10}$ -2
	18.0	52.351	1.3337	2.5295	0.9233	1.2300	1.181	0.841	0.273	+7.9841$_{10}$ -2
	20.0	56.797	1.3925	2.8445	0.8928	1.1127	1.091	0.781	0.284	+1.1345$_{10}$ -1
δ =	21.2448	61.366	1.4477	3.1465	0.8618	1.0000	1.000	0.734	0.285	+1.4870$_{10}$ -1
δ =	21.5426	64.753	1.4848	3.3518	0.8403	0.9221	0.934	0.707	0.279	+1.7398$_{10}$ -1
	20.0	71.896	1.5503	3.7189	0.8015	0.7765	0.804	0.665	0.242	+2.2122$_{10}$ -1
	18.0	75.333	1.5750	3.8584	0.7869	0.7177	0.749	0.651	0.212	+2.3971$_{10}$ -1
	16.0	77.796	1.5898	3.9419	0.7781	0.6811	0.714	0.644	0.185	+2.5090$_{10}$ -1
	14.0	79.816	1.5999	3.9997	0.7721	0.6551	0.689	0.638	0.159	+2.5868$_{10}$ -1
	12.0	81.577	1.6074	4.0419	0.7677	0.6357	0.670	0.635	0.135	+2.6438$_{10}$ -1
	10.0	83.170	1.6129	4.0733	0.7644	0.6210	0.655	0.632	0.111	+2.6864$_{10}$ -1
	8.0	84.648	1.6170	4.0967	0.7620	0.6099	0.645	0.630	0.089	+2.7182$_{10}$ -1
	6.0	86.047	1.6200	4.1137	0.7602	0.6017	0.636	0.628	0.066	+2.7413$_{10}$ -1
	4.0	87.392	1.6220	4.1252	0.7590	0.5961	0.631	0.627	0.044	+2.7570$_{10}$ -1
	2.0	88.704	1.6232	4.1319	0.7583	0.5928	0.628	0.627	0.022	+2.7662$_{10}$ -1
	1.0	89.353	1.6235	4.1336	0.7582	0.5920	0.627	0.627	0.011	+2.7684$_{10}$ -1
	0.0	90.000	1.6236	4.1341	0.7581	0.5918	0.627	0.627	0.000	+2.7691$_{10}$ -1
1.94 (1.605)	0.0	31.028	1.0000	1.0000	1.0000	1.9400	1.605	1.605	0.000	0.0000
	1.0	31.857	1.0158	1.0565	1.0000	1.9044	1.588	1.543	0.027	+1.6974$_{10}$ -5
	2.0	32.710	1.0318	1.1156	0.9999	1.8691	1.571	1.485	0.052	+1.3348$_{10}$ -4
	4.0	34.494	1.0641	1.2416	0.9990	1.7983	1.535	1.376	0.096	+1.0325$_{10}$ -3
	6.0	36.390	1.0972	1.3788	0.9966	1.7284	1.498	1.277	0.134	+3.3736$_{10}$ -3
	8.0	38.411	1.1313	1.5283	0.9923	1.6572	1.458	1.188	0.167	+7.7533$_{10}$ -3
	10.0	40.579	1.1669	1.6913	0.9854	1.5843	1.416	1.107	0.195	+1.4712$_{10}$ -2
	12.0	42.922	1.2043	1.8696	0.9755	1.5090	1.370	1.034	0.220	+2.4767$_{10}$ -2
	14.0	45.484	1.2439	2.0658	0.9623	1.4300	1.320	0.967	0.241	+3.8478$_{10}$ -2
	16.0	48.342	1.2868	2.2843	0.9450	1.3454	1.263	0.904	0.259	+5.6573$_{10}$ -2
	18.0	51.653	1.3346	2.5341	0.9228	1.2516	1.196	0.845	0.275	+8.0304$_{10}$ -2
	20.0	55.828	1.3915	2.8390	0.8933	1.1393	1.112	0.787	0.286	+1.1283$_{10}$ -1
δ_* =	21.6206	61.392	1.4606	3.2175	0.8544	1.0000	1.000	0.729	0.289	+1.5735$_{10}$ -1
δ_m =	21.9106	64.728	1.4977	3.4239	0.8327	0.9226	0.934	0.702	0.282	+1.8308$_{10}$ -1
	20.0	72.607	1.5703	3.8319	0.7896	0.7620	0.790	0.658	0.239	+2.3618$_{10}$ -1
	18.0	75.762	1.5927	3.9586	0.7764	0.7087	0.740	0.646	0.210	+2.5314$_{10}$ -1
	16.0	78.105	1.6066	4.0377	0.7681	0.6743	0.707	0.639	0.183	+2.6381$_{10}$ -1
	14.0	80.052	1.6164	4.0932	0.7624	0.6494	0.683	0.634	0.158	+2.7134$_{10}$ -1
	12.0	81.761	1.6235	4.1340	0.7581	0.6307	0.665	0.630	0.134	+2.7690$_{10}$ -1
	10.0	83.313	1.6289	4.1647	0.7550	0.6164	0.651	0.628	[illegible]	+2.8108$_{10}$ -1
	8.0	84.756	1.6329	4.1875	0.7526	0.6056	0.640	0.626	[illegible]	+2.8420$_{10}$ -1
	6.0	86.125	1.6358	4.2041	0.7509	0.5977	0.683	0.625	[illegible]	+2.8648$_{10}$ -1
	4.0	87.443	1.6378	4.2155	0.7497	0.5923	0.627	0.624	0.044	+2.8803$_{10}$ -1
	2.0	83.729	1.6390	4.2220	0.7491	0.5891	0.624	0.623	0.022	+2.8893$_{10}$ -1
	1.0	89.365	1.6393	4.2237	0.7489	0.5883	0.623	0.623	0.011	+2.8915$_{10}$ -1
	0.0	90.000	1.6394	4.2242	0.7488	0.5880	0.623	0.623	0.000	+2.8922$_{10}$ -1
1.96 (1.615)	0.0	30.677	1.0000	1.0000	1.0000	1.9600	1.615	1.615	0.000	0.0000
	1.0	31.500	1.0159	1.0569	1.0000	1.9243	1.598	1.552	0.027	+1.7310$_{10}$ -5
	2.0	32.347	1.0320	1.1164	0.9999	1.8888	1.581	1.493	0.052	+1.3609$_{10}$ -4
	4.0	34.117	1.0645	1.2433	0.9989	1.8182	1.545	1.382	0.097	+1.0522$_{10}$ -3
	6.0	35.996	1.0978	1.3815	0.9966	1.7475	1.508	1.283	0.135	+3.4358$_{10}$ -3
	8.0	38.000	1.1322	1.5321	0.9921	1.6761	1.469	1.193	0.168	+7.8905$_{10}$ -3
	10.0	40.145	1.1680	1.6963	0.9852	1.6032	1.427	1.112	0.196	+1.4958$_{10}$ -2
	12.0	42.458	1.2055	1.8757	0.9752	1.5279	1.382	1.038	0.221	+2.5150$_{10}$ -2
	14.0	44.981	1.2453	2.0728	0.9617	1.4491	1.332	0.970	0.242	+3.9008$_{10}$ -2
	16.0	47.782	1.2832	2.2915	0.9444	1.3651	1.276	0.908	0.260	+5.7215$_{10}$ -2
	18.0	50.996	1.3357	2.5399	0.9223	1.2726	1.212	0.849	0.276	+8.0893$_{10}$ -2
	20.0	54.956	1.3912	2.8375	0.8935	1.1642	1.131	0.792	0.283	+1.1266$_{10}$ -1
	22.0	61.485	1.4744	3.2938	0.8464	0.9984	0.999	0.723	0.292	+1.6675$_{10}$ -1
δ_* =	21.9893	61.421	1.4736	3.2896	0.8469	1.0000	1.000	0.723	0.292	+1.6622$_{10}$ -1
δ_m =	22.2717	64.705	1.5108	3.4970	0.8250	0.9231	0.935	0.698	0.286	+1.9239$_{10}$ -1
	22.0	67.775	1.5424	3.6740	0.8063	0.8559	0.876	0.678	0.274	+2.1533$_{10}$ -1
	20.0	73.226	1.5898	3.9419	0.7781	0.7494	0.778	0.651	0.237	+2.5090$_{10}$ -1
	18.0	76.158	1.6103	4.0586	0.7659	0.7004	0.732	0.641	0.208	+2.6665$_{10}$ -1
	16.0	78.395	1.6235	4.1338	0.7582	0.6678	0.701	0.634	0.182	+2.7687$_{10}$ -1
	14.0	80.275	1.6329	4.1873	0.7526	0.6440	0.678	0.630	0.157	+2.8417$_{10}$ -1
	12.0	81.935	1.6398	4.2270	0.7486	0.6259	0.660	0.626	0.133	+2.8961$_{10}$ -1
	10.0	83.448	1.6451	4.2568	0.7455	0.6121	0.647	0.624	0.110	+2.9370$_{10}$ -1
	8.0	84.860	1.6490	4.2792	0.7432	0.6016	0.636	0.622	0.087	+2.9678$_{10}$ -1
	6.0	86.200	1.6518	4.2955	0.7415	0.5938	0.629	0.621	0.065	+2.9902$_{10}$ -1
	4.0	87.491	1.6538	4.3066	0.7404	0.5885	0.623	0.620	0.043	+3.0055$_{10}$ -1
	2.0	88.753	1.6549	4.3131	0.7398	0.5854	0.620	0.620	0.022	+3.0143$_{10}$ -1
	1.0	89.377	1.6552	4.3147	0.7396	0.5846	0.620	0.619	0.011	+3.0165$_{10}$ -1
	0.0	90.000	1.6553	4.3152	0.7395	0.5844	0.619	0.619	0.000	+3.0173$_{10}$ -1
1.98 (1.624)	0.0	30.335	1.0000	1.0000	1.0000	1.9800	1.624	1.624	0.000	0.0000
	1.0	31.152	1.0161	1.0573	1.0000	1.9441	1.607	1.561	0.027	+1.7655$_{10}$ -5
	2.0	31.992	1.0322	1.1172	0.9999	1.9084	1.590	1.500	0.052	+1.3877$_{10}$ -4
	4.0	33.749	1.0649	1.2450	0.9989	1.8375	1.555	1.389	0.097	+1.0724$_{10}$ -3
	6.0	35.614	1.0985	1.3843	0.9965	1.7666	1.518	1.289	0.135	+3.4999$_{10}$ -3
	8.0	37.599	1.1331	1.5360	0.9920	1.6950	1.480	1.198	0.168	+8.0318$_{10}$ -3
	10.0	39.723	1.1690	1.7014	0.9849	1.6219	1.438	1.116	0.197	+1.5212$_{10}$ -2
	12.0	42.009	1.2068	1.8819	0.9748	1.5466	1.393	1.041	0.221	+2.5549$_{10}$ -2
	14.0	44.496	1.2468	2.0800	0.9612	1.4679	1.344	0.973	0.243	+3.9567$_{10}$ -2
	16.0	47.245	1.2897	2.2993	0.9437	1.3844	1.289	0.911	0.261	+5.7909$_{10}$ -2
	18.0	50.375	1.3370	2.5468	0.9216	1.2931	1.226	0.852	0.277	+8.1590$_{10}$ -2
	20.0	54.160	1.3915	2.8390	0.8933	1.1877	1.149	0.796	0.290	+1.1283$_{10}$ -1
	22.0	59.744	1.4657	3.2459	0.8514	1.0421	1.035	0.733	0.296	+1.6083$_{10}$ -1
δ_* =	22.3511	61.452	1.4867	3.3625	0.8392	1.0000	1.000	0.718	0.295	+1.7534$_{10}$ -1
δ_m =	22.6260	64.686	1.5240	3.5709	0.8172	0.9237	0.935	0.693	0.289	+2.0191$_{10}$ -1
	22.0	69.278	1.5708	3.8345	0.7894	0.8239	0.847	0.665	0.269	+2.3653$_{10}$ -1
	20.0	73.776	1.6088	4.0501	0.7668	0.7380	0.768	0.645	0.235	+2.6549$_{10}$ -1
	18.0	76.523	1.6279	4.1587	0.7556	0.6927	0.725	0.636	0.207	+2.8026$_{10}$ -1
	16.0	78.666	1.6405	4.2305	0.7482	0.6616	0.695	0.630	0.181	+2.9009$_{10}$ -1
	14.0	80.485	1.6495	4.2822	0.7429	0.6387	0.673	0.626	0.156	+2.9718$_{10}$ -1

M_1	δ	θ	T_2/T_1	p_2/p_1	p_{02}/p_{01}	M_2	q_2/q_*	u_2/q_*	v_2/q_*	$(s_2 - s_1)/R$
	12.0	82.100	1.6562	4.3207	0.7390	0.6213	0.656	0.622	0.132	+3.0249$_{10}$ -1
	10.0	83.577	1.6613	4.3499	0.7360	0.6078	0.643	0.620	0.109	+3.0651$_{10}$ -1
	8.0	84.958	1.6652	4.3718	0.7338	0.5976	0.632	0.619	0.087	+3.0953$_{10}$ -1
	6.0	86.271	1.6679	4.3878	0.7322	0.5900	0.625	0.617	0.065	+3.1174$_{10}$ -1
	4.0	87.538	1.6698	4.3987	0.7311	0.5849	0.620	0.616	0.043	+3.1325$_{10}$ -1
	2.0	83.776	1.6710	4.4050	0.7304	0.5818	0.617	0.616	0.022	+3.1413$_{10}$ -1
	1.0	89.389	1.6712	4.4066	0.7303	0.5811	0.616	0.616	0.011	+3.1435$_{10}$ -1
	0.0	90.000	1.6713	4.4071	0.7302	0.5808	0.616	0.616	0.000	+3.1442$_{10}$ -1
2.00 (1.633)	0.0	30.000	1.0000	1.0000	1.0000	2.0000	1.633	1.633	0.000	0.0000
	1.0	30.811	1.0162	1.0577	1.0000	1.9639	1.616	1.569	0.027	+1.8008$_{10}$ -5
	2.0	31.646	1.0324	1.1180	0.9999	1.9281	1.600	1.508	0.053	+1.4152$_{10}$ -4
	4.0	33.390	1.0654	1.2467	0.9989	1.8568	1.565	1.395	0.098	+1.0932$_{10}$ -3
	6.0	35.241	1.0991	1.3871	0.9964	1.7856	1.528	1.294	0.136	+3.5656$_{10}$ -3
	8.0	37.210	1.1339	1.5400	0.9919	1.7137	1.490	1.202	0.169	+8.1774$_{10}$ -3
	10.0	39.314	1.1702	1.7066	0.9846	1.6405	1.449	1.120	0.197	+1.5475$_{10}$ -2
	12.0	41.575	1.2081	1.8884	0.9744	1.5651	1.405	1.045	0.222	+2.5963$_{10}$ -2
	14.0	44.029	1.2483	2.0876	0.9606	1.4866	1.356	0.976	0.243	+4.0152$_{10}$ -2
	16.0	46.731	1.2913	2.3075	0.9430	1.4034	1.302	0.914	0.262	+5.8650$_{10}$ -2
	18.0	49.785	1.3384	2.5546	0.9209	1.3131	1.240	0.856	0.278	+8.2382$_{10}$ -2
	20.0	53.423	1.3922	2.8429	0.8929	1.2102	1.166	0.800	0.291	+1.1327$_{10}$ -1
	22.0	58.457	1.4616	3.2228	0.8539	1.0760	1.062	0.741	0.299	+1.5799$_{10}$ -1
	$\delta_* =$ 22.7060	61.485	1.5000	3.4365	0.8314	1.0000	1.000	0.713	0.298	+1.8468$_{10}$ -1
	$\delta_m =$ 22.9735	64.669	1.5373	3.6458	0.8093	0.9243	0.936	0.689	0.292	+2.1164$_{10}$ -1
	22.0	70.332	1.5950	3.9713	0.7750	0.8017	0.827	0.656	0.265	+2.5485$_{10}$ -1
	20.0	74.270	1.6276	4.1570	0.7558	0.7278	0.758	0.639	0.233	+2.8004$_{10}$ -1
	18.0	76.861	1.6454	4.2589	0.7453	0.6854	0.718	0.631	0.205	+2.9398$_{10}$ -1
	16.0	78.921	1.6575	4.3277	0.7383	0.6558	0.689	0.625	0.179	+3.0345$_{10}$ -1
	14.0	80.684	1.6662	4.3777	0.7332	0.6337	0.668	0.622	0.155	+3.1035$_{10}$ -1
	12.0	82.257	1.6727	4.4153	0.7294	0.6168	0.651	0.619	0.132	+3.1555$_{10}$ -1
	10.0	83.700	1.6777	4.4438	0.7265	0.6037	0.638	0.617	0.109	+3.1950$_{10}$ -1
	8.0	85.052	1.6815	4.4653	0.7244	0.5937	0.629	0.615	0.086	+3.2247$_{10}$ -1
	6.0	86.339	1.6842	4.4810	0.7228	0.5864	0.621	0.614	0.065	+3.2465$_{10}$ -1
	4.0	87.582	1.6861	4.4917	0.7217	0.5813	0.616	0.613	0.043	+3.2614$_{10}$ -1
	2.0	83.798	1.6871	4.4979	0.7211	0.5783	0.613	0.613	0.021	+3.2701$_{10}$ -1
	1.0	89.400	1.6874	4.4995	0.7209	0.5776	0.613	0.612	0.011	+3.2722$_{10}$ -1
	0.0	90.000	1.6875	4.5000	0.7209	0.5774	0.612	0.612	0.000	+3.2729$_{10}$ -1
2.02 (1.642)	0.0	29.673	1.0000	1.0000	1.0000	2.0200	1.642	1.642	0.000	0.0000
	1.0	30.479	1.0163	1.0581	1.0000	1.9837	1.626	1.577	0.028	+1.8369$_{10}$ -5
	2.0	31.308	1.0326	1.1188	0.9999	1.9477	1.609	1.516	0.053	+1.4433$_{10}$ -4
	4.0	33.040	1.0658	1.2485	0.9989	1.8761	1.574	1.402	0.098	+1.1144$_{10}$ -3
	6.0	34.878	1.0998	1.3899	0.9964	1.8046	1.538	1.299	0.137	+3.6331$_{10}$ -3
	8.0	36.831	1.1348	1.5440	0.9917	1.7325	1.500	1.207	0.170	+8.3269$_{10}$ -3
	10.0	38.916	1.1713	1.7119	0.9844	1.6591	1.460	1.123	0.198	+1.5746$_{10}$ -2
	12.0	41.154	1.2095	1.8950	0.9740	1.5836	1.416	1.048	0.223	+2.6392$_{10}$ -2
	14.0	43.577	1.2498	2.0954	0.9601	1.5051	1.368	0.979	0.244	+4.0762$_{10}$ -2
	16.0	46.236	1.2930	2.3162	0.9423	1.4222	1.315	0.917	0.263	+5.9434$_{10}$ -2
	18.0	49.223	1.3401	2.5632	0.9201	1.3327	1.254	0.858	0.279	+8.3258$_{10}$ -2
	20.0	52.736	1.3932	2.8486	0.8923	1.2318	1.182	0.803	0.292	+1.1391$_{10}$ -1
	22.0	57.389	1.4594	3.2111	0.8551	1.1054	1.085	0.746	0.302	+1.5656$_{10}$ -1
	$\delta_* =$ 23.0542	61.521	1.5134	3.5114	0.8235	1.0000	1.000	0.708	0.301	+1.9424$_{10}$ -1
	$\delta_m =$ 23.3144	64.655	1.5508	3.7215	0.8013	0.9248	0.936	0.684	0.295	+2.2156$_{10}$ -1
	22.0	71.171	1.6172	4.0979	0.7619	0.7841	0.811	0.648	0.262	+2.7198$_{10}$ -1
	20.0	74.719	1.6462	4.2632	0.7449	0.7184	0.749	0.634	0.231	+2.9457$_{10}$ -1
	18.0	77.177	1.6630	4.3593	0.7351	0.6785	0.711	0.626	0.204	+3.0781$_{10}$ -1
	16.0	79.162	1.6745	4.4255	0.7284	0.6502	0.684	0.621	0.178	+3.1696$_{10}$ -1
	14.0	80.873	1.6830	4.4740	0.7235	0.6288	0.663	0.618	0.154	+3.2368$_{10}$ -1
	12.0	82.406	1.6894	4.5107	0.7198	0.6124	0.647	0.615	0.131	+3.2877$_{10}$ -1
	10.0	83.817	1.6942	4.5386	0.7170	0.5997	0.635	0.613	0.108	+3.3265$_{10}$ -1
	8.0	85.142	1.6979	4.5597	0.7149	0.5900	0.625	0.611	0.086	+3.3559$_{10}$ -1
	6.0	86.404	1.7006	4.5751	0.7134	0.5828	0.618	0.610	0.064	+3.3773$_{10}$ -1
	4.0	87.625	1.7024	4.5856	0.7123	0.5778	0.613	0.610	0.043	+3.3920$_{10}$ -1
	2.0	88.819	1.7035	4.5918	0.7117	0.5749	0.610	0.609	0.021	+3.4006$_{10}$ -1
	1.0	89.410	1.7037	4.5933	0.7116	0.5742	0.609	0.609	0.011	+3.4027$_{10}$ -1
	0.0	90.000	1.7038	4.5938	0.7115	0.5740	0.609	0.609	0.000	+3.4034$_{10}$ -1
2.04 (1.651)	0.0	29.353	1.0000	1.0000	1.0000	2.0400	1.651	1.651	0.000	0.0000
	1.0	30.154	1.0164	1.0585	1.0000	2.0035	1.635	1.585	0.028	+1.8739$_{10}$ -5
	2.0	30.979	1.0329	1.1196	0.9999	1.9673	1.618	1.523	0.053	+1.4721$_{10}$ -4
	4.0	32.699	1.0663	1.2503	0.9989	1.8953	1.584	1.408	0.098	+1.1362$_{10}$ -3
	6.0	34.524	1.1005	1.3928	0.9963	1.8235	1.548	1.304	0.137	+3.7024$_{10}$ -3

M_1	δ	θ	T_2/T_1	p_2/p_1	p_{02}/p_{01}	M_2	q_2/q_*	u_2/q_*	v_2/q_*	$(s_2 - s_1)/R$
	8.0	36.463	1.1358	1.5481	0.9916	1.7512	1.510	1.211	0.170	+8.4805$_{10}$ -3
	10.0	38.530	1.1724	1.7173	0.9841	1.6775	1.470	1.127	0.199	+1.6025$_{10}$ -2
	12.0	40.746	1.2109	1.9018	0.9735	1.6020	1.427	1.051	0.223	+2.6834$_{10}$ -2
	14.0	43.140	1.2514	2.1035	0.9594	1.5235	1.379	0.982	0.245	+4.1395$_{10}$ -2
	16.0	45.760	1.2948	2.3253	0.9415	1.4408	1.327	0.919	0.264	+6.0260$_{10}$ -2
	18.0	48.687	1.3418	2.5725	0.9192	1.3519	1.267	0.861	0.280	+8.4209$_{10}$ -2
	20.0	52.093	1.3946	2.8558	0.8916	1.2528	1.197	0.806	0.293	+1.1472$_{10}$ -1
	22.0	56.459	1.4586	3.2062	0.8556	1.1319	1.106	0.751	0.303	+1.5596$_{10}$ -1
	$\delta_* =$ 23.3957	61.559	1.5269	3.5873	0.8154	1.0000	1.000	0.703	0.304	+2.0403$_{10}$ -1
	$\delta_m =$ 23.6483	64.643	1.5644	3.7981	0.7932	0.9254	0.937	0.680	0.298	+2.3169$_{10}$ -1
	22.0	71.877	1.6384	4.2187	0.7494	0.7693	0.797	0.641	0.259	+2.8848$_{10}$ -1
	20.0	75.131	1.6646	4.3688	0.7341	0.7097	0.741	0.629	0.229	+3.0912$_{10}$ -1
	18.0	77.473	1.6806	4.4601	0.7249	0.6720	0.705	0.622	0.202	+3.2176$_{10}$ -1
	16.0	79.389	1.6917	4.5239	0.7185	0.6448	0.679	0.617	0.177	+3.3061$_{10}$ -1
	14.0	81.052	1.6999	4.5711	0.7138	0.6242	0.659	0.614	0.153	+3.3718$_{10}$ -1
	12.0	82.549	1.7061	4.6069	0.7102	0.6082	0.643	0.611	0.130	+3.4217$_{10}$ -1
	10.0	83.929	1.7108	4.6342	0.7075	0.5958	0.631	0.609	0.107	+3.4598$_{10}$ -1
	8.0	85.227	1.7144	4.6549	0.7055	0.5863	0.621	0.608	0.085	+3.4887$_{10}$ -1
	6.0	86.467	1.7171	4.6701	0.7040	0.5793	0.614	0.607	0.064	+3.5099$_{10}$ -1
	4.0	87.665	1.7189	4.6805	0.7030	0.5745	0.609	0.606	0.042	+3.5244$_{10}$ -1
	2.0	83.839	1.7199	4.6865	0.7024	0.5716	0.607	0.606	0.021	+3.5329$_{10}$ -1
	1.0	89.420	1.7202	4.6880	0.7022	0.5709	0.606	0.606	0.011	+3.5350$_{10}$ -1
	0.0	90.000	1.7203	4.6885	0.7022	0.5707	0.606	0.606	0.000	+3.5357$_{10}$ -1
2.06 (1.660)	0.0	29.041	1.0000	1.0000	1.0000	2.0600	1.660	1.660	0.000	0.0000
	1.0	29.837	1.0165	1.0589	1.0000	2.0233	1.643	1.593	0.028	+1.9117$_{10}$ -5
	2.0	30.656	1.0331	1.1205	0.9998	1.9869	1.627	1.530	0.053	+1.5015$_{10}$ -4
	4.0	32.366	1.0667	1.2521	0.9988	1.9146	1.593	1.414	0.099	+1.1585$_{10}$ -3
	6.0	34.179	1.1011	1.3958	0.9962	1.8424	1.558	1.309	0.138	+3.7733$_{10}$ -3
	8.0	36.104	1.1367	1.5523	0.9914	1.7698	1.520	1.215	0.171	+8.6381$_{10}$ -3
	10.0	38.155	1.1736	1.7229	0.9838	1.6959	1.480	1.131	0.199	+1.6311$_{10}$ -2
	12.0	40.350	1.2123	1.9087	0.9731	1.6202	1.437	1.054	0.224	+2.7290$_{10}$ -2
	14.0	42.718	1.2531	2.1118	0.9588	1.5417	1.390	0.985	0.246	+4.2052$_{10}$ -2
	16.0	45.301	1.2966	2.3348	0.9407	1.4592	1.339	0.922	0.264	+6.1123$_{10}$ -2
	18.0	48.174	1.3437	2.5825	0.9183	1.3709	1.280	0.864	0.281	+8.5230$_{10}$ -2
	20.0	51.485	1.3962	2.8644	0.8908	1.2731	1.212	0.809	0.294	+1.1569$_{10}$ -1
	22.0	55.626	1.4585	3.2060	0.8556	1.1566	1.125	0.755	0.305	+1.5594$_{10}$ -1
	$\delta_* =$ 23.7307	61.598	1.5406	3.6641	0.8073	1.0000	1.000	0.698	0.307	+2.1403$_{10}$ -1
	$\delta_m =$ 23.9768	64.634	1.5781	3.8756	0.7851	0.9260	0.937	0.676	0.301	+2.4200$_{10}$ -1
	22.0	72.490	1.6589	4.3360	0.7374	0.7564	0.785	0.635	0.257	+3.0460$_{10}$ -1
	20.0	75.510	1.6830	4.4743	0.7235	0.7016	0.733	0.624	0.227	+3.2372$_{10}$ -1
	18.0	77.750	1.6982	4.5613	0.7148	0.6658	0.699	0.618	0.201	+3.3582$_{10}$ -1
	16.0	79.604	1.7089	4.6230	0.7086	0.6396	0.674	0.613	0.176	+3.4441$_{10}$ -1
	14.0	81.222	1.7169	4.6689	0.7041	0.6197	0.654	0.610	0.152	+3.5082$_{10}$ -1
	12.0	82.684	1.7229	4.7039	0.7007	0.6042	0.639	0.608	0.129	+3.5572$_{10}$ -1
	10.0	84.036	1.7276	4.7308	0.6980	0.5921	0.627	0.606	0.107	+3.5947$_{10}$ -1
	8.0	85.309	1.7311	4.7511	0.6961	0.5828	0.618	0.605	0.085	+3.6232$_{10}$ -1
	6.0	86.526	1.7337	4.7660	0.6946	0.5759	0.611	0.604	0.063	+3.6441$_{10}$ -1
	4.0	87.704	1.7355	4.7763	0.6936	0.5712	0.606	0.603	0.042	+3.6584$_{10}$ -1
	2.0	83.858	1.7365	4.7822	0.6930	0.5684	0.603	0.603	0.021	+3.6668$_{10}$ -1
	1.0	89.430	1.7368	4.7837	0.6929	0.5677	0.603	0.603	0.011	+3.6689$_{10}$ -1
	0.0	90.000	1.7369	4.7842	0.6928	0.5675	0.603	0.603	0.000	+3.6696$_{10}$ -1
2.08 (1.668)	0.0	28.736	1.0000	1.0000	1.0000	2.0800	1.668	1.668	0.000	0.0000
	1.0	29.527	1.0166	1.0593	1.0000	2.0431	1.652	1.601	0.028	+1.9503$_{10}$ -5
	2.0	30.341	1.0333	1.1213	0.9998	2.0064	1.636	1.537	0.054	+1.5317$_{10}$ -4
	4.0	32.041	1.0672	1.2540	0.9988	1.9338	1.602	1.420	0.099	+1.1813$_{10}$ -3
	6.0	33.842	1.1018	1.3987	0.9962	1.8613	1.567	1.314	0.138	+3.8458$_{10}$ -3
	8.0	35.754	1.1376	1.5566	0.9912	1.7883	1.530	1.219	0.171	+8.7995$_{10}$ -3
	10.0	37.789	1.1748	1.7285	0.9835	1.7143	1.490	1.134	0.200	+1.6605$_{10}$ -2
	12.0	39.965	1.2137	1.9158	0.9726	1.6383	1.448	1.057	0.225	+2.7759$_{10}$ -2
	14.0	42.309	1.2548	2.1203	0.9582	1.5598	1.401	0.987	0.246	+4.2730$_{10}$ -2
	16.0	44.858	1.2985	2.3446	0.9399	1.4774	1.350	0.924	0.265	+6.2023$_{10}$ -2
	18.0	47.682	1.3457	2.5930	0.9173	1.3895	1.293	0.865	0.281	+8.6314$_{10}$ -2
	20.0	50.911	1.3979	2.8741	0.8898	1.2929	1.226	0.811	0.295	+1.1678$_{10}$ -1
	22.0	54.868	1.4591	3.2093	0.8553	1.1798	1.143	0.759	0.306	+1.5633$_{10}$ -1
	24.0	61.279	1.5497	3.7152	0.8019	1.0091	1.008	0.696	0.310	+2.2074$_{10}$ -1
	$\delta_* =$ 24.0592	61.639	1.5544	3.7418	0.7991	1.0000	1.000	0.693	0.309	+2.2424$_{10}$ -1
	$\delta_m =$ 24.2985	64.626	1.5919	3.9539	0.7768	0.9267	0.938	0.672	0.303	+2.5251$_{10}$ -1
	24.0	67.795	1.6281	4.1599	0.7555	0.8535	0.873	0.653	0.291	+2.8043$_{10}$ -1
	22.0	73.033	1.6790	4.4510	0.7258	0.7449	0.774	0.629	0.254	+3.2049$_{10}$ -1
	20.0	75.862	1.7014	4.5797	0.7129	0.6941	0.726	0.620	0.226	+3.3837$_{10}$ -1

M_1	δ	θ	T_2/T_1	p_2/p_1	p_{o2}/p_{o1}	M_2	q_2/q_*	u_2/q_*	v_2/q_*	$(s_2 - s_1)/R$
	18.0	78.011	1.7158	4.6630	0.7047	0.6599	0.693	0.614	0.199	+3.5000D -1
	16.0	79.807	1.7262	4.7227	0.6983	0.6347	0.659	0.610	0.175	+3.5835D -1
	14.0	81.385	1.7340	4.7575	0.6945	0.6153	0.650	0.607	0.151	+3.6462D -1
	12.0	82.814	1.7399	4.8018	0.6911	0.6002	0.635	0.605	0.128	+3.6943D -1
	10.0	84.139	1.7445	4.8282	0.6886	0.5834	0.623	0.603	0.106	+3.7312D -1
	8.0	85.383	1.7479	4.8482	0.6867	0.5794	0.614	0.601	0.085	+3.7593D -1
	6.0	86.584	1.7505	4.8629	0.6852	0.5726	0.608	0.601	0.063	+3.7799D -1
	4.0	87.742	1.7522	4.8730	0.6843	0.5680	0.603	0.600	0.042	+3.7941D -1
	2.0	88.877	1.7533	4.8789	0.6837	0.5652	0.600	0.600	0.021	+3.8024D -1
	1.0	89.439	1.7535	4.8803	0.6836	0.5646	0.600	0.599	0.010	+3.8044D -1
	0.0	90.000	1.7536	4.8308	0.6835	0.5643	0.599	0.599	0.000	+3.8051D -1
2.10 (1.677)	0.0	28.437	1.0000	1.0000	1.0000	2.1000	1.677	1.677	0.000	0.0000
	1.0	29.224	1.0167	1.0597	1.0000	2.0628	1.661	1.609	0.028	+1.9898D -5
	2.0	30.033	1.0335	1.1222	0.9998	2.0260	1.645	1.544	0.054	+1.5624D -4
	4.0	31.723	1.0676	1.2558	0.9988	1.9530	1.611	1.426	0.100	+1.2046D -3
	6.0	33.513	1.1025	1.4018	0.9961	1.8301	1.576	1.319	0.139	+3.9201D -3
	8.0	35.412	1.1386	1.5609	0.9911	1.8069	1.540	1.223	0.172	+8.9648D -3
	10.0	37.433	1.1760	1.7342	0.9832	1.7325	1.500	1.137	0.201	+1.6906D -2
	12.0	39.592	1.2152	1.9231	0.9722	1.6564	1.458	1.060	0.225	+2.8241D -2
	14.0	41.912	1.2565	2.1291	0.9575	1.5777	1.412	0.990	0.247	+4.3430D -2
	16.0	44.431	1.3004	2.3547	0.9390	1.4954	1.362	0.926	0.266	+6.2958D -2
	18.0	47.210	1.3477	2.6040	0.9163	1.4078	1.305	0.868	0.282	+8.7456D -2
	20.0	50.365	1.3999	2.8847	0.8887	1.3122	1.240	0.814	0.296	+1.1799D -1
	22.0	54.169	1.4602	3.2152	0.8547	1.2019	1.160	0.762	0.308	+1.5705D -1
	24.0	59.768	1.5424	3.6740	0.8063	1.0493	1.041	0.704	0.313	+2.1533D -1
δ_*	24.3815	61.681	1.5683	3.8205	0.7908	1.0000	1.000	0.688	0.312	+2.3466D -1
δ_m	24.6140	64.621	1.6058	4.0332	0.7686	0.9273	0.938	0.668	0.306	+2.6321D -1
	24.0	69.103	1.6558	4.3238	0.7387	0.8245	0.847	0.643	0.286	+3.0291D -1
	22.0	73.521	1.6987	4.5644	0.7145	0.7345	0.764	0.624	0.252	+3.3624D -1
	20.0	76.190	1.7197	4.6852	0.7025	0.6870	0.719	0.615	0.224	+3.5309D -1
	18.0	78.258	1.7336	4.7652	0.6947	0.6543	0.688	0.610	0.198	+3.6430D -1
	16.0	80.001	1.7436	4.8232	0.6891	0.6299	0.664	0.606	0.174	+3.7243D -1
	14.0	81.539	1.7512	4.8670	0.6848	0.6111	0.646	0.603	0.150	+3.7857D -1
	12.0	82.938	1.7570	4.9006	0.6816	0.5964	0.631	0.601	0.128	+3.8329D -1
	10.0	84.237	1.7615	4.9264	0.6791	0.5849	0.620	0.600	0.106	+3.8693D -1
	8.0	85.463	1.7649	4.9461	0.6773	0.5760	0.611	0.598	0.084	+3.8970D -1
	6.0	86.639	1.7674	4.9606	0.6759	0.5694	0.604	0.597	0.063	+3.9173D -1
	4.0	87.778	1.7691	4.9706	0.6749	0.5648	0.600	0.597	0.042	+3.9314D -1
	2.0	88.894	1.7701	4.9764	0.6744	0.5622	0.597	0.596	0.021	+3.9395D -1
	1.0	89.448	1.7704	4.9779	0.6742	0.5615	0.597	0.596	0.010	+3.9416D -1
	0.0	90.000	1.7705	4.9783	0.6742	0.5613	0.596	0.596	0.000	+3.9422D -1
2.12 (1.685)	0.0	28.145	1.0000	1.0000	1.0000	2.1200	1.685	1.685	0.000	0.0000
	1.0	28.927	1.0168	1.0601	1.0000	2.0826	1.669	1.616	0.028	+2.0302D -5
	2.0	29.732	1.0338	1.1230	0.9998	2.0456	1.653	1.551	0.054	+1.5939D -4
	4.0	31.413	1.0681	1.2577	0.9988	1.9721	1.620	1.431	0.100	+1.2284D -3
	6.0	33.192	1.1032	1.4048	0.9960	1.8989	1.586	1.324	0.139	+3.9960D -3
	8.0	35.080	1.1395	1.5652	0.9909	1.8253	1.549	1.227	0.172	+9.1340D -3
	10.0	37.087	1.1772	1.7400	0.9829	1.7507	1.510	1.140	0.201	+1.7215D -2
	12.0	39.228	1.2167	1.9304	0.9717	1.6743	1.468	1.062	0.226	+2.8736D -2
	14.0	41.527	1.2583	2.1380	0.9568	1.5955	1.423	0.992	0.247	+4.4150D -2
	16.0	44.017	1.3024	2.3651	0.9381	1.5132	1.373	0.928	0.265	+6.3925D -2
	18.0	46.755	1.3499	2.6156	0.9152	1.4260	1.317	0.870	0.283	+8.8653D -2
	20.0	49.844	1.4020	2.8963	0.8875	1.3311	1.253	0.816	0.297	+1.1930D -1
	22.0	53.518	1.4616	3.2232	0.8538	1.2230	1.175	0.764	0.309	+1.5804D -1
	24.0	58.617	1.5390	3.6548	0.8083	1.0814	1.066	0.710	0.316	+2.1282D -1
δ_* =	24.6975	61.724	1.5824	3.9001	0.7825	1.0000	1.000	0.684	0.314	+2.4529D -1
δ_m =	24.9234	64.618	1.6199	4.1133	0.7603	0.9279	0.939	0.664	0.308	+2.7408D -1
	24.0	70.055	1.6817	4.4667	0.7242	0.8036	0.828	0.635	0.283	+3.2266D -1
	22.0	73.965	1.7182	4.6707	0.7033	0.7251	0.756	0.619	0.250	+3.5192D -1
	20.0	76.496	1.7380	4.7909	0.6922	0.6803	0.713	0.611	0.223	+3.6789D -1
	18.0	78.490	1.7514	4.8580	0.6847	0.6489	0.683	0.606	0.197	+3.7872D -1
	16.0	80.185	1.7611	4.9244	0.6793	0.6254	0.660	0.603	0.173	+3.8664D -1
	14.0	81.687	1.7685	4.9672	0.6753	0.6071	0.642	0.600	0.150	+3.9266D -1
	12.0	83.056	1.7742	5.0002	0.6721	0.5927	0.620	0.598	0.127	+3.9730D -1
	10.0	84.330	1.7786	5.0256	0.6697	0.5814	0.616	0.596	0.105	+4.0088D -1
	8.0	85.536	1.7820	5.0450	0.6679	0.5727	0.608	0.595	0.084	+4.0361D -1
	6.0	86.691	1.7844	5.0593	0.6666	0.5663	0.601	0.594	0.062	+4.0563D -1
	4.0	87.812	1.7861	5.0692	0.6656	0.5618	0.597	0.594	0.042	+4.0701D -1
	2.0	88.911	1.7871	5.0749	0.6651	0.5592	0.594	0.593	0.021	+4.0782D -1
	1.0	89.456	1.7874	5.0763	0.6650	0.5585	0.594	0.593	0.010	+4.0802D -1
	0.0	90.000	1.7875	5.0768	0.6649	0.5583	0.593	0.593	0.000	+4.0809D -1

M_1	δ	θ	T_2/T_1	p_2/p_1	p_{o2}/p_{o1}	M_2	q_2/q_*	u_2/q_*	v_2/q_*	$(s_2 - s_1)/R$
2.14 (1.694)	0.0	27.859	1.0000	1.0000	1.0000	2.1400	1.694	1.694	0.000	0.0000
	1.0	28.637	1.0169	1.0605	1.0000	2.1024	1.678	1.624	0.028	+2.0713D -5
	2.0	29.438	1.0340	1.1239	0.9998	2.0651	1.662	1.558	0.054	+1.6260D -4
	4.0	31.109	1.0685	1.2596	0.9987	1.9913	1.629	1.437	0.100	+1.2528D -3
	6.0	32.879	1.1040	1.4079	0.9959	1.9177	1.595	1.328	0.140	+4.0736D -3
	8.0	34.755	1.1405	1.5697	0.9907	1.8437	1.558	1.231	0.173	+9.3069D -3
	10.0	36.749	1.1785	1.7459	0.9826	1.7688	1.520	1.143	0.202	+1.7531D -2
	12.0	38.875	1.2182	1.9379	0.9712	1.6922	1.478	1.065	0.226	+2.9243D -2
	14.0	41.154	1.2601	2.1472	0.9561	1.6132	1.433	0.994	0.248	+4.4890D -2
	16.0	43.617	1.3045	2.3759	0.9371	1.5309	1.384	0.930	0.267	+6.4924D -2
	18.0	46.317	1.3522	2.6275	0.9140	1.4438	1.329	0.872	0.283	+8.9901D -2
	20.0	49.347	1.4043	2.9086	0.8863	1.3497	1.266	0.818	0.298	+1.2071D -1
	22.0	52.909	1.4634	3.2330	0.8528	1.2435	1.190	0.767	0.310	+1.5924D -1
	24.0	57.652	1.5375	3.6466	0.8092	1.1093	1.089	0.714	0.318	+2.1175D -1
δ_* =	25.0074	61.769	1.5966	3.9807	0.7741	1.0000	1.000	0.679	0.317	+2.5611D -1
δ_m =	25.2269	64.616	1.6341	4.1943	0.7519	0.9286	0.939	0.660	0.311	+2.8514D -1
	24.0	70.825	1.7049	4.5998	0.7109	0.7868	0.813	0.628	0.279	+3.4118D -1
	22.0	74.372	1.7376	4.7885	0.6924	0.7163	0.747	0.615	0.248	+3.6755D -1
	20.0	76.783	1.7564	4.8969	0.6820	0.6739	0.707	0.607	0.221	+3.8277D -1
	18.0	78.711	1.7693	4.9714	0.6749	0.6438	0.678	0.603	0.196	+3.9325D -1
	16.0	80.360	1.7787	5.0264	0.6697	0.6210	0.655	0.599	0.172	+4.0099D -1
	14.0	81.828	1.7860	5.0682	0.6657	0.6031	0.638	0.597	0.149	+4.0688D -1
	12.0	83.170	1.7916	5.1006	0.6627	0.5891	0.624	0.595	0.126	+4.1145D -1
	10.0	84.420	1.7959	5.1257	0.6604	0.5781	0.613	0.593	0.105	+4.1498D -1
	8.0	85.605	1.7992	5.1448	0.6586	0.5696	0.605	0.592	0.083	+4.1768D -1
	6.0	86.742	1.8016	5.1589	0.6573	0.5632	0.598	0.591	0.062	+4.1967D -1
	4.0	87.845	1.8033	5.1686	0.6564	0.5588	0.594	0.591	0.041	+4.2104D -1
	2.0	88.928	1.8043	5.1743	0.6558	0.5562	0.591	0.591	0.021	+4.2184D -1
	1.0	89.465	1.8045	5.1757	0.6557	0.5556	0.591	0.590	0.010	+4.2204D -1
	0.0	90.000	1.8046	5.1762	0.6557	0.5554	0.590	0.590	0.000	+4.2210D -1
2.16 (1.702)	0.0	27.578	1.0000	1.0000	1.0000	2.1600	1.702	1.702	0.000	0.0000
	1.0	28.353	1.0171	1.0610	1.0000	2.1222	1.686	1.631	0.028	+2.1133D -5
	2.0	29.150	1.0342	1.1248	0.9998	2.0847	1.670	1.565	0.055	+1.6587D -4
	4.0	30.812	1.0690	1.2615	0.9987	2.0104	1.638	1.442	0.101	+1.2776D -3
	6.0	32.573	1.1047	1.4110	0.9959	1.9364	1.604	1.332	0.140	+4.1528D -3
	8.0	34.438	1.1415	1.5741	0.9906	1.8621	1.567	1.234	0.173	+9.4836D -3
	10.0	36.420	1.1798	1.7519	0.9823	1.7869	1.529	1.146	0.202	+1.7854D -2
	12.0	38.531	1.2198	1.9455	0.9707	1.7100	1.488	1.067	0.227	+2.9762D -2
	14.0	40.791	1.2619	2.1565	0.9554	1.6308	1.443	0.996	0.248	+4.5650D -2
	16.0	43.229	1.3066	2.3868	0.9362	1.5484	1.395	0.932	0.267	+6.5954D -2
	18.0	45.895	1.3545	2.6399	0.9128	1.4615	1.340	0.873	0.284	+9.1198D -2
	20.0	48.871	1.4067	2.9216	0.8850	1.3679	1.278	0.819	0.298	+1.2220D -1
	22.0	52.335	1.4654	3.2442	0.8516	1.2632	1.205	0.769	0.311	+1.6062D -1
	24.0	56.807	1.5372	3.6452	0.8093	1.1347	1.108	0.718	0.320	+2.1156D -1
δ_* =	25.3113	61.814	1.6109	4.0622	0.7656	1.0000	1.000	0.675	0.319	+2.6713D -1
δ_m =	25.5245	64.616	1.6485	4.2762	0.7435	0.9292	0.940	0.656	0.313	+2.9637D -1
	24.0	71.479	1.7270	4.7273	0.6984	0.7725	0.800	0.622	0.277	+3.5899D -1
	22.0	74.747	1.7569	4.8998	0.6817	0.7082	0.740	0.610	0.247	+3.8318D -1
	20.0	77.053	1.7748	5.0033	0.6718	0.6679	0.701	0.604	0.220	+3.9774D -1
	18.0	78.920	1.7872	5.0755	0.6650	0.6389	0.673	0.599	0.195	+4.0791D -1
	16.0	80.527	1.7965	5.1291	0.6600	0.6167	0.651	0.596	0.171	+4.1546D -1
	14.0	81.963	1.8035	5.1701	0.6562	0.5993	0.634	0.594	0.148	+4.2125D -1
	12.0	83.278	1.8090	5.2020	0.6533	0.5856	0.621	0.592	0.126	+4.2574D -1
	10.0	84.507	1.8133	5.2267	0.6510	0.5748	0.610	0.590	0.104	+4.2922D -1
	8.0	85.672	1.8165	5.2455	0.6493	0.5665	0.602	0.589	0.083	+4.3189D -1
	6.0	86.791	1.8189	5.2595	0.6480	0.5602	0.595	0.589	0.062	+4.3386D -1
	4.0	87.877	1.8206	5.2691	0.6471	0.5559	0.591	0.588	0.041	+4.3521D -1
	2.0	88.944	1.8216	5.2747	0.6466	0.5534	0.588	0.588	0.021	+4.3600D -1
	1.0	89.472	1.8218	5.2761	0.6465	0.5528	0.588	0.588	0.010	+4.3620D -1
	0.0	90.000	1.8219	5.2765	0.6464	0.5525	0.588	0.588	0.000	+4.3626D -1
2.18 (1.710)	0.0	27.304	1.0000	1.0000	1.0000	2.1800	1.710	1.710	0.000	0.0000
	1.0	28.075	1.0172	1.0614	1.0000	2.1419	1.694	1.639	0.029	+2.1562D -5
	2.0	28.868	1.0345	1.1257	0.9998	2.1042	1.679	1.571	0.055	+1.6921D -4
	4.0	30.522	1.0695	1.2634	0.9987	2.0295	1.646	1.447	0.101	+1.3030D -3
	6.0	32.273	1.1054	1.4141	0.9958	1.9551	1.612	1.337	0.140	+4.2337D -3
	8.0	34.129	1.1425	1.5787	0.9904	1.8804	1.577	1.237	0.174	+9.6641D -3
	10.0	36.099	1.1810	1.7580	0.9820	1.8049	1.538	1.149	0.203	+1.8185D -2
	12.0	38.195	1.2213	1.9533	0.9702	1.7277	1.498	1.069	0.227	+3.0294D -2
	14.0	40.438	1.2637	2.1660	0.9546	1.6483	1.453	0.998	0.249	+4.6429D -2
	16.0	42.854	1.3087	2.3981	0.9352	1.5658	1.405	0.933	0.268	+6.7013D -2
	18.0	45.487	1.3569	2.6527	0.9116	1.4790	1.351	0.875	0.284	+9.2540D -2

M_1	δ	θ	T_2/T_1	p_2/p_1	p_{o2}/p_{o1}	M_2	q_2/q_*	u_2/q_*	v_2/q_*	$(s_2 - s_1)/R$
	20.0	48.414	1.4092	2.9352	0.8836	1.3859	1.290	0.821	0.299	+1.2376$_{10}$ -1
	22.0	51.793	1.4677	3.2567	0.8503	1.2825	1.219	0.771	0.311	+1.6216$_{10}$ -1
	24.0	56.048	1.5378	3.6484	0.8090	1.1583	1.127	0.721	0.321	+2.1198$_{10}$ -1
δ_* =	25.6093	61.861	1.6254	4.1446	0.7570	1.0000	1.000	0.671	0.321	+2.7834$_{10}$ -1
δ_m =	25.8165	64.617	1.6629	4.3590	0.7351	0.9299	0.941	0.652	0.316	+3.0777$_{10}$ -1
	24.0	72.050	1.7485	4.8512	0.6864	0.7601	0.788	0.616	0.274	+3.7636$_{10}$ -1
	22.0	75.095	1.7761	5.0110	0.671	0.7007	0.732	0.606	0.245	+3.9882$_{10}$ -1
	20.0	77.308	1.7932	5.1102	0.6618	0.6622	0.696	0.600	0.218	+4.1279$_{10}$ -1
	18.0	79.119	1.8053	5.1802	0.6553	0.6341	0.668	0.596	0.194	+4.2267$_{10}$ -1
	16.0	80.686	1.8143	5.2326	0.6505	0.6126	0.647	0.593	0.170	+4.3006$_{10}$ -1
	14.0	82.092	1.8212	5.2728	0.6468	0.5957	0.631	0.591	0.147	+4.3574$_{10}$ -1
	12.0	83.383	1.8266	5.3042	0.6439	0.5822	0.617	0.589	0.125	+4.4017$_{10}$ -1
	10.0	84.590	1.8308	5.3285	0.6417	0.5717	0.607	0.588	0.104	+4.4360$_{10}$ -1
	8.0	85.736	1.8340	5.3471	0.6400	0.5635	0.599	0.586	0.082	+4.4623$_{10}$ -1
	6.0	86.838	1.8364	5.3609	0.6388	0.5573	0.592	0.586	0.062	+4.4818$_{10}$ -1
	4.0	87.908	1.8380	5.3704	0.6379	0.5531	0.583	0.585	0.041	+4.4952$_{10}$ -1
	2.0	88.959	1.8390	5.3760	0.6374	0.5506	0.586	0.585	0.020	+4.5030$_{10}$ -1
	1.0	89.480	1.8392	5.3773	0.6373	0.5500	0.585	0.585	0.010	+4.5050$_{10}$ -1
	0.0	90.000	1.8393	5.3778	0.6373	0.5498	0.585	0.585	0.000	+4.5056$_{10}$ -1
2.20	0.0	27.036	1.0000	1.0000	1.0000	2.2000	1.718	1.718	0.000	0.0000
(1.718)	1.0	27.803	1.0173	1.0618	1.0000	2.1617	1.703	1.646	0.029	+2.1999$_{10}$ -5
	2.0	28.592	1.0347	1.1266	0.9998	2.1237	1.687	1.578	0.055	+1.7262$_{10}$ -4
	4.0	30.238	1.0700	1.2654	0.9987	2.0485	1.655	1.453	0.102	+1.3288$_{10}$ -3
	6.0	31.981	1.1061	1.4173	0.9957	1.9738	1.621	1.341	0.141	+4.3162$_{10}$ -3
	8.0	33.827	1.1435	1.5832	0.9902	1.8987	1.585	1.241	0.174	+9.8483$_{10}$ -3
	10.0	35.785	1.1823	1.7641	0.9816	1.8228	1.548	1.151	0.203	+1.8522$_{10}$ -2
	12.0	37.869	1.2229	1.9611	0.9696	1.7454	1.507	1.071	0.228	+3.0837$_{10}$ -2
	14.0	40.095	1.2656	2.1756	0.9539	1.6657	1.463	0.999	0.249	+4.7226$_{10}$ -2
	16.0	42.489	1.3109	2.4095	0.9342	1.5831	1.415	0.935	0.268	+6.8100$_{10}$ -2
	18.0	45.092	1.3593	2.6658	0.9103	1.4963	1.362	0.876	0.285	+9.3927$_{10}$ -2
	20.0	47.976	1.4118	2.9494	0.8821	1.4035	1.302	0.822	0.299	+1.2540$_{10}$ -1
	22.0	51.277	1.4701	3.2704	0.8489	1.3012	1.232	0.772	0.312	+1.6385$_{10}$ -1
	24.0	55.356	1.5390	3.6552	0.8083	1.1805	1.144	0.723	0.322	+2.1287$_{10}$ -1
	26.0	62.695	1.6512	4.2918	0.7419	0.9795	0.983	0.661	0.322	+2.9850$_{10}$ -1
δ_* =	25.9016	61.908	1.6400	4.2279	0.7485	1.0000	1.000	0.666	0.324	+2.8973$_{10}$ -1
δ_m =	26.1028	64.620	1.6775	4.4426	0.7266	0.9305	0.941	0.649	0.318	+3.1933$_{10}$ -1
	26.0	66.480	1.7015	4.5807	0.7128	0.8849	0.901	0.638	0.311	+3.3852$_{10}$ -1
	24.0	72.559	1.7695	4.9728	0.6747	0.7490	0.778	0.611	0.272	+3.9344$_{10}$ -1
	22.0	75.420	1.7953	5.1222	0.6607	0.6936	0.726	0.602	0.243	+4.1449$_{10}$ -1
	20.0	77.549	1.8117	5.2175	0.6519	0.6568	0.690	0.597	0.217	+4.2793$_{10}$ -1
	18.0	79.308	1.8234	5.2856	0.6456	0.6296	0.664	0.593	0.193	+4.3755$_{10}$ -1
	16.0	80.839	1.8323	5.3369	0.6410	0.6086	0.643	0.590	0.169	+4.4478$_{10}$ -1
	14.0	82.216	1.8391	5.3764	0.6374	0.5921	0.627	0.588	0.147	+4.5037$_{10}$ -1
	12.0	83.483	1.8444	5.4073	0.6346	0.5789	0.614	0.586	0.125	+4.5472$_{10}$ -1
	10.0	84.670	1.8485	5.4313	0.6325	0.5686	0.604	0.585	0.103	+4.5812$_{10}$ -1
	8.0	85.798	1.8517	5.4497	0.6308	0.5605	0.596	0.584	0.082	+4.6072$_{10}$ -1
	6.0	86.883	1.8540	5.4633	0.6296	0.5545	0.590	0.583	0.061	+4.6264$_{10}$ -1
	4.0	87.938	1.8556	5.4727	0.6288	0.5503	0.585	0.582	0.041	+4.6397$_{10}$ -1
	2.0	88.973	1.8565	5.4782	0.6283	0.5479	0.583	0.582	0.020	+4.6474$_{10}$ -1
	1.0	89.487	1.8568	5.4795	0.6282	0.5473	0.582	0.582	0.010	+4.6493$_{10}$ -1
	0.0	90.000	1.8569	5.4800	0.6281	0.5471	0.582	0.582	0.000	+4.6500$_{10}$ -1
2.30	0.0	25.771	1.0000	1.0000	1.0000	2.3000	1.756	1.756	0.000	0.0000
(1.756)	1.0	26.522	1.0179	1.0640	1.0000	2.2604	1.741	1.680	0.029	+2.4312$_{10}$ -5
	2.0	27.294	1.0359	1.1311	0.9998	2.2212	1.726	1.608	0.056	+1.9066$_{10}$ -4
	4.0	28.906	1.0724	1.2753	0.9985	2.1437	1.695	1.477	0.103	+1.4657$_{10}$ -3

M_1	δ	θ	T_2/T_1	p_2/p_1	p_{o2}/p_{o1}	M_2	q_2/q_*	u_2/q_*	v_2/q_*	$(s_2 - s_1)/R$
	6.0	30.611	1.1099	1.4336	0.9953	2.0667	1.663	1.360	0.143	+4.7532$_{10}$ -3
	8.0	32.415	1.1487	1.6068	0.9892	1.9896	1.628	1.256	0.176	+1.0825$_{10}$ -2
	10.0	34.326	1.1890	1.7959	0.9799	1.9117	1.592	1.163	0.205	+2.0313$_{10}$ -2
	12.0	36.354	1.2311	2.0019	0.9668	1.8325	1.553	1.080	0.230	+3.3727$_{10}$ -2
	14.0	38.510	1.2755	2.2261	0.9498	1.7514	1.510	1.006	0.251	+5.1484$_{10}$ -2
	16.0	40.815	1.3224	2.4700	0.9287	1.6677	1.464	0.940	0.270	+7.3942$_{10}$ -2
	18.0	43.299	1.3724	2.7360	0.9035	1.5804	1.414	0.881	0.286	+1.0146$_{10}$ -1
	20.0	46.007	1.4261	3.0276	0.8741	1.4885	1.357	0.827	0.301	+1.3452$_{10}$ -1
	22.0	49.026	1.4847	3.3514	0.8403	1.3894	1.293	0.778	0.314	+1.7394$_{10}$ -1
	24.0	52.536	1.5508	3.7216	0.8013	1.2788	1.216	0.732	0.326	+2.2158$_{10}$ -1
	26.0	57.077	1.6319	4.1819	0.7532	1.1425	1.115	0.685	0.334	+2.8343$_{10}$ -1
δ_* =	27.2803	62.151	1.7150	4.6582	0.7052	1.0000	1.000	0.647	0.333	+3.4932$_{10}$ -1
δ_m =	27.4541	64.653	1.7524	4.8739	0.6842	0.9338	0.944	0.631	0.328	+3.7955$_{10}$ -1
	26.0	71.264	1.8377	5.3683	0.6381	0.7743	0.801	0.601	0.293	+4.4922$_{10}$ -1
	24.0	74.512	1.8714	5.5649	0.6206	0.7060	0.738	0.591	0.263	+4.7699$_{10}$ -1
	22.0	76.770	1.8915	5.6818	0.6105	0.6635	0.697	0.585	0.236	+4.9349$_{10}$ -1
	20.0	78.583	1.9054	5.7032	0.6035	0.6328	0.667	0.581	0.211	+5.0498$_{10}$ -1
	18.0	80.133	1.9158	5.8238	0.5984	0.6092	0.644	0.578	0.183	+5.1353$_{10}$ -1
	16.0	81.510	1.9238	5.8705	0.5945	0.5906	0.626	0.576	0.165	+5.2011$_{10}$ -1
	14.0	82.764	1.9301	5.9071	0.5914	0.5757	0.611	0.574	0.143	+5.2527$_{10}$ -1
	12.0	83.928	1.9350	5.9360	0.5890	0.5638	0.599	0.573	0.122	+5.2934$_{10}$ -1
	10.0	85.026	1.9389	5.9586	0.5871	0.5543	0.589	0.571	0.101	+5.3253$_{10}$ -1
	8.0	86.074	1.9419	5.9761	0.5857	0.5469	0.582	0.571	0.080	+5.3499$_{10}$ -1
	6.0	87.085	1.9441	5.9890	0.5846	0.5413	0.576	0.570	0.060	+5.3682$_{10}$ -1
	4.0	88.070	1.9456	5.9980	0.5839	0.5374	0.572	0.570	0.040	+5.3808$_{10}$ -1
	2.0	89.039	1.9465	6.0033	0.5834	0.5352	0.570	0.569	0.020	+5.3882$_{10}$ -1
	1.0	89.520	1.9467	6.0046	0.5833	0.5346	0.570	0.569	0.010	+5.3900$_{10}$ -1
	0.0	90.000	1.9468	6.0050	0.5833	0.5344	0.569	0.569	0.000	+5.3906$_{10}$ -1
2.40	0.0	24.624	1.0000	1.0000	1.0000	2.4000	1.792	1.792	0.000	0.0000
(1.792)	1.0	25.361	1.0185	1.0662	1.0000	2.3590	1.778	1.712	0.030	+2.6840$_{10}$ -5
	2.0	26.119	1.0371	1.1357	0.9998	2.3184	1.763	1.637	0.057	+2.1038$_{10}$ -4
	4.0	27.702	1.0749	1.2856	0.9984	2.2383	1.733	1.498	0.105	+1.6154$_{10}$ -3
	6.0	29.377	1.1138	1.4505	0.9948	2.1589	1.701	1.376	0.145	+5.2313$_{10}$ -3
	8.0	31.149	1.1540	1.6314	0.9882	2.0795	1.668	1.268	0.178	+1.1894$_{10}$ -2
	10.0	33.023	1.1959	1.8292	0.9780	1.9994	1.633	1.172	0.207	+2.2274$_{10}$ -2
	12.0	35.007	1.2398	2.0450	0.9638	1.9181	1.595	1.087	0.231	+3.6898$_{10}$ -2
	14.0	37.112	1.2860	2.2798	0.9454	1.8350	1.554	1.011	0.252	+5.6171$_{10}$ -2
	16.0	39.352	1.3348	2.5351	0.9227	1.7497	1.509	0.944	0.271	+8.0412$_{10}$ -2
	18.0	41.748	1.3866	2.8128	0.8959	1.6613	1.461	0.884	0.287	+1.0990$_{10}$ -1
	20.0	44.336	1.4421	3.1155	0.8650	1.5689	1.407	0.830	0.302	+1.4497$_{10}$ -1
	22.0	47.174	1.5021	3.4480	0.8302	1.4709	1.346	0.781	0.315	+1.8615$_{10}$ -1
	24.0	50.371	1.5682	3.8196	0.7909	1.3644	1.276	0.736	0.328	+2.3454$_{10}$ -1
	26.0	54.184	1.6442	4.2522	0.7460	1.2426	1.190	0.693	0.338	+2.9306$_{10}$ -1
	28.0	59.656	1.7462	4.8383	0.6876	1.0779	1.064	0.647	0.344	+3.7454$_{10}$ -1
δ_* =	28.5313	62.400	1.7933	5.1109	0.6617	1.0000	1.000	0.629	0.342	+4.1289$_{10}$ -1
δ_m =	28.6814	64.710	1.8305	5.3269	0.6419	0.9370	0.947	0.616	0.337	+4.4337$_{10}$ -1
	28.0	69.291	1.8968	5.7130	0.6078	0.8201	0.843	0.595	0.316	+4.9790$_{10}$ -1
	26.0	73.399	1.9468	6.0048	0.5833	0.7260	0.756	0.581	0.283	+5.3904$_{10}$ -1
	24.0	75.888	1.9722	6.1539	0.5712	0.6751	0.708	0.574	0.256	+5.6000$_{10}$ -1
	22.0	77.802	1.9892	6.2534	0.5633	0.6397	0.674	0.570	0.230	+5.7396$_{10}$ -1
	20.0	79.402	2.0016	6.3260	0.5576	0.6129	0.648	0.567	0.206	+5.8415$_{10}$ -1
	18.0	80.801	2.0111	6.3816	0.5533	0.5919	0.627	0.565	0.184	+5.9192$_{10}$ -1
	16.0	82.060	2.0185	6.4251	0.5499	0.5751	0.610	0.563	0.161	+5.9801$_{10}$ -1
	14.0	83.217	2.0244	6.4596	0.5473	0.5615	0.597	0.562	0.140	+6.0283$_{10}$ -1
	12.0	84.299	2.0290	6.4870	0.5452	0.5505	0.586	0.561	0.119	+6.0667$_{10}$ -1
	10.0	85.324	2.0327	6.4087	0.5435	0.5416	0.577	0.560	0.099	+6.0969$_{10}$ -1
	8.0	86.306	2.0356	6.5254	0.5423	0.5348	0.570	0.559	0.079	+6.1203$_{10}$ -1
	6.0	87.255	2.0377	6.5379	0.5413	0.5296	0.564	0.559	0.059	+6.1377$_{10}$ -1
	4.0	88.182	2.0392	6.5466	0.5407	0.5260	0.561	0.558	0.039	+6.1498$_{10}$ -1
	2.0	89.095	2.0400	6.5517	0.5403	0.5238	0.559	0.558	0.019	+6.1569$_{10}$ -1
	1.0	89.548	2.0403	6.5529	0.5402	0.5233	0.558	0.558	0.010	+6.1586$_{10}$ -1
	0.0	90.000	2.0403	6.5533	0.5401	0.5231	0.558	0.558	0.000	+6.1592$_{10}$ -1

TABLE 6 PROPAGATION OF A NON–STEEP WAVE IN AIR WITHOUT FRICTION IN A CONSTANT AREA DUCT

Flow and fluid parameters are calculated with the pressure ratio (p/p_1) across the wave-front as the independent variable for the range, $p/p_1 = 0 \rightarrow (0{\cdot}001) \rightarrow 0{\cdot}05 \rightarrow 0{\cdot}01 \rightarrow 3 \rightarrow (0{\cdot}05) \rightarrow 5 \rightarrow (0{\cdot}1) \rightarrow 10$.

The summary of equations used is given in Section 8·5. The values of the following parameters are tabulated:

- $(p/p_1)^{1/7}$ (for convenience in calculations)
- T/T_1 static temperature ratio across wave-front
- ρ/ρ_1 static density ratio across wave-front
- c/a_1 ratio of wave propagation velocity to the speed of sound of air ahead of wave
- q/a_1 ratio of fluid particle velocity to the speed of sound of air ahead of wave
- $M = q/a$, Mach number of induced flow behind wave-front

(Note: directional significance is given by the movement of A1 in Figure 8·6)

p/p_1	$(p/p_1)^{1/7}$	T/T_1	ρ/ρ_1	c/a_1	q/a_1	$M = q/a$
0	0	0	0	−5	−5	∞
0.001	0.3728	0.1389	0.0072	-2.7634	-3.1362	8.4135
0.002	0.4116	0.1694	0.0118	-2.5306	-2.9422	7.1489
0.003	0.4361	0.1902	0.0158	-2.3834	-2.8195	6.4652
0.004	0.4544	0.2065	0.0194	-2.2736	-2.7280	6.0036
0.005	0.4691	0.2201	0.0227	-2.1853	-2.6544	5.6583
0.006	0.4815	0.2318	0.0259	-2.1110	-2.5925	5.3843
0.007	0.4922	0.2423	0.0289	-2.0467	-2.5389	5.1581
0.008	0.5017	0.2517	0.0318	-1.9898	-2.4915	4.9662
0.009	0.5102	0.2603	0.0346	-1.9387	-2.4490	4.7999
0.010	0.5179	0.2683	0.0373	-1.8923	-2.4103	4.6535
0.011	0.5250	0.2757	0.0399	-1.8497	-2.3748	4.5229
0.012	0.5316	0.2826	0.0425	-1.8103	-2.3419	4.4053
0.013	0.5377	0.2892	0.0450	-1.7736	-2.3114	4.2984
0.014	0.5435	0.2953	0.0474	-1.7393	-2.2827	4.2004
0.015	0.5488	0.3012	0.0498	-1.7070	-2.2558	4.1102
0.016	0.5539	0.3068	0.0521	-1.6765	-2.2304	4.0266
0.017	0.5587	0.3122	0.0545	-1.6476	-2.2063	3.9483
0.018	0.5633	0.3173	0.0567	-1.6201	-2.1834	3.8760
0.019	0.5677	0.3223	0.0590	-1.5939	-2.1616	3.8077
0.020	0.5719	0.3270	0.0612	-1.5688	-2.1407	3.7434
0.021	0.5759	0.3316	0.0633	-1.5448	-2.1207	3.6827
0.022	0.5797	0.3361	0.0655	-1.5218	-2.1015	3.6252
0.023	0.5834	0.3403	0.0676	-1.4996	-2.0830	3.5706
0.024	0.5870	0.3445	0.0697	-1.4783	-2.0652	3.5186
0.025	0.5904	0.3486	0.0717	-1.4577	-2.0481	3.4691
0.026	0.5937	0.3525	0.0738	-1.4378	-2.0315	3.4218
0.027	0.5969	0.3563	0.0758	-1.4185	-2.0154	3.3765
0.028	0.6000	0.3600	0.0778	-1.3999	-1.9999	3.3331
0.029	0.6030	0.3637	0.0797	-1.3818	-1.9848	3.2914
0.030	0.6060	0.3672	0.0817	-1.3642	-1.9702	3.2513
0.031	0.6088	0.3706	0.0836	-1.3472	-1.9560	3.2128
0.032	0.6116	0.3740	0.0856	-1.3305	-1.9421	3.1756
0.033	0.6143	0.3773	0.0875	-1.3144	-1.9287	3.1397
0.034	0.6169	0.3806	0.0893	-1.2986	-1.9155	3.1051
0.035	0.6195	0.3837	0.0912	-1.2833	-1.9027	3.0716
0.036	0.6220	0.3868	0.0931	-1.2683	-1.8902	3.0392
0.037	0.6244	0.3899	0.0949	-1.2536	-1.8780	3.0078
0.038	0.6268	0.3928	0.0967	-1.2393	-1.8661	2.9773
0.039	0.6291	0.3958	0.0985	-1.2254	-1.8545	2.9478
0.040	0.6314	0.3986	0.1003	-1.2117	-1.8431	2.9191
0.041	0.6336	0.4015	0.1021	-1.1983	-1.8319	2.8912
0.042	0.6358	0.4042	0.1039	-1.1852	-1.8210	2.8641
0.043	0.6379	0.4070	0.1057	-1.1723	-1.8103	2.8377
0.044	0.6400	0.4097	0.1074	-1.1598	-1.7998	2.8120
0.045	0.6421	0.4123	0.1091	-1.1474	-1.7895	2.7870

p/p_1	$(p/p_1)^{1/7}$	T/T_1	ρ/ρ_1	c/a_1	q/a_1	$M = q/a$
0.046	0.6441	0.4149	0.1109	-1.1353	-1.7794	2.7626
0.047	0.6461	0.4174	0.1126	-1.1234	-1.7695	2.7387
0.048	0.6480	0.4200	0.1143	-1.1117	-1.7598	2.7155
0.049	0.6500	0.4224	0.1160	-1.1003	-1.7502	2.6928
0.05	0.6518	0.4249	0.1177	-1.0890	-1.7408	2.6706
0.06	0.6690	0.4476	0.1340	-0.9858	-1.6548	2.4734
0.07	0.6839	0.4678	0.1496	-0.8964	-1.5803	2.3107
0.08	0.6971	0.4860	0.1646	-0.8174	-1.5145	2.1725
0.09	0.7089	0.5026	0.1791	-0.7464	-1.4553	2.0528
0.10	0.7197	0.5179	0.1931	-0.6819	-1.4016	1.9475
0.11	0.7296	0.5322	0.2067	-0.6227	-1.3522	1.8535
0.12	0.7387	0.5456	0.2199	-0.5679	-1.3066	1.7689
0.13	0.7472	0.5583	0.2329	-0.5170	-1.2641	1.6919
0.14	0.7551	0.5702	0.2455	-0.4693	-1.2244	1.6214
0.15	0.7626	0.5816	0.2579	-0.4244	-1.1870	1.5565
0.16	0.7697	0.5924	0.2701	-0.3820	-1.1517	1.4963
0.17	0.7764	0.6027	0.2820	-0.3418	-1.1182	1.4403
0.18	0.7827	0.6127	0.2938	-0.3036	-1.0864	1.3879
0.19	0.7888	0.6222	0.3054	-0.2672	-1.0560	1.3388
0.20	0.7946	0.6314	0.3168	-0.2324	-1.0270	1.2925
0.21	0.8002	0.6402	0.3280	-0.1991	-0.9992	1.2488
0.22	0.8055	0.6488	0.3391	-0.1671	-0.9725	1.2074
0.23	0.8106	0.6571	0.3500	-0.1363	-0.9469	1.1681
0.24	0.8156	0.6651	0.3608	-0.1066	-0.9222	1.1307
0.25	0.8203	0.6730	0.3715	-0.0780	-0.8983	1.0951
0.26	0.8249	0.6805	0.3821	-0.0503	-0.8753	1.0610
0.27	0.8294	0.6879	0.3925	-0.0236	-0.8530	1.0284
0.2791	0.8333	0.6944	0.4019	0.0000	-0.8333	1.0000
0.28	0.8337	0.6951	0.4028	+0.0023	-0.8314	0.9972
0.29	0.8379	0.7021	0.4130	+0.0275	-0.8104	0.9672
0.30	0.8420	0.7089	0.4232	+0.0519	-0.7901	0.9384
0.31	0.8459	0.7156	0.4332	+0.0756	-0.7703	0.9106
0.32	0.8498	0.7221	0.4431	+0.0987	-0.7511	0.8839
0.33	0.8535	0.7285	0.4530	+0.1212	-0.7324	0.8581
0.34	0.8572	0.7347	0.4627	+0.1430	-0.7141	0.8331
0.35	0.8607	0.7409	0.4724	+0.1644	-0.6964	0.8090
0.36	0.8642	0.7468	0.4820	+0.1852	-0.6790	0.7857
0.37	0.8676	0.7527	0.4916	+0.2055	-0.6621	0.7631
0.38	0.8709	0.7585	0.5010	+0.2254	-0.6455	0.7412
0.39	0.8741	0.7641	0.5104	+0.2448	-0.6293	0.7199

p/p_1	$(p/p_1)^{1/7}$	T/T_1	ρ/ρ_1	c/a_1	q/a_1	$M = q/a$
0.40	0.8773	0.7697	0.5197	+0.2638	-0.6135	0.6993
0.41	0.8804	0.7751	0.5290	+0.2824	-0.5980	0.6792
0.42	0.8834	0.7805	0.5381	+0.3007	-0.5828	0.6597
0.43	0.8864	0.7857	0.5473	+0.3185	-0.5679	0.6407
0.44	0.8893	0.7909	0.5563	+0.3360	-0.5533	0.6222
0.45	0.8922	0.7960	0.5653	+0.3532	-0.5390	0.6042
0.46	0.8950	0.8010	0.5743	+0.3700	-0.5250	0.5866
0.47	0.8978	0.8060	0.5832	+0.3865	-0.5112	0.5695
0.48	0.9005	0.8108	0.5920	+0.4027	-0.4977	0.5527
0.49	0.9031	0.8156	0.6008	+0.4187	-0.4844	0.5364
0.50	0.9057	0.8203	0.6095	+0.4343	-0.4714	0.5204
0.51	0.9083	0.8250	0.6182	+0.4497	-0.4586	0.5049
0.52	0.9108	0.8296	0.6268	+0.4649	-0.4459	0.4896
0.53	0.9133	0.8341	0.6354	+0.4798	-0.4335	0.4747
0.54	0.9157	0.8386	0.6440	+0.4944	-0.4213	0.4601
0.55	0.9181	0.8430	0.6524	+0.5088	-0.4093	0.4458
0.56	0.9205	0.8473	0.6609	+0.5230	-0.3975	0.4318
0.57	0.9228	0.8516	0.6693	+0.5370	-0.3858	0.4181
0.58	0.9251	0.8559	0.6777	+0.5508	-0.3743	0.4046
0.59	0.9274	0.8601	0.6860	+0.5644	-0.3630	0.3914
0.60	0.9296	0.8642	0.6943	+0.5777	-0.3519	0.3785
0.61	0.9318	0.8683	0.7025	+0.5909	-0.3409	0.3658
0.62	0.9340	0.8723	0.7107	+0.6039	-0.3301	0.3534
0.63	0.9361	0.8763	0.7189	+0.6168	-0.3194	0.3412
0.64	0.9382	0.8803	0.7270	+0.6294	-0.3088	0.3292
0.65	0.9403	0.8842	0.7351	+0.6419	-0.2984	0.3174
0.66	0.9424	0.8881	0.7432	+0.6542	-0.2882	0.3058
0.67	0.9444	0.8919	0.7512	+0.6664	-0.2780	0.2944
0.68	0.9464	0.8957	0.7592	+0.6784	-0.2680	0.2832
0.69	0.9484	0.8994	0.7672	+0.6902	-0.2581	0.2722
0.70	0.9503	0.9031	0.7751	+0.7019	-0.2484	0.2614
0.71	0.9523	0.9068	0.7830	+0.7135	-0.2387	0.2507
0.72	0.9542	0.9104	0.7909	+0.7249	-0.2292	0.2402
0.73	0.9560	0.9140	0.7987	+0.7362	-0.2198	0.2299
0.74	0.9579	0.9176	0.8065	+0.7474	-0.2105	0.2198
0.75	0.9597	0.9211	0.8143	+0.7584	-0.2013	0.2098
0.76	0.9616	0.9246	0.8220	+0.7693	-0.1922	0.1999
0.77	0.9634	0.9280	0.8297	+0.7801	-0.1832	0.1902
0.78	0.9651	0.9315	0.8374	+0.7908	-0.1744	0.1807
0.79	0.9669	0.9349	0.8450	+0.8013	-0.1656	0.1712
0.80	0.9686	0.9382	0.8527	+0.8118	-0.1569	0.1620
0.81	0.9703	0.9416	0.8603	+0.8221	-0.1483	0.1528
0.82	0.9720	0.9449	0.8678	+0.8323	-0.1398	0.1438
0.83	0.9737	0.9482	0.8754	+0.8424	-0.1313	0.1349
0.84	0.9754	0.9514	0.8829	+0.8524	-0.1230	0.1261
0.85	0.9771	0.9546	0.8904	+0.8623	-0.1147	0.1174
0.86	0.9787	0.9578	0.8979	+0.8721	-0.1066	0.1089
0.87	0.9803	0.9610	0.9053	+0.8818	-0.0985	0.1005
0.88	0.9819	0.9641	0.9127	+0.8914	-0.0905	0.0921
0.89	0.9835	0.9673	0.9201	+0.9009	-0.0825	0.0839
0.90	0.9851	0.9703	0.9275	+0.9104	-0.0747	0.0758
0.91	0.9866	0.9734	0.9349	+0.9197	-0.0669	0.0678
0.92	0.9882	0.9765	0.9422	+0.9290	-0.0592	0.0599
0.93	0.9897	0.9795	0.9495	+0.9381	-0.0516	0.0521
0.94	0.9912	0.9825	0.9568	+0.9472	-0.0440	0.0444

p/p_1	$(p/p_1)^{1/7}$	T/T_1	ρ/ρ_1	c/a_1	q/a_1	$M = q/a$
0.95	0.9927	0.9855	0.9640	+0.9562	-0.0365	0.0368
0.96	0.9942	0.9884	0.9713	+0.9651	-0.0291	0.0292
0.97	0.9957	0.9913	0.9785	+0.9739	-0.0217	0.0218
0.98	0.9971	0.9942	0.9857	+0.9827	-0.0144	0.0145
0.99	0.9986	0.9971	0.9928	+0.9914	-0.0072	0.0072
1	1	1	1	+1	0	0
1.01	1.0014	1.0028	1.0071	+1.0085	+0.0071	0.0071
1.02	1.0028	1.0057	1.0142	+1.0170	+0.0142	0.0141
1.03	1.0042	1.0085	1.0213	+1.0254	+0.0212	0.0211
1.04	1.0056	1.0113	1.0284	+1.0337	+0.0281	0.0279
1.05	1.0070	1.0140	1.0355	+1.0420	+0.0350	0.0347
1.06	1.0084	1.0168	1.0425	+1.0502	+0.0418	0.0414
1.07	1.0097	1.0195	1.0495	+1.0583	+0.0486	0.0481
1.08	1.0111	1.0222	1.0565	+1.0663	+0.0553	0.0547
1.09	1.0124	1.0249	1.0635	+1.0743	+0.0619	0.0612
1.10	1.0137	1.0276	1.0704	+1.0823	+0.0685	0.0676
1.11	1.0150	1.0303	1.0774	+1.0901	+0.0751	0.0740
1.12	1.0163	1.0329	1.0843	+1.0979	+0.0816	0.0803
1.13	1.0176	1.0355	1.0912	+1.1057	+0.0881	0.0865
1.14	1.0189	1.0381	1.0981	+1.1134	+0.0945	0.0927
1.15	1.0202	1.0407	1.1050	+1.1210	+0.1008	0.0988
1.16	1.0214	1.0433	1.1118	+1.1286	+0.1071	0.1049
1.17	1.0227	1.0459	1.1187	+1.1361	+0.1134	0.1109
1.18	1.0239	1.0484	1.1255	+1.1436	+0.1196	0.1168
1.19	1.0252	1.0510	1.1323	+1.1510	+0.1258	0.1227
1.20	1.0264	1.0535	1.1391	+1.1583	+0.1319	0.1285
1.21	1.0276	1.0560	1.1459	+1.1656	+0.1380	0.1343
1.22	1.0288	1.0585	1.1526	+1.1729	+0.1441	0.1400
1.23	1.0300	1.0609	1.1594	+1.1801	+0.1501	0.1457
1.24	1.0312	1.0634	1.1661	+1.1872	+0.1560	0.1513
1.25	1.0324	1.0658	1.1728	+1.1943	+0.1620	0.1569
1.26	1.0336	1.0683	1.1795	+1.2014	+0.1678	0.1624
1.27	1.0347	1.0707	1.1862	+1.2084	+0.1737	0.1678
1.28	1.0359	1.0731	1.1928	+1.2154	+0.1795	0.1733
1.29	1.0370	1.0755	1.1995	+1.2223	+0.1852	0.1786
1.30	1.0382	1.0778	1.2061	+1.2292	+0.1910	0.1839
1.31	1.0393	1.0802	1.2127	+1.2360	+0.1966	0.1892
1.32	1.0405	1.0826	1.2193	+1.2428	+0.2023	0.1944
1.33	1.0416	1.0849	1.2259	+1.2495	+0.2079	0.1996
1.34	1.0427	1.0872	1.2325	+1.2562	+0.2135	0.2047
1.35	1.0438	1.0895	1.2391	+1.2628	+0.2190	0.2098
1.36	1.0449	1.0918	1.2456	+1.2694	+0.2245	0.2149
1.37	1.0460	1.0941	1.2522	+1.2760	+0.2300	0.2199
1.38	1.0471	1.0964	1.2587	+1.2825	+0.2354	0.2248
1.39	1.0482	1.0987	1.2652	+1.2890	+0.2408	0.2298
1.40	1.0492	1.1009	1.2717	+1.2954	+0.2462	0.2347
1.41	1.0503	1.1031	1.2782	+1.3019	+0.2515	0.2395
1.42	1.0514	1.1054	1.2846	+1.3082	+0.2568	0.2443
1.43	1.0524	1.1076	1.2911	+1.3145	+0.2621	0.2491
1.44	1.0535	1.1098	1.2975	+1.3208	+0.2674	0.2538
1.45	1.0545	1.1120	1.3040	+1.3271	+0.2726	0.2585
1.46	1.0556	1.1142	1.3104	+1.3333	+0.2778	0.2631
1.47	1.0566	1.1164	1.3168	+1.3395	+0.2829	0.2678
1.48	1.0576	1.1185	1.3232	+1.3456	+0.2880	0.2723
1.49	1.0586	1.1207	1.3295	+1.3517	+0.2931	0.2769

p/p_1	$(p/p_1)^{1/7}$	T/T_1	ρ/ρ_1	c/a_1	q/a_1	$M = q/a$
1.50	1.0596	1.1228	1.3359	+1.3578	+0.2982	0.2814
1.51	1.0606	1.1250	1.3423	+1.3638	+0.3032	0.2859
1.52	1.0616	1.1271	1.3486	+1.3698	+0.3082	0.2903
1.53	1.0626	1.1292	1.3549	+1.3758	+0.3132	0.2947
1.54	1.0636	1.1313	1.3613	+1.3818	+0.3181	0.2991
1.55	1.0646	1.1334	1.3676	+1.3877	+0.3230	0.3034
1.56	1.0656	1.1355	1.3739	+1.3935	+0.3279	0.3078
1.57	1.0666	1.1376	1.3802	+1.3994	+0.3328	0.3120
1.58	1.0675	1.1396	1.3864	+1.4052	+0.3376	0.3163
1.59	1.0685	1.1417	1.3927	+1.4109	+0.3425	0.3205
1.60	1.0694	1.1437	1.3989	+1.4167	+0.3472	0.3247
1.61	1.0704	1.1458	1.4052	+1.4224	+0.3520	0.3289
1.62	1.0713	1.1478	1.4114	+1.4281	+0.3567	0.3330
1.63	1.0723	1.1498	1.4176	+1.4337	+0.3615	0.3371
1.64	1.0732	1.1518	1.4238	+1.4394	+0.3661	0.3412
1.65	1.0742	1.1538	1.4300	+1.4450	+0.3708	0.3452
1.66	1.0751	1.1558	1.4362	+1.4505	+0.3754	0.3492
1.67	1.0760	1.1578	1.4424	+1.4561	+0.3801	0.3532
1.68	1.0769	1.1598	1.4486	+1.4616	+0.3846	0.3572
1.69	1.0778	1.1617	1.4547	+1.4671	+0.3892	0.3611
1.70	1.0788	1.1637	1.4609	+1.4725	+0.3938	0.3650
1.71	1.0797	1.1657	1.4670	+1.4779	+0.3983	0.3689
1.72	1.0806	1.1676	1.4731	+1.4833	+0.4028	0.3727
1.73	1.0815	1.1695	1.4792	+1.4887	+0.4073	0.3766
1.74	1.0823	1.1715	1.4853	+1.4940	+0.4117	0.3804
1.75	1.0832	1.1734	1.4914	+1.4994	+0.4161	0.3842
1.76	1.0841	1.1753	1.4975	+1.5047	+0.4205	0.3879
1.77	1.0850	1.1772	1.5036	+1.5099	+0.4249	0.3917
1.78	1.0859	1.1791	1.5096	+1.5152	+0.4293	0.3954
1.79	1.0867	1.1810	1.5157	+1.5204	+0.4337	0.3990
1.80	1.0876	1.1829	1.5217	+1.5256	+0.4380	0.4027
1.81	1.0885	1.1847	1.5278	+1.5307	+0.4423	0.4063
1.82	1.0893	1.1866	1.5338	+1.5359	+0.4466	0.4100
1.83	1.0902	1.1885	1.5398	+1.5410	+0.4508	0.4135
1.84	1.0910	1.1903	1.5458	+1.5461	+0.4551	0.4171
1.85	1.0919	1.1922	1.5518	+1.5512	+0.4593	0.4207
1.86	1.0927	1.1940	1.5578	+1.5562	+0.4635	0.4242
1.87	1.0935	1.1958	1.5638	+1.5612	+0.4677	0.4277
1.88	1.0944	1.1977	1.5697	+1.5662	+0.4719	0.4312
1.89	1.0952	1.1995	1.5757	+1.5712	+0.4760	0.4346
1.90	1.0960	1.2013	1.5816	+1.5762	+0.4801	0.4381
1.91	1.0969	1.2031	1.5876	+1.5811	+0.4843	0.4415
1.92	1.0977	1.2049	1.5935	+1.5860	+0.4883	0.4449
1.93	1.0985	1.2067	1.5994	+1.5909	+0.4924	0.4483
1.94	1.0993	1.2085	1.6054	+1.5958	+0.4965	0.4516
1.95	1.1001	1.2102	1.6113	+1.6006	+0.5005	0.4550
1.96	1.1009	1.2120	1.6172	+1.6054	+0.5045	0.4583
1.97	1.1017	1.2138	1.6231	+1.6102	+0.5085	0.4616
1.98	1.1025	1.2155	1.6289	+1.6150	+0.5125	0.4649
1.99	1.1033	1.2173	1.6348	+1.6198	+0.5165	0.4681
2.00	1.1041	1.2190	1.6407	+1.6245	+0.5204	0.4714
2.01	1.1049	1.2208	1.6465	+1.6293	+0.5244	0.4746
2.02	1.1057	1.2225	1.6524	+1.6340	+0.5283	0.4778
2.03	1.1064	1.2242	1.6582	+1.6386	+0.5322	0.4810
2.04	1.1072	1.2259	1.6640	+1.6433	+0.5361	0.4842

p/p_1	$(p/p_1)^{1/7}$	T/T_1	ρ/ρ_1	c/a_1	q/a_1	$M = q/a$
2.05	1.1080	1.2276	1.6699	+1.6479	+0.5400	0.4873
2.06	1.1088	1.2294	1.6757	+1.6526	+0.5438	0.4905
2.07	1.1095	1.2311	1.6815	+1.6572	+0.5476	0.4936
2.08	1.1103	1.2328	1.6873	+1.6618	+0.5515	0.4967
2.09	1.1111	1.2344	1.6931	+1.6663	+0.5553	0.4998
2.10	1.1118	1.2361	1.6989	+1.6709	+0.5591	0.5028
2.11	1.1126	1.2378	1.7046	+1.6754	+0.5628	0.5059
2.12	1.1133	1.2395	1.7104	+1.6799	+0.5666	0.5089
2.13	1.1141	1.2411	1.7162	+1.6844	+0.5703	0.5119
2.14	1.1148	1.2428	1.7219	+1.6889	+0.5741	0.5149
2.15	1.1156	1.2445	1.7277	+1.6933	+0.5778	0.5179
2.16	1.1163	1.2461	1.7334	+1.6978	+0.5815	0.5209
2.17	1.1170	1.2478	1.7391	+1.7022	+0.5852	0.5239
2.18	1.1178	1.2494	1.7448	+1.7066	+0.5888	0.5268
2.19	1.1185	1.2510	1.7505	+1.7110	+0.5925	0.5297
2.20	1.1192	1.2527	1.7563	+1.7154	+0.5961	0.5326
2.21	1.1200	1.2543	1.7620	+1.7197	+0.5998	0.5355
2.22	1.1207	1.2559	1.7676	+1.7240	+0.6034	0.5384
2.23	1.1214	1.2575	1.7733	+1.7284	+0.6070	0.5413
2.24	1.1221	1.2591	1.7790	+1.7327	+0.6105	0.5441
2.25	1.1228	1.2607	1.7847	+1.7369	+0.6141	0.5469
2.26	1.1235	1.2623	1.7903	+1.7412	+0.6177	0.5498
2.27	1.1242	1.2639	1.7960	+1.7455	+0.6212	0.5526
2.28	1.1250	1.2655	1.8016	+1.7497	+0.6248	0.5554
2.29	1.1257	1.2671	1.8073	+1.7539	+0.6283	0.5581
2.30	1.1264	1.2687	1.8129	+1.7581	+0.6318	0.5609
2.31	1.1271	1.2702	1.8185	+1.7623	+0.6353	0.5637
2.32	1.1277	1.2718	1.8242	+1.7665	+0.6387	0.5664
2.33	1.1284	1.2734	1.8298	+1.7707	+0.6422	0.5691
2.34	1.1291	1.2749	1.8354	+1.7748	+0.6457	0.5718
2.35	1.1298	1.2765	1.8410	+1.7789	+0.6491	0.5745
2.36	1.1305	1.2780	1.8466	+1.7830	+0.6525	0.5772
2.37	1.1312	1.2796	1.8522	+1.7871	+0.6559	0.5799
2.38	1.1319	1.2811	1.8577	+1.7912	+0.6594	0.5825
2.39	1.1325	1.2827	1.8633	+1.7953	+0.6627	0.5852
2.40	1.1332	1.2842	1.8689	+1.7993	+0.6661	0.5878
2.41	1.1339	1.2857	1.8744	+1.8034	+0.6695	0.5904
2.42	1.1346	1.2872	1.8800	+1.8074	+0.6728	0.5930
2.43	1.1352	1.2888	1.8855	+1.8114	+0.6762	0.5956
2.44	1.1359	1.2903	1.8911	+1.8154	+0.6795	0.5982
2.45	1.1366	1.2918	1.8966	+1.8194	+0.6828	0.6008
2.46	1.1372	1.2933	1.9021	+1.8234	+0.6861	0.6033
2.47	1.1379	1.2948	1.9076	+1.8273	+0.6894	0.6059
2.48	1.1385	1.2963	1.9132	+1.8313	+0.6927	0.6084
2.49	1.1392	1.2978	1.9187	+1.8352	+0.6960	0.6110
2.50	1.1399	1.2993	1.9242	+1.8391	+0.6993	0.6135
2.51	1.1405	1.3007	1.9297	+1.8430	+0.7025	0.6160
2.52	1.1412	1.3022	1.9352	+1.8469	+0.7058	0.6185
2.53	1.1418	1.3037	1.9406	+1.8508	+0.7090	0.6209
2.54	1.1424	1.3052	1.9461	+1.8546	+0.7122	0.6234
2.55	1.1431	1.3066	1.9516	+1.8585	+0.7154	0.6259
2.56	1.1437	1.3081	1.9570	+1.8623	+0.7186	0.6283
2.57	1.1444	1.3096	1.9625	+1.8661	+0.7218	0.6307
2.58	1.1450	1.3110	1.9680	+1.8700	+0.7250	0.6332
2.59	1.1456	1.3125	1.9734	+1.8738	+0.7281	0.6356

p/p_1	$(p/p_1)^{1/7}$	T/T_1	ρ/ρ_1	c/a_1	q/a_1	$M = q/a$
2.60	1.1463	1.3139	1.9783	+1.8775	+0.7313	0.6380
2.61	1.1469	1.3153	1.9843	+1.8813	+0.7344	0.6404
2.62	1.1475	1.3168	1.9897	+1.8851	+0.7376	0.6427
2.63	1.1481	1.3182	1.9951	+1.8888	+0.7407	0.6451
2.64	1.1483	1.3196	2.0005	+1.8926	+0.7438	0.6475
2.65	1.1494	1.3211	2.0059	+1.8963	+0.7469	0.6498
2.66	1.1500	1.3225	2.0113	+1.9000	+0.7500	0.6522
2.67	1.1506	1.3239	2.0167	+1.9037	+0.7531	0.6545
2.68	1.1512	1.3253	2.0221	+1.9074	+0.7562	0.6568
2.69	1.1518	1.3267	2.0275	+1.9111	+0.7592	0.6591
2.70	1.1525	1.3281	2.0329	+1.9147	+0.7623	0.6614
2.71	1.1531	1.3296	2.0383	+1.9184	+0.7653	0.6637
2.72	1.1537	1.3310	2.0436	+1.9220	+0.7683	0.6660
2.73	1.1543	1.3323	2.0490	+1.9256	+0.7714	0.6683
2.74	1.1549	1.3337	2.0544	+1.9293	+0.7744	0.6705
2.75	1.1555	1.3351	2.0597	+1.9329	+0.7774	0.6728
2.76	1.1561	1.3365	2.0651	+1.9365	+0.7804	0.6750
2.77	1.1567	1.3379	2.0704	+1.9401	+0.7834	0.6773
2.78	1.1573	1.3393	2.0757	+1.9436	+0.7864	0.6795
2.79	1.1579	1.3407	2.0811	+1.9472	+0.7893	0.6817
2.80	1.1585	1.3420	2.0864	+1.9507	+0.7923	0.6839
2.81	1.1590	1.3434	2.0917	+1.9543	+0.7952	0.6861
2.82	1.1596	1.3448	2.0970	+1.9578	+0.7982	0.6883
2.83	1.1602	1.3461	2.1023	+1.9613	+0.8011	0.6905
2.84	1.1608	1.3475	2.1077	+1.9648	+0.8040	0.6926
2.85	1.1614	1.3483	2.1129	+1.9683	+0.8069	0.6948
2.86	1.1620	1.3502	2.1182	+1.9718	+0.8099	0.6970
2.87	1.1626	1.3515	2.1235	+1.9753	+0.8128	0.6991
2.88	1.1631	1.3529	2.1288	+1.9788	+0.8156	0.7012
2.89	1.1637	1.3542	2.1341	+1.9822	+0.8185	0.7034
2.90	1.1643	1.3555	2.1394	+1.9857	+0.8214	0.7055
2.91	1.1649	1.3569	2.1446	+1.9891	+0.8243	0.7076
2.92	1.1654	1.3582	2.1499	+1.9925	+0.8271	0.7097
2.93	1.1660	1.3595	2.1551	+1.9959	+0.8300	0.7118
2.94	1.1666	1.3609	2.1604	+1.9994	+0.8328	0.7139
2.95	1.1671	1.3622	2.1656	+2.0028	+0.8356	0.7160
2.96	1.1677	1.3635	2.1709	+2.0061	+0.8384	0.7180
2.97	1.1683	1.3648	2.1761	+2.0095	+0.8413	0.7201
2.98	1.1688	1.3661	2.1814	+2.0129	+0.8441	0.7222
2.99	1.1694	1.3674	2.1866	+2.0162	+0.8469	0.7242
3.00	1.1699	1.3687	2.1918	+2.0196	+0.8497	0.7262
3.05	1.1727	1.3752	2.2178	+2.0362	+0.8635	0.7363
3.10	1.1754	1.3816	2.2437	+2.0525	+0.8771	0.7462
3.15	1.1781	1.3880	2.2695	+2.0687	+0.8906	0.7559
3.20	1.1808	1.3942	2.2952	+2.0846	+0.9038	0.7655
3.25	1.1834	1.4004	2.3208	+2.1003	+0.9169	0.7748
3.30	1.1860	1.4065	2.3462	+2.1158	+0.9298	0.7840
3.35	1.1885	1.4126	2.3715	+2.1311	+0.9426	0.7931
3.40	1.1910	1.4186	2.3968	+2.1462	+0.9552	0.8020
3.45	1.1935	1.4245	2.4219	+2.1611	+0.9676	0.8107
3.50	1.1960	1.4304	2.4469	+2.1759	+0.9799	0.8193
3.55	1.1984	1.4362	2.4718	+2.1904	+0.9920	0.8278
3.60	1.2008	1.4419	2.4967	+2.2048	+1.0040	0.8361
3.65	1.2032	1.4476	2.5214	+2.2190	+1.0159	0.8443
3.70	1.2055	1.4533	2.5460	+2.2331	+1.0276	0.8524

p/p_1	$(p/p_1)^{1/7}$	T/T_1	ρ/ρ_1	c/a_1	q/a_1	$M = q/a$
3.75	1.2078	1.4588	2.5705	+2.2470	+1.0391	0.8603
3.80	1.2101	1.4644	2.5950	+2.2607	+1.0506	0.8682
3.85	1.2124	1.4699	2.6193	+2.2743	+1.0619	0.8759
3.90	1.2146	1.4753	2.6436	+2.2877	+1.0731	0.8835
3.95	1.2168	1.4807	2.6677	+2.3010	+1.0841	0.8909
4.00	1.2190	1.4860	2.6918	+2.3141	+1.0951	0.8983
4.05	1.2212	1.4913	2.7158	+2.3271	+1.1059	0.9056
4.10	1.2233	1.4965	2.7397	+2.3399	+1.1166	0.9128
4.15	1.2254	1.5017	2.7635	+2.3526	+1.1272	0.9198
4.20	1.2275	1.5069	2.7873	+2.3652	+1.1377	0.9268
4.25	1.2296	1.5120	2.8109	+2.3777	+1.1481	0.9337
4.30	1.2317	1.5170	2.8345	+2.3900	+1.1584	0.9405
4.35	1.2337	1.5220	2.8580	+2.4023	+1.1685	0.9472
4.40	1.2357	1.5270	2.8814	+2.4143	+1.1786	0.9538
4.45	1.2377	1.5320	2.9048	+2.4263	+1.1886	0.9603
4.50	1.2397	1.5369	2.9281	+2.4382	+1.1985	0.9668
4.55	1.2417	1.5417	2.9513	+2.4499	+1.2083	0.9731
4.60	1.2436	1.5465	2.9744	+2.4616	+1.2180	0.9794
4.65	1.2455	1.5513	2.9975	+2.4731	+1.2276	0.9856
4.70	1.2474	1.5561	3.0204	+2.4845	+1.2371	0.9917
4.75	1.2493	1.5608	3.0434	+2.4959	+1.2466	0.9978
4.7684	1.2500	1.5625	3.0518	+2.5000	+1.2500	1.0000
4.80	1.2512	1.5655	3.0662	+2.5071	+1.2559	1.0038
4.85	1.2530	1.5701	3.0890	+2.5182	+1.2652	1.0097
4.90	1.2549	1.5747	3.1117	+2.5292	+1.2744	1.0155
4.95	1.2567	1.5793	3.1343	+2.5402	+1.2835	1.0213
5.00	1.2585	1.5838	3.1569	+2.5510	+1.2925	1.0270
5.10	1.2621	1.5928	3.2019	+2.5724	+1.3103	1.0382
5.20	1.2656	1.6017	3.2466	+2.5934	+1.3279	1.0492
5.30	1.2690	1.6104	3.2911	+2.6141	+1.3451	1.0599
5.40	1.2724	1.6190	3.3353	+2.6345	+1.3621	1.0705
5.50	1.2758	1.6275	3.3793	+2.6545	+1.3788	1.0807
5.60	1.2790	1.6359	3.4231	+2.6742	+1.3952	1.0908
5.70	1.2823	1.6442	3.4667	+2.6937	+1.4114	1.1007
5.80	1.2855	1.6524	3.5100	+2.7128	+1.4273	1.1104
5.90	1.2886	1.6605	3.5531	+2.7317	+1.4431	1.1199
6.00	1.2917	1.6685	3.5960	+2.7503	+1.4585	1.1292
6.10	1.2948	1.6764	3.6387	+2.7686	+1.4738	1.1383
6.20	1.2978	1.6842	3.6812	+2.7866	+1.4889	1.1472
6.30	1.3007	1.6919	3.7236	+2.8045	+1.5037	1.1560
6.40	1.3037	1.6996	3.7657	+2.8220	+1.5184	1.1647
6.50	1.3066	1.7071	3.8076	+2.8394	+1.5328	1.1732
6.60	1.3094	1.7146	3.8494	+2.8565	+1.5471	1.1815
6.70	1.3122	1.7220	3.8909	+2.8734	+1.5612	1.1897
6.80	1.3150	1.7293	3.9323	+2.8901	+1.5751	1.1978
6.90	1.3178	1.7365	3.9735	+2.9065	+1.5888	1.2057
7.00	1.3205	1.7436	4.0146	+2.9228	+1.6023	1.2135
7.10	1.3231	1.7507	4.0555	+2.9389	+1.6157	1.2211
7.20	1.3258	1.7577	4.0962	+2.9548	+1.6290	1.2287
7.30	1.3284	1.7647	4.1367	+2.9705	+1.6420	1.2361
7.40	1.3310	1.7715	4.1771	+2.9860	+1.6550	1.2434
7.50	1.3335	1.7784	4.2174	+3.0013	+1.6577	1.2506
7.60	1.3361	1.7851	4.2575	+3.0164	+1.6804	1.2577
7.70	1.3386	1.7918	4.2974	+3.0314	+1.6929	1.2647
7.80	1.3410	1.7984	4.3372	+3.0462	+1.7052	1.2716
7.90	1.3435	1.8049	4.3769	+3.0609	+1.7174	1.2783

p/p_1	$(p/p_1)^{1/7}$	T/T_1	ρ/ρ_1	c/a_1	q/a_1	$M = q/a$
8.00	1.3459	1.8114	4.4164	+3.0754	+1.7295	1.2850
8.10	1.3483	1.8179	4.4557	+3.0897	+1.7415	1.2916
8.20	1.3507	1.8243	4.4949	+3.1039	+1.7533	1.2981
8.30	1.3530	1.8306	4.5340	+3.1180	+1.7650	1.3045
8.40	1.3553	1.8369	4.5730	+3.1319	+1.7766	1.3108
8.50	1.3576	1.8431	4.6118	+3.1456	+1.7880	1.3170
8.60	1.3599	1.8493	4.6505	+3.1593	+1.7994	1.3232
8.70	1.3621	1.8554	4.6891	+3.1728	+1.8106	1.3293
8.80	1.3644	1.8615	4.7275	+3.1861	+1.8218	1.3353
8.90	1.3666	1.8675	4.7658	+3.1993	+1.8328	1.3412

p/p_1	$(p/p_1)^{1/7}$	T/T_1	ρ/ρ_1	c/a_1	q/a_1	$M = q/a$
9.00	1.3687	1.8734	4.8040	+3.2124	+1.8437	1.3470
9.10	1.3709	1.8794	4.8421	+3.2254	+1.8545	1.3528
9.20	1.3730	1.8852	4.8800	+3.2383	+1.8652	1.3585
9.30	1.3752	1.8911	4.9178	+3.2510	+1.8758	1.3641
9.40	1.3773	1.8969	4.9555	+3.2636	+1.8863	1.3696
9.50	1.3794	1.9026	4.9931	+3.2761	+1.8968	1.3751
9.60	1.3814	1.9083	5.0306	+3.2885	+1.9071	1.3805
9.70	1.3835	1.9140	5.0680	+3.3008	+1.9173	1.3859
9.80	1.3855	1.9196	5.1053	+3.3129	+1.9275	1.3912
9.90	1.3875	1.9252	5.1424	+3.3250	+1.9375	1.3964
10.00	1.3895	1.9307	5.1795	+3.3370	+1.9475	1.4016

TABLE 7 PROPAGATION OF A STEEP–FRONTED COMPRESSION (SHOCK) WAVE WITHOUT FRICTION IN A CONSTANT AREA DUCT

Flow and fluid parameters are calculated with the pressure ratio (p/p_1) across the wave-front as the independent variable for the range, $p/p_1 = 1 \rightarrow (0{\cdot}05) \rightarrow 5 \rightarrow (0{\cdot}1) \rightarrow 10$. The summary of equations used is given in Section 8·5. The values of the following parameters are tabulated:

T/T_1 static temperature ratio across wave-front (normal shock)
ρ/ρ_1 static density ratio across wave-front (normal shock)
c/a_1 ratio of wave propagation velocity to the speed of sound in air ahead of wave
q/a_1 ratio of fluid particle velocity to the speed of sound in air ahead of wave
$M = q/a$, Mach number of induced flow behind wave-front
$(s - s_1)/R$ parameter for entropy change across the wave-front (normal shock)

(Note: directional significance is given by the movement of A1 in Figure 8·7)

p/p_1	T/T_1	ρ/ρ_1	c/a_1	q/a_1	$M = q/a$	$(s - s_1)/R$
1	1	1	+1	0	0	0
1.05	1.0140	1.0355	+1.0212	+0.0350	0.0347	+1.1851$_{10}$ -5
1.10	1.0276	1.0704	+1.0420	+0.0686	0.0676	+8.8328$_{10}$ -5
1.15	1.0408	1.1049	+1.0623	+0.1009	0.0989	+2.7845$_{10}$ -4
1.20	1.0537	1.1389	+1.0823	+0.1320	0.1286	+6.1794$_{10}$ -4
1.25	1.0662	1.1724	+1.1019	+0.1621	0.1569	+1.1324$_{10}$ -3
1.30	1.0784	1.2055	+1.1212	+0.1911	0.1840	+1.8399$_{10}$ -3
1.35	1.0904	1.2381	+1.1402	+0.2193	0.2100	+2.7521$_{10}$ -3
1.40	1.1021	1.2703	+1.1588	+0.2466	0.2349	+3.8767$_{10}$ -3
1.45	1.1137	1.3020	+1.1772	+0.2731	0.2587	+5.2174$_{10}$ -3
1.50	1.1250	1.3333	+1.1952	+0.2988	0.2817	+6.7755$_{10}$ -3
1.55	1.1362	1.3642	+1.2130	+0.3239	0.3038	+8.5502$_{10}$ -3
1.60	1.1472	1.3947	+1.2306	+0.3483	0.3252	+1.0539$_{10}$ -2
1.65	1.1580	1.4248	+1.2479	+0.3721	0.3457	+1.2738$_{10}$ -2
1.70	1.1687	1.4545	+1.2649	+0.3953	0.3656	+1.5144$_{10}$ -2
1.75	1.1793	1.4839	+1.2817	+0.4180	0.3849	+1.7750$_{10}$ -2
1.80	1.1898	1.5128	+1.2984	+0.4401	0.4035	+2.0551$_{10}$ -2
1.85	1.2002	1.5414	+1.3148	+0.4618	0.4215	+2.3542$_{10}$ -2
1.90	1.2105	1.5696	+1.3310	+0.4830	0.4390	+2.6717$_{10}$ -2
1.95	1.2207	1.5975	+1.3470	+0.5038	0.4560	+3.0068$_{10}$ -2
2.00	1.2308	1.6250	+1.3628	+0.5241	0.4725	+3.3591$_{10}$ -2
2.05	1.2408	1.6522	+1.3784	+0.5441	0.4885	+3.7278$_{10}$ -2
2.10	1.2507	1.6790	+1.3939	+0.5637	0.5040	+4.1123$_{10}$ -2
2.15	1.2606	1.7055	+1.4092	+0.5829	0.5192	+4.5121$_{10}$ -2
2.20	1.2704	1.7317	+1.4243	+0.6018	0.5339	+4.9266$_{10}$ -2
2.25	1.2802	1.7576	+1.4392	+0.6204	0.5483	+5.3551$_{10}$ -2
2.30	1.2899	1.7831	+1.4541	+0.6386	0.5623	+5.7972$_{10}$ -2
2.35	1.2995	1.8084	+1.4687	+0.6565	0.5759	+6.2522$_{10}$ -2
2.40	1.3091	1.8333	+1.4832	+0.6742	0.5893	+6.7197$_{10}$ -2
2.45	1.3186	1.8580	+1.4976	+0.6916	0.6023	+7.1990$_{10}$ -2
2.50	1.3281	1.8824	+1.5119	+0.7087	0.6149	+7.6898$_{10}$ -2
2.55	1.3376	1.9064	+1.5260	+0.7255	0.6273	+8.1915$_{10}$ -2
2.60	1.3470	1.9302	+1.5399	+0.7421	0.6394	+8.7037$_{10}$ -2
2.65	1.3564	1.9538	+1.5538	+0.7585	0.6513	+9.2259$_{10}$ -2
2.70	1.3657	1.9770	+1.5675	+0.7747	0.6629	+9.7577$_{10}$ -2
2.75	1.3750	2.0000	+1.5811	+0.7906	0.6742	+1.0299$_{10}$ -1
2.80	1.3843	2.0227	+1.5946	+0.8063	0.6853	+1.0848$_{10}$ -1
2.85	1.3935	2.0452	+1.6080	+0.8218	0.6961	+1.1407$_{10}$ -1
2.90	1.4027	2.0674	+1.6213	+0.8371	0.7068	+1.1973$_{10}$ -1
2.95	1.4119	2.0894	+1.6345	+0.8522	0.7172	+1.2547$_{10}$ -1

p/p_1	T/T_1	ρ/ρ_1	c/a_1	q/a_1	$M = q/a$	$(s - s_1)/R$
3.00	1.4211	2.1111	+1.6475	+0.8671	0.7274	+1.3128$_{10}$ -1
3.05	1.4302	2.1326	+1.6605	+0.8819	0.7374	+1.3716$_{10}$ -1
3.10	1.4393	2.1538	+1.6733	+0.8964	0.7472	+1.4311$_{10}$ -1
3.15	1.4484	2.1749	+1.6861	+0.9108	0.7568	+1.4913$_{10}$ -1
3.20	1.4574	2.1957	+1.6987	+0.9251	0.7663	+1.5520$_{10}$ -1
3.25	1.4665	2.2162	+1.7113	+0.9391	0.7755	+1.6133$_{10}$ -1
3.30	1.4755	2.2366	+1.7238	+0.9531	0.7846	+1.6752$_{10}$ -1
3.35	1.4845	2.2567	+1.7362	+0.9668	0.7935	+1.7376$_{10}$ -1
3.40	1.4935	2.2766	+1.7485	+0.9804	0.8023	+1.8005$_{10}$ -1
3.45	1.5024	2.2963	+1.7607	+0.9939	0.8109	+1.8639$_{10}$ -1
3.50	1.5114	2.3158	+1.7728	+1.0073	0.8193	+1.9278$_{10}$ -1
3.55	1.5203	2.3351	+1.7849	+1.0205	0.8276	+1.9921$_{10}$ -1
3.60	1.5292	2.3542	+1.7968	+1.0336	0.8358	+2.0568$_{10}$ -1
3.65	1.5381	2.3731	+1.8087	+1.0465	0.8438	+2.1219$_{10}$ -1
3.70	1.5470	2.3918	+1.8205	+1.0594	0.8517	+2.1874$_{10}$ -1
3.75	1.5559	2.4103	+1.8323	+1.0721	0.8595	+2.2532$_{10}$ -1
3.80	1.5647	2.4286	+1.8439	+1.0847	0.8671	+2.3194$_{10}$ -1
3.85	1.5735	2.4467	+1.8555	+1.0971	0.8746	+2.3859$_{10}$ -1
3.90	1.5824	2.4646	+1.8670	+1.1095	0.8820	+2.4527$_{10}$ -1
3.95	1.5912	2.4824	+1.8784	+1.1217	0.8893	+2.5198$_{10}$ -1
4.00	1.6000	2.5000	+1.8898	+1.1339	0.8964	+2.5872$_{10}$ -1
4.05	1.6088	2.5174	+1.9011	+1.1459	0.9035	+2.6548$_{10}$ -1
4.10	1.6176	2.5347	+1.9124	+1.1579	0.9104	+2.7227$_{10}$ -1
4.15	1.6264	2.5517	+1.9235	+1.1697	0.9172	+2.7908$_{10}$ -1
4.20	1.6351	2.5686	+1.9346	+1.1815	0.9239	+2.8591$_{10}$ -1
4.25	1.6439	2.5854	+1.9457	+1.1931	0.9306	+2.9276$_{10}$ -1
4.30	1.6526	2.6019	+1.9567	+1.2047	0.9371	+2.9963$_{10}$ -1
4.35	1.6613	2.6184	+1.9676	+1.2161	0.9435	+3.0652$_{10}$ -1
4.40	1.6701	2.6346	+1.9785	+1.2275	0.9499	+3.1343$_{10}$ -1
4.45	1.6788	2.6507	+1.9893	+1.2388	0.9561	+3.2035$_{10}$ -1
4.50	1.6875	2.6667	+2.0000	+1.2500	0.9623	+3.2729$_{10}$ -1
4.55	1.6962	2.6825	+2.0107	+1.2611	0.9683	+3.3424$_{10}$ -1
4.60	1.7049	2.6981	+2.0213	+1.2722	0.9743	+3.4121$_{10}$ -1
4.65	1.7136	2.7136	+2.0319	+1.2831	0.9802	+3.4818$_{10}$ -1
4.70	1.7223	2.7290	+2.0424	+1.2940	0.9860	+3.5517$_{10}$ -1
4.75	1.7309	2.7442	+2.0529	+1.3048	0.9917	+3.6217$_{10}$ -1
4.80	1.7396	2.7593	+2.0633	+1.3155	0.9974	+3.6917$_{10}$ -1
4.8231	1.7436	2.7662	+2.0681	+1.3205	1.0000	+3.7250$_{10}$ -1

p/p_1	T/T_1	ρ/ρ_1	c/a_1	q/a_1	$M = q/a$	$(s - s_1)/R$
4.85	1.7483	2.7742	+2.0736	+1.3262	1.0030	+3.7619$_{10}$ -1
4.90	1.7569	2.7890	+2.0840	+1.3367	1.0085	+3.8321$_{10}$ -1
4.95	1.7656	2.8037	+2.0942	+1.3473	1.0139	+3.9024$_{10}$ -1
5.00	1.7742	2.8182	+2.1044	+1.3577	1.0193	+3.9727$_{10}$ -1
5.10	1.7915	2.8468	+2.1247	+1.3784	1.0298	+4.1136$_{10}$ -1
5.20	1.8087	2.8750	+2.1448	+1.3988	1.0401	+4.2546$_{10}$ -1
5.30	1.8259	2.9027	+2.1647	+1.4189	1.0501	+4.3958$_{10}$ -1
5.40	1.8431	2.9298	+2.1844	+1.4388	1.0598	+4.5370$_{10}$ -1
5.50	1.8603	2.9565	+2.2039	+1.4585	1.0693	+4.6782$_{10}$ -1
5.60	1.8775	2.9828	+2.2233	+1.4779	1.0786	+4.8195$_{10}$ -1
5.70	1.8946	3.0085	+2.2424	+1.4971	1.0876	+4.9607$_{10}$ -1
5.80	1.9117	3.0339	+2.2615	+1.5161	1.0965	+5.1018$_{10}$ -1
5.90	1.9288	3.0588	+2.2804	+1.5349	1.1051	+5.2427$_{10}$ -1
6.00	1.9459	3.0833	+2.2991	+1.5534	1.1136	+5.3836$_{10}$ -1
6.10	1.9630	3.1074	+2.3176	+1.5718	1.1218	+5.5243$_{10}$ -1
6.20	1.9801	3.1311	+2.3361	+1.5900	1.1299	+5.6647$_{10}$ -1
6.30	1.9972	3.1545	+2.3543	+1.6080	1.1378	+5.8050$_{10}$ -1
6.40	2.0142	3.1774	+2.3725	+1.6258	1.1455	+5.9450$_{10}$ -1
6.50	2.0313	3.2000	+2.3905	+1.6434	1.1531	+6.0848$_{10}$ -1
6.60	2.0483	3.2222	+2.4083	+1.6609	1.1605	+6.2242$_{10}$ -1
6.70	2.0653	3.2441	+2.4260	+1.6782	1.1678	+6.3634$_{10}$ -1
6.80	2.0823	3.2656	+2.4437	+1.6954	1.1749	+6.5023$_{10}$ -1
6.90	2.0993	3.2868	+2.4611	+1.7123	1.1818	+6.6408$_{10}$ -1
7.00	2.1163	3.3077	+2.4785	+1.7292	1.1886	+6.7790$_{10}$ -1
7.10	2.1333	3.3282	+2.4957	+1.7459	1.1953	+6.9168$_{10}$ -1
7.20	2.1502	3.3485	+2.5128	+1.7624	1.2019	+7.0542$_{10}$ -1
7.30	2.1672	3.3684	+2.5298	+1.7788	1.2083	+7.1913$_{10}$ -1
7.40	2.1841	3.3881	+2.5467	+1.7950	1.2146	+7.3280$_{10}$ -1
7.50	2.2011	3.4074	+2.5635	+1.8112	1.2208	+7.4643$_{10}$ -1
7.60	2.2180	3.4265	+2.5801	+1.8271	1.2268	+7.6001$_{10}$ -1
7.70	2.2350	3.4453	+2.5967	+1.8430	1.2328	+7.7356$_{10}$ -1
7.80	2.2519	3.4638	+2.6132	+1.8587	1.2386	+7.8706$_{10}$ -1
7.90	2.2688	3.4820	+2.6295	+1.8743	1.2444	+8.0052$_{10}$ -1
8.00	2.2857	3.5000	+2.6458	+1.8898	1.2500	+8.1393$_{10}$ -1
8.10	2.3026	3.5177	+2.6619	+1.9052	1.2555	+8.2730$_{10}$ -1
8.20	2.3195	3.5352	+2.6780	+1.9204	1.2610	+8.4063$_{10}$ -1
8.30	2.3364	3.5524	+2.6939	+1.9356	1.2663	+8.5391$_{10}$ -1
8.40	2.3533	3.5694	+2.7098	+1.9506	1.2715	+8.6714$_{10}$ -1
8.50	2.3702	3.5862	+2.7255	+1.9655	1.2767	+8.8033$_{10}$ -1
8.60	2.3871	3.6027	+2.7412	+1.9803	1.2818	+8.9347$_{10}$ -1
8.70	2.4039	3.6190	+2.7568	+1.9951	1.2867	+9.0657$_{10}$ -1
8.80	2.4208	3.6351	+2.7723	+2.0097	1.2916	+9.1962$_{10}$ -1
8.90	2.4377	3.6510	+2.7877	+2.0242	1.2965	+9.3262$_{10}$ -1
9.00	2.4545	3.6667	+2.8031	+2.0386	1.3012	+9.4557$_{10}$ -1
9.10	2.4714	3.6821	+2.8183	+2.0529	1.3059	+9.5848$_{10}$ -1
9.20	2.4883	3.6974	+2.8335	+2.0671	1.3104	+9.7133$_{10}$ -1
9.30	2.5051	3.7124	+2.8486	+2.0813	1.3150	+9.8414$_{10}$ -1
9.40	2.5220	3.7273	+2.8636	+2.0953	1.3194	+9.9691$_{10}$ -1
9.50	2.5388	3.7419	+2.8785	+2.1092	1.3238	+1.0096$_{10}$ 0
9.60	2.5556	3.7564	+2.8933	+2.1231	1.3281	+1.0223$_{10}$ 0
9.70	2.5725	3.7707	+2.9081	+2.1369	1.3323	+1.0349$_{10}$ 0
9.80	2.5893	3.7848	+2.9228	+2.1506	1.3365	+1.0475$_{10}$ 0
9.90	2.6061	3.7987	+2.9374	+2.1642	1.3406	+1.0600$_{10}$ 0
10.00	2.6230	3.8125	+2.9520	+2.1777	1.3446	+1.0725$_{10}$ 0

PROBLEMS (with answers)

PROBLEM 1. Figure P1 shows a thin-walled conical diffuser attached to the bottom of a large water tank, into which water is fed, as shown, at the rate of M kg/s. The upper and lower sections of the cone are respectively A_1 and A_2 m^2 and the flow in the cone is frictionless. If the water density is ϱ kg/m^3 show that the net liquid force on the cone is upward and has a magnitude (newtons) of

$$\frac{M^2A_2}{2\varrho}\left(\frac{1}{A_2}-\frac{1}{A_1}\right)^2$$

(U. Birmingham 1963)

Fig. P1

PROBLEM 2. Figure P2 shows a control valve for dividing the flow of oil entering at A into two streams—one leaving at B and the other, after passing through the hollow piston, at C. Axial positioning of the piston is controlled hydraulically by means of an independent oil supply to the chamber D. In the position shown the axial widths of the left and right port openings are 4·8 mm and 1·6 mm respectively and the associated flows to B and C are 0·015 m^3/s and 0·003 m^3/s. The piston diameter is 25 mm, the specific gravity of the oil is 0·8 and the oil enters at A with a static pressure of $13{\cdot}3\times10^5$ N/m^2.

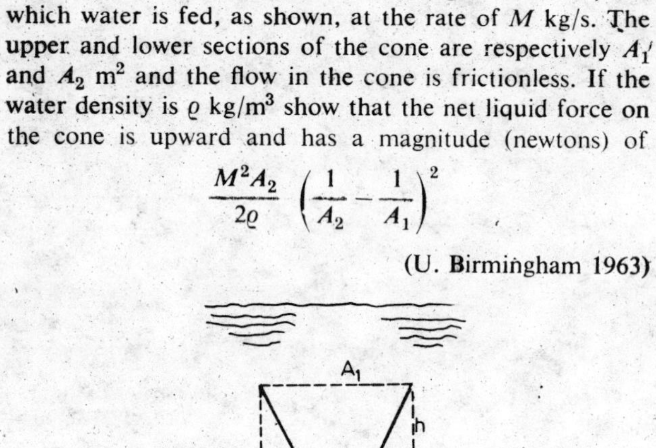

Fig. P2

Assuming that just as it leaves the bore the oil flows at 70° to the piston axis and neglecting the effects of oil viscosity, find the pressure at D.

(U. Birmingham 1962)

Answer: pressure at D, $p_D = 20{\cdot}94\times10^5$ N/m^2

PROBLEM 3. The nozzle of a rocket motor is designed for complete expansion of the gases from combustion pressure to atmospheric pressure. Assuming that the gases are perfect and the flow is isentropic show that the rocket thrust per mass flow rate ($F/\dot{m}$) is given by,

$$\frac{F}{\dot{m}} = \left[RT_1\frac{2\gamma}{\gamma-1}\left(1-\left(\frac{p_a}{p_1}\right)^{\frac{\gamma-1}{\gamma}}\right)\right]^{0\cdot5}$$

where (p_a) is the atmospheric pressure and (R), (T_1) and (p_1) are respectively the gas constant, combustion temperature and combustion pressure.

At lift-off 50% of the entire mass of a certain rocket is fuel. The rocket is designed to fire for one minute at a constant combustion temperature and pressure of 3 000 K and 2·0 MN/m^2, using all its fuel in the process. Determine the acceleration at lift-off when the rocket is fired vertically from sea-level and the nozzle design back pressure of 0·101 MN/m^2 prevails. Take the average relative molecular mass of the gases to be 33·5 and $\gamma = 1{\cdot}2$. The universal gas constant is 8314 Nm/mole K.

(U. London 1969)

Answer: acceleration = 15·6 m/s^2

PROBLEM 4. Express the compressible form of the energy equation in terms of

(a) the limit particle velocity, c,
(b) the throat or sonic velocity, q_*.

Hence show that the Mach number is given by

$$M^2 = 5/(6(q_*/q)^2 - 1)$$

Figure P4 shows air flowing at uniform speed (q) in a duct of constant cross-section. At points A and B, 1 m apart, are located receiver/transmitter sources which are used to propagate and record sound waves.

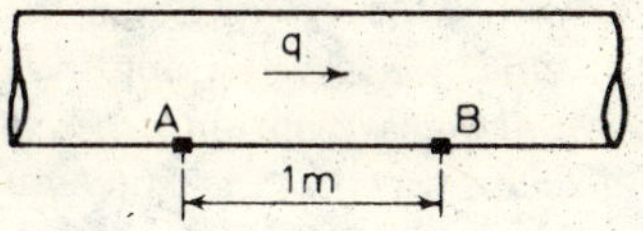

Fig. P4

If the time taken for a sound wave to be transmitted from A to B (in direction of flow) is 1·049 ms, and from B to A is 9·440 ms, determine the flow speed (q) and the Mach number (M). If the total head density, $\varrho_0 = 1{\cdot}2$ kg/m^3, calculate the static pressure in the duct.

What is the obvious limitation of the above method for determining the flow Mach number?

(U. Leeds 1967)

Answer: Flow speed, $q = 423{\cdot}7$ m/s

Mach number, $M = 0{\cdot}8$

PROBLEM 5. An aircraft is flying at a speed of 264 m/s at an altitude at which the speed of sound is 330 m/s. A bullet is fired at the aircraft from a position directly ahead of it, and passes close to the wing surface. The maximum local Mach number in this region is 1·5 and at this point the bow (shock) wave of the bullet just vanishes.

Calculate the velocity of the bullet.

Answer: Velocity of bullet = 688·1 m/s.

(U. Leeds 1964)

PROBLEM 6. In the critical flow valve (Figure P6), air ($\gamma = 1{\cdot}4$) enters via A at 14×10^5 N/m^2 and 116°C and divides into two streams which leave at B and C respectively. Simultaneous control of the two valves is effected by axial movement of the differential piston, F, the conditions at the two annular throats D and E being sonic. With the valve in

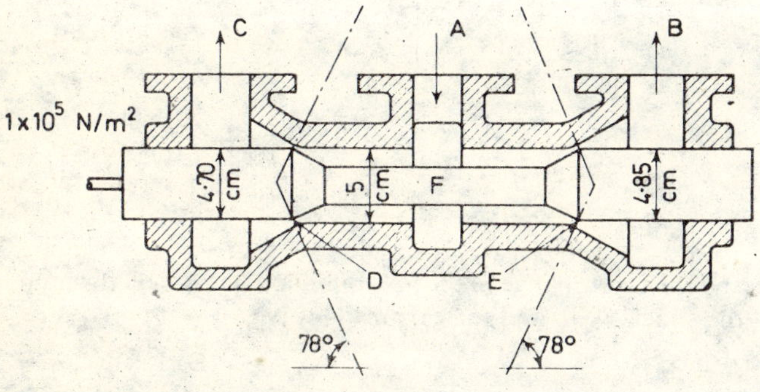

Fig. P6

the position shown, the velocity at each throat may be assumed to be at an angle of 12° with the piston axis; without calculation, the throat areas D and E, measured at right angles to the sonic flow, may be taken as 2 cm^2 and 1 cm^2 respectively. Assuming steady frictionless adiabatic flow as far as the throats, neglecting the kinetic energy at A and taking atmospheric pressure as 1×10^5 N/m^2, find the net axial thrust on the piston.

(U. Birmingham 1961)

Answer: 112·05 N →

PROBLEM 7. Air flows subsonically through the diverging passage of a diffuser. The static pressures in the passage at sections of area 96 cm^2 and 156 cm^2 are $1{\cdot}38 \times 10^5$ N/m^2 and $2{\cdot}07 \times 10^5$ N/m^2 respectively.

Calculate the Mach number of the flow at the larger section assuming isentropic flow. Find the mass flow per second of air if the total head temperature of the flow is 40°C.

Determine the resultant thrust on the diffuser if the ambient uniform pressure is 1×10^5 N/m^2.

(U. Leeds 1970)

Answers: $M_2 = 0{\cdot}384$, Mass flow 4·95 kg/s, Thrust 134·5 N (in direction of flow).

PROBLEM 8 For adiabatic frictionless flow in a non-parallel duct, show that the Mach number (M) and the cross-sectional area (A) at any two sections are related by the following expression:

$$\left(\frac{A_2 M_2}{A_1 M_1}\right)^2 = \left(\frac{1+\dfrac{\gamma-1}{2}M_2^2}{1+\dfrac{\gamma-1}{2}M_1^2}\right)^{\frac{\gamma+1}{\gamma-1}}$$

An aero-engine nacelle has a frontal opening 0·15 m^2 in area, while the engine lies in a parallel portion of the nacelle 0·4 m^2 in cross-sectional area. At a certain operating condition at 3000 m altitude, air enters the nacelle at a speed of 300 m/s. Estimate the mean Mach number of the flow immediately upstream of the engine, assuming adiabatic frictionless flow in the nacelle. What is the speed of air immediately upstream of the engine?

At 3000 m altitude, take the pressure of the air as $0{\cdot}7 \times 10^5$ N/m^2 and the density as 0·8 kg/m^3, γ being uniformly 1·4.

(U. Leeds 1955)

Answers: $M_2 = 0{\cdot}219$, $q_2 = 81{\cdot}5$ m/s.

PROBLEM 9. Assuming steady, one-dimensional, adiabatic, reversible flow of a perfect gas along a duct of slowly varying cross-section, derive expressions for the pressure (p), temperature (T) and density (ϱ) of the gas in terms of the Mach number (M), the ratio of the principal specific heat capacities and the stagnation conditions p_0, T_0 and ϱ_0.

Show that sonic velocity may be obtained only at a minimum cross-section in such a flow.

A supersonic wind tunnel is designed to operate at $M = 2$ using air as the working fluid. Conditions in the settling chamber are maintained constant such that $p_0 = 0{\cdot}5$ MN/m^2 and $T_0 = 16$°C. Estimate the Reynolds number for a model having a characteristic dimension of 100 mm mounted in the test section, assuming that the viscosity (μ) is given by,

$$(\mu/(171 \times 10^{-5}\ \text{mg/ms})) = (T/273\ \text{K})^{0{\cdot}76}$$

(U. London 1970)

Answer: Reynolds number, $R_l = ql\varrho/\mu = 6\,153\,000$

PROBLEM 10. Establish the Rankine–Hugoniot relationship between the absolute pressure (p_1) and density (ϱ_1) upstream of a plane normal shock wave in a perfect gas and the corresponding downstream quantities (p_2, ϱ_2).

Hence show that if $h = p_2/p_1 - 1$, then,

$$\frac{\varrho_2}{\varrho_1} = 1 = \frac{h}{\gamma} - \frac{(\gamma-1)}{2}\left(\frac{h}{\gamma}\right)^2 + \left(\frac{\gamma-1}{2}\right)^2\left(\frac{h}{\gamma}\right)^3 \text{ etc.}$$

Establish a similar relationship for an isentropic compression through the same pressure ratio and show that, if fourth and higher powers of (h/γ) are negligible in comparison with unity, the density ratios for the isentropic and shock processes differ by $\left(\frac{\gamma^2-1}{12}\right)\left(\frac{h}{\gamma}\right)^3$.

(U. Birmingham 1960)

PROBLEM 11. Figure P11 shows two parallel-sided ducts connected by a transition section with flanges at A and B. The supersonic air flow is uniform at sections (1) and (2) before and after the change in direction of 60°. At section (2) the duct area $A_2 = 0{\cdot}0225$ m² and the flow properties are pressure, $p_2 = 0{\cdot}1\times10^5$ N/m², velocity $q_2 = 1101$ m/s and the local speed of sound $a_2 = 300$ m/s.

Determine the Mach number (M_1) at section (1) and the angular extent (α) of the expansion fan.

Assuming that (p_1) and (p_2) are the uniform pressures of the air flow at flanges A and B, use the momentum equation to determine the resultant thrust on the transition section.

(U. Leeds 1967)

Answer: $M_1 = 1{\cdot}09$

$\alpha = 110{\cdot}74°$

resultant thrust = 4067·8 N

at angle to the horizontal = 47·93° ↗

PROBLEM 13. Gas flows along a duct of constant circular cross-section, diameter D; heat energy is transferred in such a way that the total temperature T_0 changes along the duct and the Mach number M remains constant. Assuming that the friction factor f is constant, show that

$$\frac{dT_0}{dx} = -T_0 \frac{\gamma M^2}{1+\gamma M^2} \frac{4C_f}{D}$$

where x is the distance measured along the duct.

1 kg/s air flows along a constant cross-section pipe of 8 cm diameter in such a way that the Mach number remains constant and equal to 0·7. The friction coefficient in the pipe is 0·01. Estimate the rate of heat energy transfer/length of the pipe in a region where the mean total temperature is 200°C.

Take $\gamma = 1{\cdot}4$ and $c_p = 1{\cdot}005$ kJ/kg K

(U. London 1967)

Answer: heat energy transfer rate = −96·71 kJ/s per metre

PROBLEM 14. Air flows isentropically, with a stagnation pressure of 7×10^5 N/m², through a two-dimensional convergent-divergent nozzle having an exit area 3·5 times the throat area. The flow leaves the nozzle with a supersonic velocity and there are no shock waves or expansion waves at the exit. A symmetrical 40° wedge is placed in the jet as shown in Figure P14. Determine (a) the angle of the oblique shock wave attached to the wedge and the pressure and Mach number immediately behind the shock wave.

The shock wave reflects from the jet edge as an expansion wave. Find (b) the pressure and Mach number behind the expansion wave, the included angle of the expansion wave, and the total change in the flow direction on passing through the shock wave and the expansion wave.

(U. Leeds 1968)

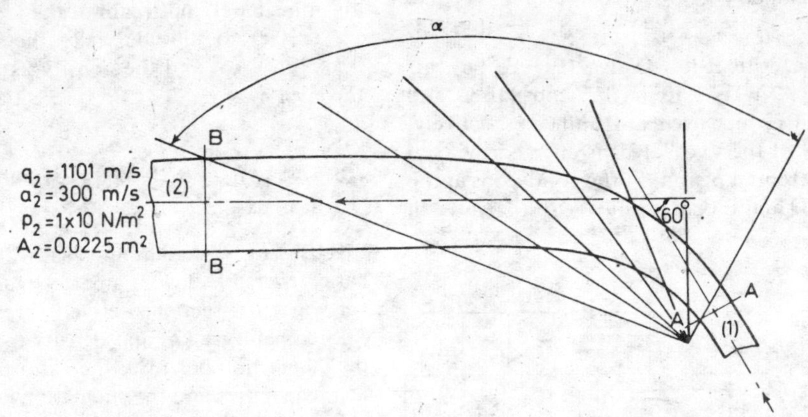

Fig. P11

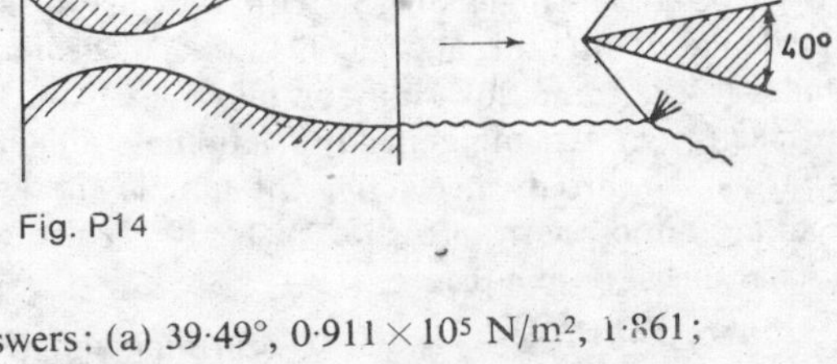

Fig. P14

Answers: (a) 39·49°, $0{\cdot}911\times10^5$ N/m², 1·861;

(b) $0{\cdot}258\times10^5$ N/m², 2·672, 31·04°, 40·53°

PROBLEM 12. A convergent–divergent nozzle of throat area 16 cm² is supplied with air from a very large reservoir in which the pressure (p_0) is 2×10^5 N/m² and the acoustic velocity (a_0) is 340 m/s. The nozzle is operated with an incorrect back pressure causing a plane normal shock wave to occur in the divergent portion with a Mach number (M_1) of 1·35 just upstream of the shock. Assuming the shock to extend wholly across the nozzle and taking the air flow elsewhere to be isentropic, find the nozzle area at the shock and the pressure (p_2) immediately downstream of it. Take $\gamma = 1{\cdot}4$ and prove any nozzle shock formulae used.

(U. Birmingham 1961)

Answer: nozzle area at shock = 0·001742 m²

pressure $p_2 = 1{\cdot}321\times10^5$ N/m²

PROBLEM 15. An airstream approaches the two-dimensional air intake, shown in Figure P15, with a Mach number of 2, a pressure of $0{\cdot}7 \times 10^5$ N/m^2 and a temperature of −6 C. The wedge has a half angle of 10° and the cross-sectional area at inlet to the duct is 0·01 m^2. If a normal shock wave forms in the duct where the cross-section area is 0·0131 m^2, determine the total reduction in stagnation pressure due to passage of flow through the oblique shock wave attached to the wedge and the normal shock wave in the duct. Calculate the mass flow through the intake.

(U. Leeds 1970)

Answers: $2{\cdot}023 \times 10^5$ N/m^2, 7·32 kg/s

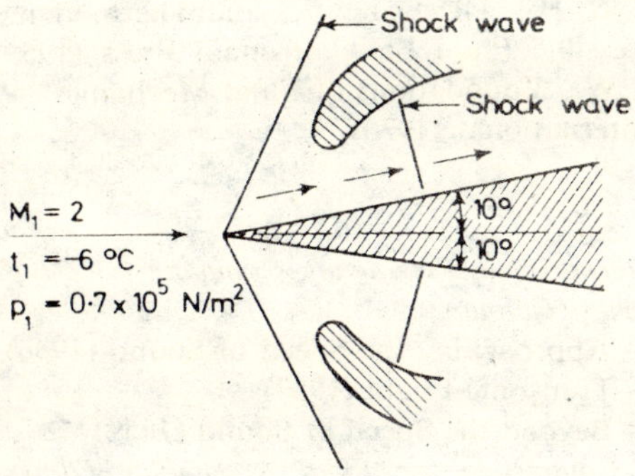

Fig. P15

PROBLEM 16. Figure P16 shows the steady adiabatic two-dimensional flow of air along the horizontal downward facing wall AB, passes under a uniform flap BC of length 10 cm and of weight W N/m run normal to the figure. The flap is freely hinged to the wall at B.

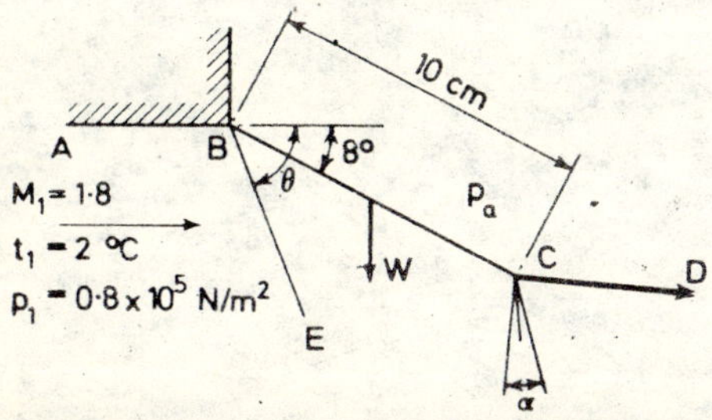

Fig. P16

For initial conditions of pressure, $p_1 = 0{\cdot}8 \times 10^5$ N/m^2, temperature, $t_1 = 2$°C and Mach number, $M_1 = 1{\cdot}80$. It is observed that the flap is at angle of 8° to the horizontal as shown. Ambient pressure $p_a = 1 \times 10^5$ N/m^2.

Using the flow tables provided, determine:

(a) the angle (θ) of the shock BE,
(b) the weight (W) of the flap,
(c) the angular extent (α) of the expansion fan centred at C,
(d) the inclination to the horizontal of the slip plane, CD,
(e) the flow speed along the slip plane.

(U. Birmingham 1958)

Answers: (a) 41·67° (b) 2 055 N/m (c) 7·4° (d) −4·3° (e) 564·8 m/s

PROBLEM 17. A frictionless pipe line is fitted with a full bore gate valve at some point along its length. With the valve fully open there is a steady flow of air of velocity 297·4 m/s and the associated static pressure and total head temperature are 7×10^5 N/m^2 and 60°C respectively. Estimate the greatest pressure difference which can occur across the valve upon closure. State any assumptions which are used.

What are the propagation velocities of the front points of the different waves?

(U. Birmingham)

Answers: $\Delta p = 19{\cdot}814 \times 10^5$ N/m^2 (assuming non-steep waves)

velocities = −42·6 m/s (i.e. upstream), +637·4 m/s (downstream)

PROBLEM 18. Atmospheric air at 1×10^5 N/m^2, having an acoustic velocity of 300 m/s, is contained within a frictionless cylinder open at one end, as shown in Figure P18. A heavy cylindrical punch is fired into the cylinder with a velocity of 150 m/s.

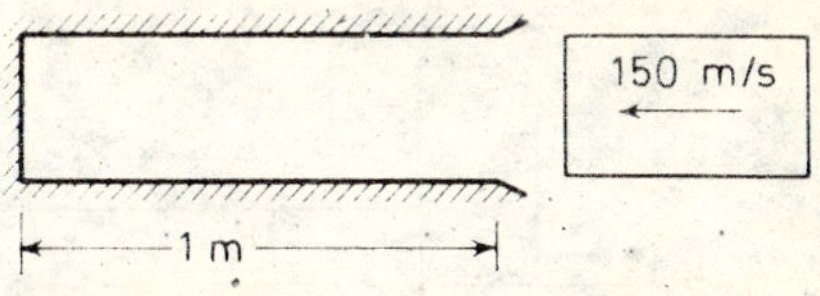

Fig. P18

In analysing the resulting events, it may be assumed

(a) that the air has constant specific heats of ratio 7/5,
(b) that the non-steep shock wave generated in the cylinder is created at the instant when the leading face of the punch just enters the cylinder,
(c) that the deceleration of the punch may be ignored in what follows.

Using otherwise the most exact methods of which one-dimensional analysis is capable, determine (a) the pressure generated at the closed end of the cylinder when the initial shock wave undergoes reflection and (b) the distance by which the punch face will have moved into the cylinder when it encounters this first reflection from the closed end.

(U. Birmingham)

Answers: (a) $3{\cdot}583 \times 10^5$ N/m^2 (b) 0·5147 m

PROBLEM 19. A certain shock tube experiment is arranged in such a way that two shock-fronted, flat-topped compression waves are generated, the second pursuing the first, with the result that, at a particular instant of time, the relevant portion of the whole process is as shown in Figure P19.

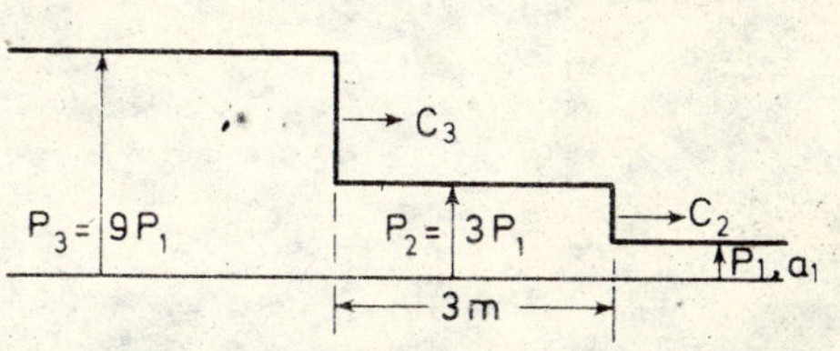

Fig. P19

(a) Determine the propagation velocities of the two shock fronts indicated.
(b) Find the distance travelled by the leading shock from the position shown when it is overtaken by the second shock.
(c) Compare the particle velocity and static temperature behind the second shock with the respective values which would arise behind a single shock of pressure ratio 9. Comment on the significance of any discrepancies.

Answers: (a) $c_2/a_1 = +1{\cdot}6475$, $c_3/a_1 \doteq +2{\cdot}8311$

(b) 4·175 m

(c) $q_3/a_1 = 1{\cdot}9008$ (2·0386);
$T_3/T_1 = 2{\cdot}0195$ (2·4545)

PROBLEM 20. A long, uniform, frictionless pipe containing quiescent air at a pressure and temperature of 1×10^5 N/m^2 and 289 K respectively, is closed at one end by a piston. The piston is given a displacement such that it generates in the air a flat crested compressive wave with an amplitude ratio, $p_2/p_1 = 7$.

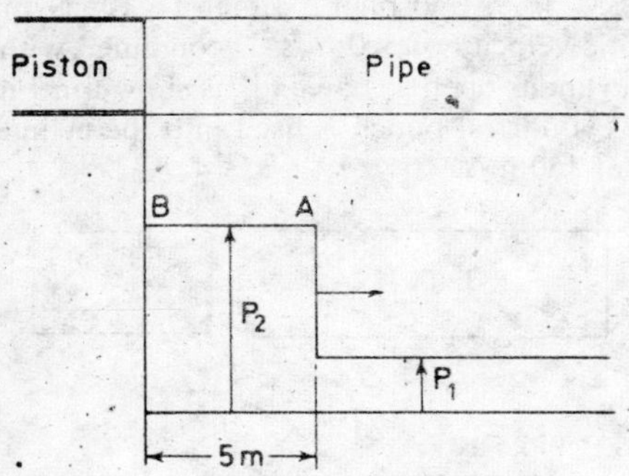

Wave diagram at time t = 0

Fig. P20

If, for the purpose of the analysis, the shape of the wave at the instant the piston comes to rest can be taken to be as shown in Figure P20, determine the pressure and temperature conditions in the wake of the wave during its passage along the pipe. Calculate the total period of time particles, situated at a distance of 10 metres from the crown of the piston at the instant is comes to rest, are affected by the wave. The length of the pipe is such that reflections from its end may be ignored.

(U. Birmingham)

Answers: The rarefaction behind the wave establishes zero particles velocity at the stationary piston. Pressure, $1{\cdot}0472 \times 10^5$ N/m^2, temperature, 355·4 K. Time for AB to pass through particles initially 10 metres from piston crown is 0·00723 s.

REFERENCES AND RESOURCES

BANNISTER, F. K., 'Pressure Waves in Gases in Pipes', *Akroyd Stuart Memorial Lectures*, University of Nottingham (1958).

BENEDICT, R. P. and STELTZ, W. G. 'Handbook of Generalised Gas Dynamics', Plenum Press Data Division (1966)

CHEERS, F., 'Elements of Compressible Flow', Wiley (1963)

HOUGHTON, E. L. and BROCK, A. B., 'Tables for the Compressible Flow of Dry Air', Arnold (1961)

JOHN, James E. A., 'Gas Dynamics', Allyn and Bacon (1969)

SHAPIRO, A. H., 'The Dynamics and Thermodynamics of Compressible Fluid Flow', Ronald Press (1953)

YUAN, S. W., 'Foundations of Fluid Mechanics', Prentice-Hall International (1970)

FILMS

Shell International Petroleum Company — High Speed Flight Series (Colour)

Part 1 — Approaching the Speed of Sound (1956)

Part 2 — Transonic Flight (1957)

Part 3 — Beyond the Speed of Sound (1959)

National Committee for Fluid Mechonics Films, U.S.A. (B/W) Channel Flow of a Compressible Fluid (Donald Coles) Waves in Fluids (Arthur F. Bryson)

INDEX